RETHINKING EVOLUTION

The Revolution That's Hiding in Plain Sight

RETHINKING EVOLUTION

The Revolution That's Hiding in Plain Sight

Gene Levinson
SmartNoter Inc., USA

World Scientific
EW JERSEY • LONDON • SINGAPORE • BEIJING • SHANGHAI • HONG KONG • TAIPEI • CHENNAI • TOKYO

Published by

World Scientific Publishing Europe Ltd.
57 Shelton Street, Covent Garden, London WC2H 9HE
Head office: 5 Toh Tuck Link, Singapore 596224
USA office: 27 Warren Street, Suite 401-402, Hackensack, NJ 07601

Library of Congress Cataloging-in-Publication Data
Names: Gene, Levinson, author.
Title: Rethinking evolution : the revolution that's hiding in plain sight /
by Gene Levinson (SmartNoter Inc, USA)
Description: New Jersey : World Scientific, [2019]
Identifiers: LCCN 2019013762 | ISBN 9781786347268 (hc)
Subjects: LCSH: Evolution (Biology)
Classification: LCC QH366.2 .G45 2019 | DDC 576.8--dc23
LC record available at https://lccn.loc.gov/2019013762

British Library Cataloguing-in-Publication Data
A catalogue record for this book is available from the British Library.

For any available supplementary material, please visit
https://www.worldscientific.com/worldscibooks/10.1142/Q0215#t=suppl

Desk Editors: Dr. Sree Meenakshi Sajani/Jennifer Brough/Shi Ying Koe

Typeset by Stallion Press
Email: enquiries@stallionpress.com

Dedication

This work would not have been possible without the tireless efforts and published contributions of literally thousands of hardworking and creative biological researchers, theorists, and communicators. This includes all of the unsung heroes—the graduate students and postdocs—whose hard work and resilience have provided so much detailed information about the inner-workings of cells.

Preface

You're an interesting species. An interesting mix. You're capable of such beautiful dreams, and such horrible nightmares. You feel so lost, so cut off, so alone, only you're not. See, in all our searching, the only thing we've found that makes the emptiness bearable, is each other.

—fictional Extraterrestrial, speaking to Earthling, in *Contact* (the film, 1997)[a]

Rethinking Evolution breaks the mold for both popular and professional science books. The conceptual destination is nothing short of a wide-ranging Updated Evolutionary Synthesis (UES), and there are many challenges along the way. To see why it's worth embarking on this journey, please take a minute or two to jump ahead to the final chapter (Chapter 17) for a comprehensive list of the principles of the UES in its initial published form. Most readers will be surprised by at least some of the entries in this list. The UES includes many new aspects of which most readers are unaware. Today's evolutionary theory is stronger than ever, because it is thoroughly supported by literally thousands of wide-ranging empirical observations and experiments. The UES incorporates the lasting core principles of classical 1859 Darwinian theory but offers a more complete and plausible scientific explanation than ever before.

Theodosius Dobzhansky paid tribute to the "Modern Synthesis" when he famously declared that "Nothing in biology makes sense except in the light of evolution".

[a]http://www.imdb.com/title/tt0118884/

That "Modern Synthesis" is, of course, no longer modern. The 21st century converse is equally illuminating: Nothing in evolution makes sense except in the light of biology.

In other words, a deep understanding of the sheer organizing power of the natural forces of evolution can be found in now well-established biological principles. These principles are based on numerous new discoveries in molecular, cellular and developmental biology.

The plausibility of the metaphorical "blind watchmaker" who creates an exquisite range of diverse structures and functions is persuasively revealed by exploring the biology of the inner-workings of living cells, and how they interact to reproduce complex organization from genetic determinants.

The stunning efficacy of complex adaptations that enable carbon-based life forms to capture energy and nutrients—and to store and reproduce hereditary information in each generation—is no longer a mystery. The UES goes far beyond Darwin's early, brilliant insights.

As with scientific theory in general, the UES is a social product. Most aspects of the UES are based on previously published, peer-reviewed scientific papers. Yet, despite numerous attempts to integrate these important new elements into a broader public awareness, evolutionary theory continues to be poorly understood among nonscientists. *Rethinking Evolution* consolidates these fragments and resolves needless confusion, while retaining both scientific rigor and improving accessibility for the general public.

Along with the contributions of other scientists, I add an important element of my own, which I call **Emergent Evolutionary Potential (EEP**; see Chapter 2).[b]

---

[b]EEP was inspired by a new perspective that may resolve some of the paradoxes of quantum physics. Although theories that pertain to the quantum realm are *not* directly applicable to the macroscopic realm of biological evolution, they do emphasize an expanded scientific view of reality where *potential events* play an important role. Ruth Kastner presented this new perspective on quantum physics in her crystal-clear popular book, *Understanding Our Unseen Reality* [17].

There has never been an attempt at reconciliation that is quite like the one you are reading now. *Rethinking Evolution* transcends familiar science book categories, and is intended to reach a much broader readership that includes working scientists, educators and communicators, as well as nonscientists who are unfamiliar with biology in particular or science in general.

21st century evolutionary theory is arguably the most interdisciplinary topic in the natural sciences. For this reason, *Rethinking Evolution* casts a wide net to bring a broad range of disparate knowledge together in one place.

In terms of practical tools for accessibility, few popular science books have leveraged the extraordinary potential of new digital media. Nonscientists will find that the UES is presented in plain language, where practical and useful web links in the footnotes provide more in-depth explanations.

Additionally, technical terms are defined in the glossary which is linked to public-domain Internet resources such as Wikipedia. The glossary can be found in printed versions of *Rethinking Evolution* as also in hyperlinked web format on the Internet at **rethinkingevolution.com**. For the natural sciences in general and for biology in particular, Wikipedia represents a treasure-trove of accessible knowledge. This is an excellent way for readers with little or no background in biology to quickly gain a working knowledge of the relevant concepts that are of greatest personal interest. Students, teachers, and biomedical professionals will find that the contents of this book are thoroughly grounded in empirical facts but are presented in an accessible way with an engaging storyline.

Why Me, and Why Now?

At the outset, such an unusual approach to the special challenges posed by the UES begs at least two questions: **First, why me? And second, why now?**

First, early in my career as a research biologist, I contributed some of the now widely-accepted discoveries concerning **DNA sequence variation** that call for rethinking the notion of "random mutations" that characterize the **Modern Synthesis** that was forged in the 1940s. My 1987 doctoral thesis described the major mechanism—**slipped-strand mispairing**, also known as **replication slippage**—that generates repetitive sequences and—in concert with several other well-known processes—expands both **repetitive** and **nonrepetitive DNA sequences** throughout the **tree of life**.

Those are just a tiny fraction of the stunning recent discoveries made by the scientific community concerning the inner workings of cells. They call for a major upgrade to the classical and **neo-Darwinian** perspectives.

The subtitle of this book refers to the revolution that is hiding in plain sight. Despite numerous laboratory discoveries and exciting new perspectives, the outdated conceptual framework of the Modern Synthesis continues to persist in textbook accounts of evolution, with a tenacity that might be metaphorically compared to the **QWERTY keyboard layout**, which also persisted long after it outlived its usefulness.

Second, although my personal interests in developmental biology and evolution extend back to my earliest undergraduate days at UC Berkeley (UCB), I encountered the same obstacles that encourage most research scientists to focus more on peer-reviewed papers and practical applications, and less on broad theoretical discussions and communication with the general public. More recently, a number of scientific, social, and personal factors have come together in a sort of "perfect storm" that compelled me to write this book. I will briefly mention three contributing factors:

(1) In addition to intriguing new perspectives on the sheer organizing power of EEP mentioned above, a number of important additions to Darwinian theory have come of age—such as **evolutionary developmental biology**[c] [15] and **facilitated variation**[d] [13, 122].
(2) Just as biological evolution can be self-accelerating, increasingly rapid empirical discoveries in molecular, cellular, and developmental biology have been driven by technology—and these provide numerous indisputable facts consistent with an updated and expanded view that goes beyond classical Natural Selection.

[c] https://en.wikipedia.org/wiki/Evolutionary_developmental_biology
[d] https://en.wikipedia.org/wiki/Facilitated_variation

(3) Having spent much of my earlier career laboring under the same institutional constraints that limit communication with the nonscientists, as well as theoretical exploration by practicing research biologists, I am no longer subject to those constraints. Today, I have more freedom to share many years of quiet theoretical exploration—and numerous empirical discoveries by others—that have culminated in my current perspectives.

A Note on What is Missing

Although *Rethinking Evolution* offers an unusually wide-ranging and interdisciplinary view of biological evolution, I have deliberately stopped short of delving into **evolutionary psychology** (evo-psych) and its implications for human happiness and survival. I do hope to return to that subject in the future. For now, I recommend an excellent popular book on the subject by Robert Wright, titled *The Moral Animal* [32], written in 1994, and *Our Political Nature: The Evolutionary Origins of What Divides Us*, by Avi Tuschman, first written in 2013 [129].

Why would I choose to defer an updated discussion on evo-psych, when its principles of evo-psych are clearly relevant to understanding human nature, and arguably more important than ever, given recent political and social developments? After all, evo-psych informs us not only about the power and limitations of human nature, but also offers a path towards overcoming those limitations and surviving our own worst tendencies.

I have deferred this discussion for two reasons:

(1) A solid understanding of the biological aspects of evolutionary theory makes it much easier to understand and appreciate evo-psych. *Rethinking Evolution* provides this prerequisite background.
(2) Evo-psych is a blend of social science and natural science, whereas a discussion of the UES should focus on objective facts that are already well-supported by empirical evidence and basic principles of the natural sciences.

The human brain is perhaps the best example of an exquisitely powerful and complex adaptation with negative tendencies. That is why evo-psych is so important, but any conclusions that we draw must be thoroughly grounded in empirical research and critical thinking. I do remain convinced that evolution has provided us with the means to overcome our own limitations. The prerequisite to making the right choices is understanding our own origins, while recognizing our potential to transform future probabilities.

Acknowledgments

Many thanks to my daughter Meghan Levinson for helpful suggestions that improved the clarity of an early draft of the manuscript.

Thanks also to Ruth Kastner for both inspiration and encouragement to write this book.

I am also grateful to the Wikipedia community of volunteers and contributors who together have created the world's most comprehensive, reliable, and accessible online encyclopedia, and made this freely available, which is freely available to all. Hundreds of Wikipedia links have added both depth and breadth to the wide-ranging subject matter of *Rethinking Evolution.*

I also thank Beth Burnside, my undergraduate mentor, who provided an early example and high expectations for attention to detail and diligence in laboratory research, George Gutman, my graduate mentor, who took me under his wing and provided so many excellent suggestions for a project that was outside of his main area of research, and my wife Patrice Levinson, who provided moral support throughout my graduate years and beyond, and also made several improvements in the current manuscript.

About the Author

Dr. Gene Levinson is a distinguished research scientist, software inventor and educator. With early interests in developmental and evolutionary biology as an undergraduate at UC Berkeley, he went on to discover a fundamental mechanism for genome expansion as a graduate student at UC Irvine. As a postdoctoral researcher at Harvard University he tutored undergraduate students in biology and developed an ongoing passion for improving science education. By combining his broad hands-on experience in biomedical research, science education and digital communication, Dr. Levinson is currently working to help close the gap between recent empirical advances in biology and a modernized understanding of evolution among the general public, which he calls the Updated Evolutionary Synthesis (UES).

Contents

Preface vii

Acknowledgments xiii

About the Author xv

How to Use Rethinking Evolution xix

PART 1 The Updated Evolutionary Synthesis (UES) 1

Chapter 1 Overview 3
Chapter 2 Emergent Evolutionary Potential 13
Chapter 3 Complex Interplay Between Empirical Discoveries and Conceptual Frameworks 27
Chapter 4 Timeless Molecular Innovations 49
Chapter 5 The Origin of Life 79
Chapter 6 Combining the Origin of Life with the Origins of Complexity 111

PART 2 From Genes to Complex Organization 121

Chapter 7 Levels of Organization That Transcend Species 123
Chapter 8 From 19th Century Natural Selection to 20th Century Mendelian Genetics 153
Chapter 9 From the "Modern Synthesis" to Molecular Genetics: What Was Missing? 173

Chapter 10	The Origins of Multicellularity and Embryonic Development	195
Chapter 11	Embryos Are Not Computers, and DNA Is Not Software	221
Chapter 12	Shape-Specific Molecular Interaction and Binding Events (SSM-IBE)	239
Chapter 13	The Genetic Algorithm (GA): Blind Production of the Combinatorial Phenotype	255
Chapter 14	The Dual Character of Complex Adaptations	267
Chapter 15	The Creative Forces Behind Evolution, from Darwin to Evo-Devo	275
Chapter 16	Repetitive Sequences, Gene Duplication, and the Varieties of Genomic Variation	301
Chapter 17	Biological Principles of the Updated Evolutionary Synthesis	321

References 333

Glossary 343

About the Cover Photograph 457

Index 459

How to Use *Rethinking Evolution*

We are currently witnessing a growing awareness among biologists that some of the foundational features of the reigning evolutionary paradigm, in particular its genetic determinism and adaptationism, require substantial revision and need to be complemented by other concepts and theories.

—Werner Callebaut, Gerd B. Müller, and Stuart A. Newman, 2007[e]

Introduction

The **Updated Evolutionary Synthesis** (**UES**) is a revolution that's been hiding in plain sight. This book presents a comprehensive 21st century update that is written with nonscientists, students, and biomedical professionals in mind. UES offers a 21st century theory of evolution that is thoroughly grounded in widely-accepted principles of molecular, cellular and developmental biology.

On balance, the core framework of Darwinian Natural Selection is incomplete. The enduring central concepts of the struggle for existence

[e] https://www.researchgate.net/profile/Gerd_Mueller/publication/286191359_The_organismic_systems_approach_Streamlining_the_naturalistic_agenda/links/56c61c4c08ae0d3b1b602329/The-organismic-systems-approach-Streamlining-the-naturalistic-agenda.pdf

and the roles of **variation** and **selection** should be updated with modern biological discoveries and principles.

The UES represents an interdisciplinary synthesis supported by literally thousands of empirical discoveries. Many crucial insights have only become available in recent years. Most of the materials in this book were made possible by the persistence and creativity of both laboratory and field biologists throughout the world. Several excellent popular accounts are referenced in this book, as well as numerous scientific publications. But *Rethinking Evolution* weaves these elements together into a unique conceptual fabric.

In many cases, the true stories of the inner-workings of nature capture the imagination in ways that are stranger and more wonderful than science fiction, yet thoroughly grounded in scientific facts. This book is sure to stimulate numerous thoughtful discussions as well as constructive debate.

Tools for accessibility and added value

Several practical tools have been provided to make this book more accessible to a broad readership of nonscientists, students, educators, and professional working scientists:

(1) Numbered references include articles and books from the scientific literature as well as peer-reviewed articles.
(2) Footnotes include supplemental materials from the Internet as well as Wikipedia articles.
(3) A list of important concepts and principles of the UES is provided at the end of the book (Chapter 17).
(4) Each chapter begins with a *Big Picture* overview of major themes, conclusions, and take-home messages.
(5) Each chapter is cross-referenced to a web-linked glossary based on Wikipedia articles and extracts. These articles extend both the depth and the breadth of the book, and readers are urged to look up at least some of these articles. The glossary can be found in printed versions of *Rethinking Evolution* as also in hyperlinked web format on the Internet at *rethinkingevolution.com*. The intention is to allow readers

unfamiliar with any aspect of biology to quickly get up to speed on the relevant topics, without disrupting the flow of the narrative.[f]

(6) Although chapters follow a deliberate progression, each is also meant to stand alone. Within each chapter, the first mention of each glossary term is highlighted with bold type.

(7) In addition to the Wikipedia-enabled glossary, several aspects of basic biology have been illustrated with images that are freely available on the Internet, thanks to the Wikimedia project.[g]

Numerous verbatim quotations and footnotes reference fascinating publications that are readily accessible on the web, yet are easily overlooked without a curated guide. Some provide historical context and help to explain the limitations of earlier evolutionary concepts as well as how to transcend them.

---

[f]Wikipedia articles in the natural sciences are a treasure-trove of accessible and reliable information that is underutilized. For readers who remain skeptical about the added value or reliability of Wikipedia, I challenge you to follow the links and decide for yourselves. Wikipedia is a curated, highly collaborative effort that provides a more comprehensive, up-to-date, and reliably vetted resource than any other encyclopedia that is out there. The extensive use of Wikipedia links in both the footnotes and the glossary is a deliberate feature of *Rethinking Evolution*—not a bug.

[g]https://www.wikimedia.org/

PART 1

The Updated Evolutionary Synthesis (UES)

Animals diverge from common ancestors through changes in their DNA. The major question, then, is, Which changes in DNA account for morphological diversity? The answers have eluded us for the half-century since the formulation of the "Modern Synthesis" and the discovery of the structure of DNA.

—Sean B. Carroll, Jennifer K. Grenier, and Scott D. Weatherbee [1]

Chapter 1
Overview

A Natural Explanation for Adaptations

Biological evolution routinely generates structures and functions that make it possible for organisms to survive and reproduce their own kind, in virtually every imaginable environmental circumstance and way of life. We refer to these diverse solutions to the struggle for existence as *niches*.

The close match and efficacy of structure or function to specific *niches*—the torpedo shapes of fishes, birds, and marine mammals, for example—appears, to the naïve human mind, to be the product of goal-seeking behavior, planning, and design. For thousands of years, humans attributed this to acts of creation by supernatural and all-powerful Gods. After all, as young children, we are taught that this is the way that human creativity works.

But Charles Darwin explained why this justification is false. He proposed that an evolutionary process, which he called **Natural Selection**, acts as a creative force. Diversity (Figure 1) arises by means of a natural process. However, no goal-seeking, planning or design are required for this process to function. In other words, a **teleological**[a] account would not suffice.

In all living things, there is a close relationship between structure and function. The natural, evolutionary process leading to the close match and efficacy of structure and function to specific *niches* is called **adaptation**, and the resulting characteristics of each species are called **adaptations**.[b]

---

[a] https://en.wikipedia.org/wiki/Teleology

[b] https://en.wikipedia.org/wiki/Adaptation

Figure 1. Photos illustrating a small sample of the extraordinary range of animal diversity. The complex structures and functions of animals, which are reproduced during development, equip them with a variety of ways to capture energy and nutrients and successfully survive long enough to reproduce their own kind, in a variety of environments and ways of life (*niches*).

Source: Page-link: https://commons.wikimedia.org/wiki/File:Animal_diversity.png. File-link: https://upload.wikimedia.org/wikipedia/commons/1/14/Animal_diversity.png. Attribution: By creator of composite, User: Medeis; original authors, as credited individually above: User: Bkmiles, User: Panda3, Hans Hillewaert (Lycaon), User: Anilocra, Rob Hanson from Welland, Ontario, Canada, User: Nhobgood Nick Hobgood, Richard Ling, A. Slotwinski as credited by uploader user: Australianplankton, User: Lviatour, User: Kevincollins123, S. Taheri, edited by Fir0002, User: Opoterser, DBCLS, TheAlphaWolf, Philippe Guillaume. Wikimedia commons.

Evolution as fact and theory

This section takes its title from enlightening discussions by Stephen J. Gould[c] [2, 3] as well as Wikipedia.[d] As with many of the footnotes and glossary entries in *Rethinking Evolution*, they provide background for nonscientists as well as greater depth and breadth of the subject matter, and are available online.

As pointed out in the Wikipedia discussion, evidence for evolution continues to be accumulated and tested, but people often talk past each other when they weigh in on the subject of whether evolution is fact or theory. Gould brings some clarity to the subject as follows:

> Well, evolution *is* a theory. It is also a fact. And facts and theories are different things, not rungs in a hierarchy of increasing certainty…
>
> The basic attack of modern creationists falls apart on two general counts before we even reach the supposed factual details of their assault against evolution. First, they play upon a vernacular misunderstanding of the word "theory" to convey the false impression that we evolutionists are covering up the rotten core of our edifice. Second, they misuse a popular philosophy of science to argue that they are behaving scientifically in attacking evolution. Yet the same philosophy demonstrates that their own belief is not science…an example of what Orwell called "newspeak".

The connection between evolution and development

In each generation and in each individual, DNA sequences act as determinants that generate the complex organization that has evolved. In this context, complex organization is a catch-all phrase that describes the astonishing range of levels of complexity found within and between all living organisms.

Complex organization emerges by a variety of mechanisms that are collectively referred to as the **development** of the organism. The interdisciplinary field describing these mechanisms is known as **developmental biology**. In other words, biological structure and function of each

---

[c] http://www.stephenjaygould.org/library/gould_fact-and-theory.html

[d] https://en.wikipedia.org/wiki/Evolution_as_fact_and_theory

individual is reproduced by virtue of a developmental process that is driven by **genetic determinants**. These genetic determinants are themselves the product of evolution. This critical developmental link between Natural Selection and biological complexity was mostly absent from Darwin's classical theory of Natural Selection. There was a good reason for that, of course: the inner-workings of cells, as well as their developmental interactions, were unknown to 19th century science.

Today, an **Updated Evolutionary Synthesis** (**UES**) must include, among other things, a deep understanding of development, because without development, hereditary information alone has little meaning or significance for cellular and organismal survival.

Today, there is a synergistic relationship between evolutionary theory and developmental biology that provides major new insights into the dual origins of biological organization. This is referred to as **Evolutionary Developmental Biology** (or **Evo-Devo** for short).

Shifting the focus from origin of species to origin of complex organization

The extraordinary organizing power of Natural Selection and efficacy of diverse adaptations is staggering. While diverse species have the *appearance* of planning or intelligent design, the naturalistic science of evolution clearly states that they are *not* planned or designed, but rather, that they arise by blind, natural forces and processes.

Recent advances in biology that expand our scientific understanding of evolution are largely unknown by the general public. If successful, the UES proposed here could go a long way towards providing a more plausible, natural explanation for the deep history of life—including the origins of our own human species—than ever before.

Today, thanks to literally thousands of dedicated biologists, there is no shortage of empirical data—that is, real-world observations and experiments that are verified and reproducible. The factual, concrete evidence for Natural Selection is far stronger than Darwin could have possibly imagined.

There is no supernatural component to biological evolution—it is a *natural* process. For many people, this is quite difficult to accept—particularly when considering our own human origins. It would seem that despite our mental capacity for rational thought, the scientific method, and acceptance of a naturalistic philosophy, there is a module in the human brain that is "wired" to prefer a religious explanation for human origins.

Numerous popular science writers—such as Carl Sagan, Richard Dawkins, Stephen J. Gould, and Ernst Mayr—have written many books [4–9, 123] that are both captivating and accessible to the general public. These books help bridge the gap between the *appearance* of design or planning, and the reality of what has historically taken place in the natural world.

Natural history increases public awareness of evolution

It is entirely fitting that Great Britain, with its rich tradition of Natural History, and birthplace and home to Charles Darwin, continues to educate and entertain the entire world with the Natural History Unit of its public broadcasting service, the BBC.[e] This is the largest wildlife documentary production house in the world. Firmly based in well-documented and scientific evolutionary theory, these wildlife documentaries are one of the best ways for the general public to gain a clear understanding of evolutionary science—through dazzling illustrations and concrete examples from real-life. Sir David Attenborough is the most prolific and well-known narrator and contributor for these extraordinary production.[f]

Charles Darwin was keenly aware of numerous examples in nature that strongly support his theory of Natural Selection. Among the most striking were examples of **sexual selection**,[g] where not only biological structures but also behaviors—such as elaborate male courtship rituals among birds to attract a mate—underscore the central importance of reproductive success as a creative force driving evolution.

The nuances of indirect empirical evidence: Has life evolved elsewhere in the universe?

As professional biologists learn more and more about the inner-workings of living cells, the gap between our expanding knowledge-base and the awareness of the general public continues to grow. For this reason, I have

---

[e] https://en.wikipedia.org/wiki/BBC_Natural_History_Unit
[f] https://en.wikipedia.org/wiki/David_Attenborough
[g] https://en.wikipedia.org/wiki/Sexual_selection

attempted to maintain a balance between a popular writing style that is accessible to a broad range of readers, while at the same time writing about information that remains thoroughly grounded in widely-accepted scientific facts.

I use the word "empirical" extensively throughout this book, because recently discovered **empirical data**—obtained by observation or experimentation[h]—plays a major role in the UES. Since most of the actual evolutionary events that we study have occurred in the distant past, many forms of empirical evidence in evolutionary biology are *indirect*.

The *nuances* of indirect empirical evidence are a common cause of confusion—particularly among people who lack formal scientific training—because they are based on rational modes of thought that take time to process.

People are wired to be reactive, and by default tend to act more on the basis of direct observation than on logic and inference. This is an understandable product of evolution, since the early survival of our human ancestors depended on the ability to quickly react to opportunities and potential threats that could be directly observed in real-time. Phrases like "show me" or "seeing is believing" reflect this predisposition in the human species.

In his introduction to *The Tangled Tree: A Radical New History of Life* [10], David Quammen states that as far as we know, life is a peculiar phenomenon that is confined to the planet Earth, and "there is zero evidence to the contrary." Presumably, what he really means is that as of August 2018, when his book was released, we had no *directly observable* evidence for the existence of life on any other planet.

Quammen goes on to say that life elsewhere in the universe is a *guess*, whereas life on Earth is a *fact*. The problem with that statement is that indirect scientific evidence is not mere guesswork. Quammen concludes that to the best of our knowledge, the evolution of life has occurred only one time in the entire universe. But that is an oversimplification. "To the best of our knowledge" lacks precision, and implies more than the literal meaning of the words. A better phrasing would be

[h] https://en.wikipedia.org/wiki/Empirical_evidence

to say it is that "we do not know for certain whether life exists elsewhere or not, because up to now, we have not been able to observe it directly."

But Nick Lane [11] and others argue that there are, in fact, sound scientific reasons to hypothesize that it is highly probable that carbon-based life forms that rely on **proton gradients** for their energy have probably evolved on numerous occasions throughout the universe, as discussed in Chapters 5 and 6.

The relevant point in the current section is that there are many kinds of empirical evidence, including indirect evidence and logical inference from well-established scientific principles. The best hypotheses, of course, are those that lead to testable predictions. Well-established principles of chemistry and physics are applicable throughout the universe, as are well-established principles of organic chemistry and biochemistry that rest on that firm foundation. Constraints that govern interactions between organic molecules should apply to any life forms that might exist elsewhere.

Some major themes of the UES

As powerful innovations evolve, they lay the groundwork for more powerful innovations to evolve in the future.

Since Natural Selection preserves innovations that are useful in the struggle for existence, some of those innovations will turn out to be tools that generate complex organization during the development of the individual.

Some of those tools may also act as facilitators that increase **evolvability**. Our presuppositions of how complexity can arise fail to appreciate the subtle ways that new structure and function can emerge when molecular entities are able to bind and interact in new ways.

Many of the real events that take place in DNA are not simply "random mutations", but rather have the capacity to generate future changes with extraordinary organizing power.

Our presuppositions of the probability of events that seem independent fail to grasp the subtle ways that actual events—that is, events that take place in a particular location and at a particular time—shift the probabilities and opportunities for *potential* events that may or may not take place in the future. The logical relationships between *actual* events and *potential* future events represent aspects of reality that transcend space and time. This is discussed in some detail in the context of Emergent Evolutionary Potential (EEP) in Chapter 2.

Some distinguishing features of the UES

The UES retains certain core elements of **classical Darwinian theory**—such as Darwin's insight concerning the preservation of innovations that facilitate survival and reproduction—while rethinking other aspects—such as gradualism, random variation, and a number of pernicious conceptual and semantic limitations from the past.

The UES goes beyond the so-called **Modern Synthesis** and **neo-Darwinism** to incorporate more recent contributions. Many of these new concepts were made possible by recent empirical discoveries that provide new insights into the creative power of the natural forces responsible for numerous innovations in the deep history of life.

The new contributions have been variously labeled as **Self-Organization, Evolutionary Developmental Biology (or Evo-Devo), Facilitated Variation, Organismic Systems Approach, Evolvability, and the Extended, Expanded, and Integrated Evolutionary Synthesis**. This confusing proliferation of labels will be simplified and consolidated in a more unified and accessible conceptual framework in the UES.

Recently, a variety of popular books on evolution have added several of these important new concepts that go beyond classical Darwinism or neo-Darwinism. Relevant examples include *The Making of a Fly, The Plausibility of Life*, *The Music of Life* and *Endless Forms Most Beautiful* [12–15].

In addition, I will be presenting some new concepts of my own. Perhaps the most interesting of these is what I call **Emergent Evolutionary Potential** (**EEP**), which redefines "reality" to include new opportunities that naturally arise during the evolutionary process. The classical concept of Darwinian Natural Selection—which is based on incremental change — simply cannot do justice to the varieties of new structures and functions that arise when formerly separate entities come together and interact in useful ways.

What is needed is an updated scientific explanation for how a completely natural process can give rise to the extraordinary diversity, complexity, and efficacy that we observe throughout the history of life on Earth. This explanation should incorporate a broad range of recent empirical discoveries in molecular, cellular, and developmental biology. It should also address several new theoretical elements that are candidates for the UES.

The resulting synthesis is a more credible and updated evolutionary theory that is consistent with our 21st century knowledge.

Chapter 2

Emergent Evolutionary Potential

We might plan to meet tomorrow for coffee at the Downtown Coffee Shop. But suppose that, unbeknownst to us, while we are making these plans, the coffee shop (actually) closes. Instantaneously and acausally, it is no longer possible for us... to have coffee at the Downtown Coffee Shop tomorrow. What is possible has been globally and acausally altered by a new actual...We simply allow that actual events can instantaneously and acausally affect what is next possible... which, in turn, influences what can next become actual, and so on.

—from *Taking Heisenberg's Potentia Seriously* [16]

The Big Picture

The concept of emergent evolutionary potential (EEP) combines scientific perspectives of emergence and potential in the context of biological evolution (BE).

Concrete examples of biological emergence go a long way towards explaining the sheer organizing power of molecular interactions within living cells. At a variety of levels of complexity, new opportunities arise from previous innovations—particularly when formerly nonexistent or separate entities are brought together in new contexts.

This new perspective updates evolutionary theory and helps to explain the origins of diverse adaptations—not only how it arises, but also how it is reproduced during development, in each generation and in every individual.

Introducing Emergent Evolutionary Potential

Considering the scope of the perceptual and conceptual shifts required to embrace it, I'm not surprised that **EEP** has not been previously integrated with Darwinian theory.

EEP was inspired in part by new perspectives in quantum physics. In the **submicroscopic realm** of the quantum, despite decades of attempts by brilliant theoreticians and philosophers, they have failed to resolve certain paradoxes in quantum mechanics, such as nonlocality, entanglement, and instantaneous collapse. I propose that even in the **macroscopic realm** of **biological organization**, where molecular and cellular interactions can be understood in classical terms, it is not possible to fully appreciate how complex organization can evolve without taking two distinct aspects of reality into account. These two levels of reality represent a central concept of EEP, even though quantum theory does not formally apply.

Returning to the submicroscopic realm governed by the laws of quantum physics, Ruth Kastner, Stuart Kauffman and Michael Epperson have written a remarkable theoretical paper [16], and Kastner has written a popular book [17], describing a novel perspective that is thoroughly grounded in logic and science. Briefly, they argue that quantum paradoxes may be resolved if we allow that reality has a dual aspect, in which *potential* events represent a hidden aspect of reality in the quantum realm that has previously been overlooked. Unlike *actual* events that have already occurred, *potential* is an aspect of reality that exists "outside of" space and time. It is therefore not bound by the same logical requirements as actual events.

Nonlocality, entanglement, and instantaneous collapse—paradoxical elements of quantum theory that would logically apply to actual events—do not pose a problem if we assume that there is a *potential* aspect of reality that precedes those actual events. This theory is known as the **Transactional Interpretation** (**TI**) or ***res potentia***. In this view, *potential* events have a real existence, but ordinary concepts of space and time do not apply.

Potential is, however, constantly changing as a consequence of the more familiar, visible events that do take place. These events often alter the potential for what may, or may not, actually occur *next*.

This concept—that actual events often alter the potential for what may occur next—has important implications for evolutionary theory. Although the *TI* and *res potentia* pertain specifically to the quantum realm, this leads me to question whether a similar perceptual shift in the macroscopic realm might help us to better understand the surprising organizing power of Natural Selection in biological evolution (BE). This would require a significant expansion and update of classical Darwinian theory, which I call EEP.

How does potential relate to actual events that take place in the real world?

The word *potential* already has a familiar meaning for the general public as well as scientists. For both groups, the concept of potential is closely allied with the familiar definition of *reality*. Reality can be defined as the state of things as they actually are. This state results from numerous prior events—actual events that have actually taken place in space and time. We can refer to the entire network of interacting objects in the universe as a **gestalt**. According to Wikipedia, a gestalt is:

> …a German word for form or shape [that] may refer to: Holism, the idea that natural systems and their properties should be viewed as wholes, not as collections of parts.[a]

Potential has several important aspects and theoretical ramifications. To begin with, potential refers to events that are possible, but have not actually taken place. If and when potential leads to one (or more) specific events, they become actual events. This does not depart from our familiar view of reality: only those events that are possible, can actually occur. Events must have the *potential to occur* in order to *actually* occur.

Yet unlike *actual* events that always occur at particular times and places, *potential* includes logical relationships between entities that are real, yet transcend—and do not fit into the conceptual framework of—space and time.

[a]https://en.wikipedia.org/wiki/Gestalt

In a popular science article summarizing the recent work in quantum physics by Kastner, Kauffman and Epperson, Tom Siegfried paraphrases a key metaphor used by the authors:

> Consider for instance that you and I agree to meet for lunch next Tuesday at the Mad Hatter restaurant (Kastner and colleagues use the example of a coffee shop, but I don't like coffee). But then on Monday, a tornado blasts the Mad Hatter to Wonderland. Meeting there is no longer on the list of *res potentia*; it's no longer possible for lunch there to become an actuality. In other words, even though an actuality can't alter a distant actuality, it can change distant potential. We could have been a thousand miles away, yet the tornado changed our possibilities for places to eat.[b]

This is a sort of thought experiment that illustrates two key aspects of potential (*res potentia*) in familiar terms:

(1) Actual events change the range of possibilities for future events;
(2) Actual events can change distant potential—that is, potential that has no direct links to local or recent events.

Future possibilities—such as a plan to meet at a restaurant or coffee shop—are obviously dependent on the continued existence of that meeting place. It is not possible to meet somewhere that no longer exists. There is a logical relationship between the plan to meet and the existence of the meeting place. If that meeting place no longer exists, the plan must be changed. However, there is no direct relationship—in space and time—between the plan that was made, and the existence of the meeting place. This aspect of reality is a logical one. Nevertheless, the tornado—an actual event that definitely took place in space and time—transformed that logical relationship, and therefore changed the *potential* for future events. Certain events can no longer take place, and other possible events—such as meeting at a new location and/or at a different time, once new plans are made—now become probable. Future possibilities and probabilities have been altered by previous events, but those changes transcend space and time.

---

[b] https://www.sciencenews.org/blog/context/quantum-mysteries-dissolve-if-possibilities-are-realities

Potential can exert a powerful effect on the future. This means that potential must be in some sense real—since it can definitely influence future realities—but that reality transcends space and time.

EEP argues that similar logical relationships also exist in the macroscopic realm of BE. For example, all cells use a **genetic code** that stores information about **protein sequences** in **DNA**. This is a logical relationship that transcends space and time.

How does the word "potential" fit into common scientific usage in physics or chemistry?

If I lift a pencil above a table, a physicist might say that the pencil has gained potential energy. That potential energy was transferred to the pencil when my muscles generated a force that opposed the pull of gravity.

A chemist might refer to a strong (covalent) chemical bond—such as the covalent bond between two hydrogen atoms in hydrogen gas (H_2)—as the optimal distance between atoms where *potential energy* is minimized, when the two **atomic nuclei** share a pair of **electrons** in a **hybrid orbital.** When this bond is made, energy is released, usually in the form of heat.

These physical and chemical definitions of potential energy are widely accepted. Potential energy is *real*, not imagined. It depends on the relative positions and properties of interacting objects. Potential energy represents the potential to do work, or to drive specific types of physical or chemical events. It does not assure, however, that actual work will be done, or that actual events will take place.

How does the word "potential" relate to biological evolution?

Although potential energy is a familiar concept in physics or chemistry, it is not commonly used in the context of **BE**. According to Wikipedia, BE can be defined as a "change in the heritable characteristics of biological populations over successive generations".[c] Since *change* refers to one or

---

[c] https://en.wikipedia.org/wiki/Evolution

more *actual* events, we can simplify matters by defining **potential biological evolution** (**PBE**) as the *potential* change in the heritable characteristics of biological populations over successive generations.

As is the case with *potential energy*, we can assert that PBE is *real*, not imagined, and that PBE depends on the relative positions and properties of interacting objects. Furthermore, PBE represents the potential to evolve, or to drive specific types of evolutionary events. It does not assure, however, that actual evolutionary events will take place.

What's the meaning and significance of the word "emergent" in EEP?

Unlike *potential*, the word *emergent* (and *emergence*) has some unfortunate fictitious connotations. However, for philosophers—including those who are quite comfortable with naturalism[d]—emergence is a down-to-Earth, real-world phenomenon that can be described and explained scientifically. According to Wikipedia:

> Emergence is a phenomenon whereby larger entities arise through interactions among smaller or simpler entities such that the larger entities exhibit properties the smaller/simpler entities do not exhibit.[e]

This simple definition applies, for example, to the formation of **water molecules**: when the covalent bonds between atoms in hydrogen and oxygen molecules break apart and the hydrogen and oxygen atoms form new covalent bonds, the resulting water molecules exhibit a broad range of emergent properties that are critical for life and BE. These properties are **emergent** simply because they are not exhibited by the smaller/simpler hydrogen or oxygen molecules.

To make this discussion both concrete and relevant to the discussion at hand, let's consider just a few of the emergent properties of water that are critical for the origin and evolution of life. Many of these emergent properties arise because water molecules, unlike hydrogen and oxygen

---

[d] https://en.wikipedia.org/wiki/Naturalism

[e] https://en.wikipedia.org/wiki/Emergence

molecules, are **polar molecules**, meaning that the **negative charges** of the outermost electrons of the molecules are not evenly distributed.

Molecules are polar when they meet two requirements: (1) they contain **polar bonds**, and (2) those polar bonds are asymmetrical.

In each water molecule, the hydrogen–oxygen bonds are polar because the electrons are shared unequally. That is simply because the larger, more positive oxygen atom nucleus attracts the negatively charged shared electrons more strongly than the smaller hydrogen atom nucleus. Oxygen has eight positively charged protons in the nucleus, but hydrogen has only one.

The polar bonds in each water molecule are asymmetrical, because the outer shell of the oxygen atom has two nonbonding pairs of electrons (**lone pairs**) that have larger **electron clouds** that repel the bonding pairs.[f]

Each asymmetrical polar bond has a negatively charged end and a positively charged end. Therefore, each polar bond creates a **dipole**. When two water molecules come close to each other, the negative end of the dipole in one water molecule will be attracted to the positive end of the dipole in another. Consequently, each water molecule will be attracted to four other water molecules. These **dipole–dipole attractions** between water molecules are unusually strong and are referred to as **hydrogen bonding**. Don't be confused by the term—hydrogen bonding simply means that hydrogen atoms are partially responsible for these **intermolecular binding forces**.

Dipole–dipole interactions between water molecules (hydrogen bonding) is the reason that water molecules—unlike hydrogen or oxygen—exist in the liquid state of matter under standard conditions of temperature and pressure.

The liquid state of water is an example of an emergent property that can be readily explained in terms of well-established scientific principles of chemistry and physics—familiar topics frequently included in introductory high school chemistry courses.

Life as we know it depends on this emergent property. Most of the inner-workings of cells require the presence of water in its liquid state.

---

[f] https://en.wikipedia.org/wiki/Lone_pair

Still other emergent properties of water are also requirements for life. At colder temperatures, hydrogen bonding causes water molecules to solidify in crystals that are less dense than liquid water. Consequently, ice floats on top of water. If that were not the case, the oceans would have frozen long ago, from the bottom up, and the evolution of fishes and marine invertebrates—and probably, the earliest origins of precellular carbon-based life on Earth (see Chapter 5) would never have taken place.

Thanks to hydrogen bonding, liquid water has a greater capacity to absorb heat than the surrounding air. When heat is absorbed, hydrogen bonds are broken and water molecules can move more freely. When the temperature of water decreases, hydrogen bonds are formed and release a considerable amount of heat. Consequently, the climate of coastal regions tends to be less extreme, because the oceans act as a heat buffer that mediates temperature extremes.

As a gas, because of dipole–dipole forces, molecules of water vapor can readily form fog or clouds of suspended liquid water droplets, which routinely redistribute water to various terrestrial environments over planet Earth in the form of rain or snow. This provides the water required for both plant and animal life on land.

Since the cellulose molecules of plants also have asymmetrical hydrogen–oxygen bonds, hydrogen bonding also causes water and cellulose to attract. Due to these attractive forces, water will travel up from the roots to the stems of plants, by **capillary action**. If that were not true, terrestrial plants that depend on roots, stems and leaves could never have evolved.

Water is commonly referred to as a "universal solvent" because a broad range of **ionic** (**charged**) and **polar** substances will readily dissolve in liquid water, due to ion-dipole and dipole–dipole attractions. Both smaller and larger molecules of biological interest—such as **proteins**, **sugars**, **DNA**, and **RNA**, readily dissolve in water. Water facilitates numerous chemical reactions, including reactions that are catalyzed by **enzymes**.

Phospholipids—the fatty molecules that form **biological membranes**—have both **polar** and **nonpolar** portions, and consequently, they spontaneously form **lipid bilayers** that are **impermeable** to water—an essential requirement for the existence of living cells.

The emergent properties of water are so important for life that astrobiologists—scientists who search for the possibility of life on other planets—usually focus only on planets where water is likely to exist or has existed in the past.

What's the relationship between potential and emergence?

The concept of EEP applies to the macroscopic realm—that is, the more familiar realm where we can observe actual events that take place. The real world represents a **gestalt**, if only because numerous interactions take place between actual objects. In the macroscopic realm where interactions take place, every imaginable actual event that can be observed—at least in theory—involves interactions between two or more objects. If a tree falls in the forest and there is no one around to hear it, it *does* make a sound, because it does create compression waves—vibrations of the surrounding air. Those events do actually occur first, *before* the sound strikes an eardrum. Of course, it is quite easy for armchair philosophers to quibble about the meaning of the word *sound*, but that defeats the purpose of understanding the nature of reality. The naturalistic, scientific view considers the events that have taken place, as well as the potential for those events to take place.

We could also ask whether the tree had the potential to fall, and in most imaginable circumstances, gravity assures that it is possible. Yet in the macroscopic realm, potential depends not only on the arrangement of nearby objects that can directly interact, but also on more indirect logical possibilities.

Emergent Properties Create Complex Levels of Biological Structure and Function

What's the connection between potential and emergence in biological evolution?

In principle, the reasons for the emergence of larger **entities** from smaller ones in living cells, organisms, or ecological interactions is no different than the emergence of water: the larger entities arise through interactions between smaller or simpler entities, and they exhibit properties not seen in the smaller/simpler parts. Unlike water, however—which arises from the interactions between two atoms of hydrogen and one atom of oxygen—emergent biological entities are usually far more complex—they involve interactions between larger numbers of entities.

One of the striking features of emergent biological structure and function is that it consists of numerous levels of organization—that is, emergent entities consisting of larger and larger numbers of smaller entities. If we arbitrarily define the interacting entities as **subunits** and the emergent entities as **units**, then virtually every subunit has its own emergent properties—and therefore represents a structural and functional unit—and at the same time, every unit consists of interacting subunits. Arthur Koestler captured this concept by coining a new word—**holon**.[g] Biologists, like all empirical scientists, perform observations and experiments to obtain better and better approximations of reality. Unlike the physical sciences, biological structure and function usually requires qualitative, rather than quantitative, descriptions. The concepts of emergence, interaction, units and subunits are all useful tools that allow us to describe, with a great deal of detail and precision, the various emergent entities—that is, structures—whose emergent properties include various *functions*.

Biological function has meaning and significance that is closely related to evolutionary theory, because functions represent the various ways through which living things are able to capture energy and nutrients, perform various molecular, cellular and developmental tasks, and reproduce their own kind, within a particular biological *niche*. Most biological functions are the product of Natural Selection.

Is Emergence a Classical Darwinian Phenomenon?

Charles Darwin did not focus on emergence in his theory of Natural Selection, because our understanding of biological emergence depends on numerous empirical discoveries—especially about the formerly hidden realm of living cells—that have not emerged until more recently.

---

[g] https://en.wikipedia.org/wiki/Holon_(philosophy)

Darwin's classical theory of Natural Selection *implicitly* acknowledged the role of *potential* in a limited sense, recognizing that in order for selection to act to preserve useful variations, those variations must have the potential to occur in the first place. Yet Darwin could only speculate about the laws and causes of variation. Today, however—again by virtue of empirical biological discoveries—we understand a great deal about the causes and laws of variation. Unfortunately, however, evolutionary theory remains fragmented. Additionally, despite many noble efforts to incorporate this broad range of knowledge into an updated evolutionary theory, bridging the gap between scientific knowledge and the understanding of the general public has thus far not been successful. Hopefully, this book will represent the beginning of change in our ability to communicate evolutionary theory to the public.

Returning to classical Darwinian theory, it is important to acknowledge that the concept of gradual accumulation of infinitesimal variations represents a limited perspective that fails to take emergence into account. Emergence represents an important source of variation. Often, emergent properties arise when a new, larger entity is created through interactions between smaller subunits. These new properties do not represent infinitesimal changes, but rather, observable and emergent changes that are qualitative in nature. Emergence creates *new* levels of structure and function and, more importantly, creates the *potential* for new levels of complexity to evolve by a variety of mechanisms, including, but not restricted to, subsequent infinitesimal variations.

Since many of those emergent properties involve the inner-workings of cells that were a "black box" to Darwin and his contemporaries, emergence did not play a significant role in classical Darwinian theory. However, as I will explore later in the book, attempts to discredit Darwinian theory with pseudoscientific ideas such as "irreducible complexity" in popular books such as *Darwin's Black Box* [18], have already been discredited by serious scientists and communicators.[h]

Although emergence can lead to sudden changes that have extraordinary future potential, emergence is often followed by long periods in which numerous infinitesimal variations in the larger functional units lead

---

[h] https://en.wikipedia.org/wiki/Irreducible_complexity

to slight modifications of structure and function. Some of those infinitesimal variations will be selected because they enhance the survival and/or reproductive capacity of those individuals that inherit them.

In other words, even though emergent units of structure and function can arise suddenly, when smaller subunits come together and interact—which is different than the classical Darwinian concept of infinitesimal variation—those same emergent units can further evolve by means of infinitesimal changes which come later. Emergence and infinitesimal variation are complementary and have greater organizing power when combined.

What's the relationship between potential and emergence in the context of EEP?

Potential is constantly changing during the evolutionary process, as a consequence of actual events. Whenever new structures and functions arise, they create opportunities for new interactions to take place. This, in turn, can lead to the emergence of new functional units—structures and functions that would not be possible if the stage had not been set by prior actual events.

This concept is entirely consistent with the idea expressed in the excerpt at the beginning of this chapter:

> ...actual events can instantaneously and acausally affect what is next possible... which, in turn, influences what can next become actual, and so on [16].

EEP is constantly changing throughout the history of each lineage of organisms, and extraordinary *new* opportunities for natural innovations become possible as a consequence of previous innovations.

As is readily confirmed by exploring the various levels of organization in living cells, organisms, and ecosystems, most evolutionary innovations arise from opportunities for new levels of interaction that were made possible by previous innovations. A common definition of evolution is "the gradual development of something, especially from a simple to a more complex form".[i] If increased complexity arises solely by means of

---

[i]https://www.google.com/search?q=evolution+definition

natural forces, then what is the source of the *organizing power* of the evolutionary process?

According to Darwin's brilliant insight—which has been echoed by proponents of the "Modern Synthesis" and neo-Darwinism—selection is a natural process which represents a necessary and sufficient explanation for the source of that organizing power. In the struggle for existence, useful variations—which occur naturally—will gradually accumulate in the population, because those individuals who inherit useful variations are more likely to survive and produce offspring in greater numbers than those who inherit less useful ones. Since the struggle for existence takes place in the real-world, useful variations will tend to match the tribulations of survival at particular points in time and in particular locations. When these variations accumulate, they represent adaptations that alter the shape and form of individuals so that they are well-equipped to survive in particular *niches*. The wings of birds, bats or insects, for example, represent adaptations that have gradually arisen by this process—a claim that is thoroughly supported by numerous lines of evidence, including fossil records.

This is the explanation for classical and neo-Darwinian views of Natural Selection—the organizing power arises from selection and accumulation of useful variants, preserved by heredity.

Based on this chapter, we can see that EEP adds to this explanation, because new qualities emerge when two or more entities interact, and these new qualities create new opportunities, or potential, for new layers of organization to subsequently arise. Emergence and gradual accumulation of infinitesimal changes are complementary aspects of the natural world, and in the deep history of life, they combine in ways that have extraordinary organizing power—power that can actually increase as a consequence of the evolutionary process *per se*.

Biological emergence is a fundamental characteristic of virtually every aspect of the living world, ranging from the simplest binding interactions between cellular molecules to the organization of entire ecosystems. Concrete examples of the various levels of biological emergence go a long way towards explaining the sheer organizing power of the living realm. Literally thousands of empirical discoveries—both in the field and in biological laboratories—provide a rich explanation for the origins of biological organization that far exceed the vision of pioneers like Charles Darwin.

Chapter 3

Complex Interplay Between Empirical Discoveries and Conceptual Frameworks

Nothing in biology makes sense except in the light of evolution.

—Theodosius Dobzhansky[a]

The Big Picture

Dobzhansky's defense of the "Modern Synthesis" stated that "nothing in biology makes sense except in the light of evolution".

In the UES, we can say:

Nothing in evolution makes sense except in the light of biology.

This emphasizes the importance of the stunning advances in laboratory technology that have revolutionized our understanding of the inner-workings and molecular interactions in living cells.

Charles Darwin's and Gregor Mendel's broad, early insights into Natural Selection and Mendelian Genetics were brilliant, yet incomplete generalizations that were limited by the limited knowledge-base and technology of the 19th century.

Today, any Updated Evolutionary Synthesis (UES) that is suitable for the 21st century must be grounded in our more detailed knowledge-base as well as the major conceptual advances that have been made.

[a] https://en.wikipedia.org/wiki/Nothing_in_Biology_Makes_Sense_Except_in_the_Light_of_Evolution

The current explanation for complex biological organization links changes in the genome to the tools that generate structure and function in each generation.

This arises during the development of the individual and is manifested in the day-to-day functions made possible by molecular interactions within living cells.

How Evolutionary Theory Evolves

Evolutionary theory is driven by complex, two-way interactions between fresh empirical discoveries and persistent conceptual frameworks. Beginning with his first edition in 1962 and with a postscript in 1969, Thomas Kuhn attempted an explanation of scientific change with *The Structure of Scientific Revolutions* [19], and his concept of **paradigm shifts** became quite popular.[b] However, the nuanced interplay between empirical discoveries and conceptual frameworks is oversimplified by such sweeping generalizations. The philosophical concepts of a **dialectical** or a **yin-yang** relationship between contradictory elements, or the concept of a **complex adaptive system**, where cause and effect become inseparable, are helpful abstractions. In the 21st century, however, we have many concrete and detailed examples that reveals the subtleties of the historical tension between well-documented empirical discoveries and persistent conceptual frameworks.

The rocky road to evolutionary biology

Both the classical 19th century Darwinian theory of **Natural Selection** and the more recent 20th and 21st century insights into the **origin of life** at **undersea alkaline hydrothermal mounds** (see Chapters 5 and 6) have been greatly influenced by both direct **geological** observations (empirical discoveries) and theories (conceptual frameworks).

[b] https://en.wikipedia.org/wiki/The_Structure_of_Scientific_Revolutions

A comprehensive website with Charles Darwin's complete publications and his private letters and manuscripts is available online.[c] As described in some detail in the Wikipedia entry on Charles Darwin, before his son became a noted naturalist, geologist, and biologist, Darwin's father sent him to Christ's College in Cambridge, England to study for a Bachelor of Arts degree, originally intended as the first step towards becoming an Anglican country parson. There he met and befriended John Stevens Henslow and other leading parson-naturalists, who saw scientific work as religious **Natural Theology**.[d]

The connection between geology and Natural Theology can be clearly seen in the works of William Buckland,[e] the English Clergyman, geologist and paleontologist who was dean of Westminster Abbey.[f] Buckland published *Geology and Mineralogy considered with reference to Natural Theology* in 1837, which is available online.[g] A quick search of those web pages for the word "design" provides some insights:

> A more minute examination traces the progress of the mineral materials of the Earth, through various stages of change and revolution, affecting the strata which compose its surface; and discloses a regular order in the superposition of these strata; recurring at distant intervals, and accompanied by a corresponding regularity in the order of succession of many extinct races of animals and vegetables, that have followed one after another during the progress of these mineral formations; arrangements like these could not have originated in chance, since they afford evidence of law and method in the disposition of mineral matter; and still stronger evidence of design in the structure of the organic remains with which the strata are interspersed.
>
> How then has it happened that a science thus important, comprehending no less than the entire physical history of our planet, and whose

[c] John van Wyhe, ed. 2002- The Complete Work of Charles Darwin Online (http://darwin-online.org.uk).
[d] https://en.wikipedia.org/wiki/Charles_Darwin
[e] https://en.wikisource.org/wiki/Author:William_Buckland
[f] https://en.wikipedia.org/wiki/William_Buckland
[g] https://en.wikisource.org/wiki/Geology_and_Mineralogy_considered_with_reference_to_Natural_Theology

> documents are co-extensive with the globe, should have been so little regarded, and almost without a name, until the commencement of the present century?

Here, we find passion as well as curiosity and a deep desire to explore and uncover the secrets of nature, in hopes that they will provide a glimpse of the sublime mind of the creator. An excerpt from the hyperlinked Table of Contents of this fascinating publication also shows how 19th century geology was deeply intertwined with studies of the fossil record:

Strata of the Tertiary Series.

- Mammalia of the Eocene Period.
- Mammalia of the Miocene Period.
- Mammalia of the Pliocene Period.

Relation of the Earth and its Inhabitants to Man.

Supposed cases of Fossil Human Bones.

General History of Fossil Organic Remains.

- Cases of Animals destroyed suddenly.

Aggregate of Animal Enjoyment increased, and that of Pain diminished by the existence of Carnivorous Races.

Proofs of Design in the Structure of Fossil Vertebrated Animals.

Fossil Mammalia—Dinotherium.

Megatherium.

Fossil Saurians.

Ichthyosaurus.

Intestinal Structure of Ichthyosaurus and of Fossil Fishes.

Plesiosaurus.

Mosasaurus, or great Animal of Maastricht.

Pterodactyle.

Megalosaurus.

Iguanodon.

Amphibious Animals allied to Crocodiles.

Fossil Tortoises or Testudinata.

Fossil Fishes.

- Sauroid Fishes in the Order Ganoid.
- Fishes in Strata of the Carboniferous Order.
- Fishes of the Magnesian Limestone, or Zechstein.
- Fishes of the Muschelkalk, Lias, and Oolite Formations.
- Fishes of the Chalk Formation.
- Fishes of the Tertiary Formations.
- Family of Sharks.
- Fossil Spines, or Ichthyodorulites.
- Fossil Rays.
- Conclusion.

Proofs of Design in the Fossil Remains of Mollusks.

Fossil Univalve and Bivalve Shells.

Fossil Remains of naked Mollusks, Pens and Ink bags of Loligo.

Proofs of Design in the Mechanism of Fossil Chambered Shells.

- Mechanical Contrivances in the Nautilus.

Ammonites.

Natural Theologists and modern biologists have more in common than meets the eye

Darwin embarked on his own most famous voyage of discovery as a naturalist and geologist aboard the HMS Beagle on its second voyage (1831–1836), which ended just before the publication by Buckland cited above. Buckland and Darwin were contemporaries and were both greatly influenced by their shared **zeitgeist**: the defining spirit, ideas and beliefs of the educated upper classes in 19th century England. But Darwin's own observations, combined with the newly emerging conceptual frameworks of his contemporaries, led him down an intellectual path that came into direct conflict with the core beliefs of the Natural Theologists.

Scientific historians often attribute Darwin's chronic physical ailments and long delay in publishing *On the Origin of Species* [20] to the anguish, soul-searching and social pressure of coming into direct conflict with the deeply held beliefs of his friends and his community.[h] In fairness to both Darwin and to science, however, we must also attribute some of the delay to the attention to detail and high standards of scholarship that are often found among the scientific community.

Darwin's theory of Natural Selection was one of the most significant developments in the history of science in general, and the history of biology in particular. Both Darwin and historical accounts have also acknowledged the contribution of Alfred Russel Wallace, who came up with a similar theory around the same time.[i] The theory of Natural Selection laid the groundwork for the passion and curiosity of the Natural Theologists to be transformed into the passion and curiosity of the natural scientists. Both groups share a common tradition of careful, detailed observations and meticulous documentation and sharing of their empirical findings. Both have a deep respect for scholarly pursuits, for philosophy, for history, for knowledge and for truth. Both share the excitement and courage of the great explorers. But where they differ is in their core beliefs regarding the source, the causes, the forces that create the natural wonders that they discover.

These core beliefs represent powerful conceptual frameworks that lead to different conclusions regarding the significance of the facts they uncover. Where the Natural Theologists attributed the wonders of nature to the designs and creations of God, Darwin's theory of Natural Selection offered a different explanation: namely, that the wonders of nature arise by natural processes, that obey the principles or laws of physics, chemistry, and biology. Although modern biologists have much in common with the Natural Theologists, they do part ways when it comes to explaining biological complexity, adaptations, and the origin of species—and in particular, the origins of human beings.

Scientific advances are limited by the empirical knowledge-base

Throughout the history of biology, but especially in recent decades, new insights have quickly emerged as a result of new techniques being applied

---

[h] http://www.actionbioscience.org/evolution/buckeridge.html

[i] https://en.wikipedia.org/wiki/Alfred_Russel_Wallace

to scientific questions. Even the most brilliant pioneers are limited by available **empirical research**[j]. Truth be told, in the hands of scientists who know how to ask the right questions, modern laboratory methods have led to more major discoveries during the past three decades than in the last 2,000 years. This includes both basic theoretical discoveries and important breakthroughs in biomedical applications.

Genuine scientific theories are grounded in naturalism

The fundamental requirement for any legitimate update of evolutionary theory is that it must be firmly grounded in **naturalism**.[k] Caltech physics professor Sean M. Carroll provides a useful working definition [21]:

> There is only one world, the natural world, exhibiting patterns we call the 'laws of nature,' and which is discoverable by the methods of science and empirical investigation… purpose and meaning in life arise through fundamentally human acts of creation…

The roots of Natural Selection have two main branches, including Natural Theology and Natural Science. Natural Theology was a central theme in 18th and 19th century Europe, and an important foundation of Deism among the "Founding Fathers" of the United States. It was also the original motivation for detailed, systematic studies of nature. It sought to glorify God by providing detailed observations and analysis of the wonders of His creation.[l]

The guiding philosophy behind Natural Theology is belief in supernatural creation by a supreme being—who created the various species that populate the Earth, and who also created Man in His own image. The assumption underlying Natural Theology is that empirical observations of natural wonders provide a better understanding of His works—a small window into the Mind of God.

---

[j] https://en.wikipedia.org/wiki/Empirical_research

[k] https://en.wikipedia.org/wiki/Naturalism

[l] https://en.wikipedia.org/wiki/Natural_theology

Natural Science,[m] on the other hand, requires that hypotheses be tested and verified by observations and experiments. When natural wonders originally thought to have supernatural origins could be fully explained in terms of natural laws (such as chemistry, geology, physics, and biology), this led to a parting of ways between the Natural Theologists and Natural Scientists, who, for the most part, conclude that naturalism, rather than Natural Theology (which has a supernatural component) is the best way to understand the natural world.

Of course, at the frontiers of physics, chemistry and biology, scientists find phenomena that cannot be currently explained in terms of empirically observable natural law. Such is the case with the frontiers of physics, and our current inability to reconcile quantum mechanics with general relativity, for example. But Natural Scientists generally are also philosophical naturalists—they continue to search for new insights that will reveal the mysteries of nature and are generally unwilling to rely on supernatural explanations for those phenomena that are not yet understood. In this respect, the greatest scientists exhibit both perseverance, optimism, and a strong belief in the ability of the human mind to transcend its inherent limitations.

For Charles Darwin, the realization that the logic of his theory of Natural Selection created a wedge between Natural Theology and Natural Science is one of the factors that delayed his publication of *On the Origin of Species* for some two decades after its original conception. Ever since Darwin, Natural Science has gone hand-in-hand with philosophical naturalism.

Ever since Darwin, the vast majority of evolutionary theorists have looked to naturalistic explanations for the origin and evolution of life, including the origin of our own species. There is no underlying presumption that involves the hand or mind of God. Natural science is based on the fundamental laws of nature. These laws of nature are presumed to have natural, rather than supernatural, causality.

Although naturalism—and evolutionary theory—inevitably come into conflict with deeply-held religious beliefs, both among scientists and the general public, this is unavoidable. Explanations based on empirically observable facts and natural laws are incompatible with supernatural

[m] https://en.wikipedia.org/wiki/Natural_science

explanations. The former is based on evidence and facts, while the latter is based on faith and belief.

The most important requirement of naturalism is the empirical approach known as the scientific method. The empirical approach relies on observations and experiments of reality—of the natural world that exists independently of human perceptions, thoughts, languages or records. Naturalism assumes that we can obtain better and better descriptions of reality and assumes that no single description will ever perfectly match the reality that exists outside of the human mind.

- ***Established scientific theories require empirical validation.***

Naturalism requires that **scientific theories** must be thoroughly grounded in empirical observation and experiments, and are constantly subject to testing, verification, and re-testing as new methods and facts emerge. Note that the phrase "scientific theory" is close to the *opposite* of the everyday use of the term, as explained in the Wikipedia entry on that topic[n]:

> The definition of a scientific theory… as used in the disciplines of science is significantly different from the common vernacular usage of the word "theory". In everyday speech, "theory" can imply that something is an unsubstantiated and speculative guess. [That is] the opposite of its meaning in science.

- ***Established theories are continually modified to reflect reality.***

Established scientific theories which may be consistent with literally thousands of lines of evidence, and which have stood up to rigorous scrutiny, must be frequently updated. When new lines of evidence are found that are inconsistent with one or more aspects of previously established theories, the established theories should be modified accordingly.

Evolution of Classical Darwinian Theory

In the light of literally thousands of new empirical facts regarding the inner-workings of cells, the classical theory of Natural Selection, as

---

[n] https://en.wikipedia.org/wiki/Scientific_theory

proposed by **Charles Darwin** (Figure 1) in 1859, requires some major updates. The theory must be modified and supplemented to remain consistent with our vastly expanded knowledge-base in biology.

Figure 1. Photograph of Charles Darwin. The frontispiece of Francis Darwin's The Life and Letters of Charles Darwin (1887) has the caption "From a Photograph (1854?) by Messrs, Maull and Fox." Engraved for Harper's Magazine, October 1884.

Source: Page-link: https://commons.wikimedia.org/wiki/File:Charles_Darwin_seated_crop.jpg. File-link: https://upload.wikimedia.org/wikipedia/commons/2/2e/Charles_Darwin_seated_crop.jpg. Attribution: By Charles_Darwin_seated.jpg: Henry Maull (1829–1914) and John Fox (1832–1907) (Maull & Fox) derivative work: Beao (Charles_Darwin_seated.jpg) [Public domain or Public domain], via Wikimedia Commons.

Classical Darwinian theory reflects 19th century knowledge and methods

In discussing **variation**, an essential part of his classical theory of **Natural Selection,**[o] Darwin freely admitted the limitations of 19th century knowledge on this subject [20]:

> Our ignorance of the laws of variation is profound. Not in one case out of a hundred can we pretend to assign any reason why this or that part has varied. But whenever we have the means of instituting a comparison, the same laws appear to have acted in producing the lesser differences between varieties of the same species, and the greater differences between species of the same genus.

In pointing out these limitations, Darwin adhered to the professional standards of transparency and honesty that are essential to scientific progress. He recognized the importance of presenting a high-level, naturalistic explanation for the origins of species and biological adaptations, while encouraging other biologists who followed to fill in the missing details.

Darwin's classical 1859 theory represented a general, high-level explanation for the origin of species as well as their diverse adaptations. While the central core of the theory has stood the test of time, other aspects require rethinking—and extensive expansion—in the light of the massive amounts of empirical data that have been discovered in the 20th and 21st centuries.

New ideas gradually gain acceptance

Although timelines are useful for identifying moments that are *opportunities* for change, the ways that paradigm shifts actually play out—both in the scientific community and public awareness—are harder to predict. Unfortunately, although logical consistency and solid empirical support are often necessary to bring about conceptual change, they are usually not sufficient. Scientific ideas—and the public awareness of those ideas—are

[o] https://en.wikipedia.org/wiki/Natural_selection

social products that are influenced by numerous psychological and cultural factors. Turning points in the acceptance of new ideas are notoriously hard to predict.

Although scientific historians can—with 20–20 hindsight—point to particular discoveries and publications that lead to major conceptual changes in the long run, it would be far more difficult to document the *actual* timing of major changes in the conceptual framework of scientists and the public-at-large. In general, it is quite common for contradictory points of view to coexist—often for decades.

Both scientists and the general public tend to be conservative and reluctant to adopt new points of view. Paradigms change slowly. One generalization that can be made, however, is that due to the QWERTY effect, earlier ideas that have already gained acceptance tend to overshadow new perspectives for prolonged periods of time.

The following timeline summarizes several milestones in evolutionary theory that are not widely known or accepted among the general public today:

(1) **1859:** Charles Darwin publishes the first edition of *On the Origin of Species* [20] mentioning Alfred Russell Wallace's closely related theory.
(2) **1915:** T.H. Morgan and colleagues publish *The Mechanism of Mendelian Heredity* [22] providing a deep understanding of the links between genes and chromosomes based on extensive observations in fruitflies.[p]
(3) **1942:** Julian Huxley publishes *Evolution: The "Modern Synthesis"*[q] [23], combining classical Darwinian theory with **Mendelian Genetics** and **Population Genetics**, linking speciation to changes in **allele** frequencies in natural populations.
(4) **1944:** Oswald Avery and colleagues prove that DNA is the genetic material[r] [24].

---

[p]Full text online at https://archive.org/details/cu31924003076639

[q]Available online at http://krishikosh.egranth.ac.in/bitstream/1/2057456/1/ANAND-52.pdf

[r]https://en.wikipedia.org/wiki/Avery-MacLeod-McCarty_experiment. Paper available online at http://jem.rupress.org/content/jem/79/2/137.full.pdf

(5) **1953:** James Watson and Francis Crick, relying on X-ray diffraction data from Rosalind Franklin, publish a brief letter in Nature[s] elegantly describing the double-helical structure of DNA that suggested a possible replication mechanism [25].

(6) **1953:** Stanley Miller publishes the Miller–Urey simulation[t] in Science [26] showing a possible way that organic compounds such as amino acids could arise on primitive Earth, catapulting the Prebiotic Soup hypothesis into the public awareness.

(7) **1957–1961:** Francis Crick and colleagues perform experiments linking DNA to messenger RNA and the "genetic code" [27].[u]

(8) **Mid-1960s:** Har Gobind Khorana and colleagues decipher the "genetic code" in which DNA and RNA triplets code for each of the 20 different amino acids of protein sequences.[v]

(9) **1970:** Susumu Ohno publishes *Evolution by Gene Duplication* [28].

(10) **1970s and 1980s:** Using powerful technical breakthroughs, including, but not limited to, recombinant DNA technology, DNA sequencing, computer databases and PCR, scientists find widespread sources of variation including transposable elements, horizontal gene transmission, repetitive sequence evolution, and more.

(11) **1975:** Mary Claire King and Alan Wilson publish "Evolution at Two Levels in Humans and Chimpanzees" in *Science* [29], introducing the critical importance of regulatory DNA sequences and control of gene expression in the evolution of complex biological organization.

(12) **1980:** Christiane Nüsslein-Volhard and Eric Wieschaus publish one of a remarkable series of papers describing the molecular and genetic determinants of fundamental aspects of animal development, such as the formation of the overall body plan and pattern formation [30].[w]

---

[s] Available online at https://www.genome.gov/edkit/pdfs/1953.pdf

[t] Available online at https://pdfs.semanticscholar.org/080b/031f91696c8cde1db41f96f03f3e5b4146c0.pdf

[u] Available online at https://www.ncbi.nlm.nih.gov/books/NBK21950/

[v] http://www.dnaftb.org/22/bio-2.html

[w] Available online at http://www.eb.tuebingen.mpg.de/fileadmin/uploads/images/Research/departments/nuesslein-volhard/Publikationen/15_Nüsslein-Volhard_Wieschaus__1980__Nature_287.pdf

(13) **1988:** Mike Russell and colleagues publish the first [31] in a series of papers proposing that life on Earth began in alkaline hydrothermal vents (Chapter 5).
(14) **1994:** Robert Wright's *The Moral Animal* [32] introduces Evolutionary Psychology to the general public, providing Darwinian perspectives on the origins of negative tendencies in the human mind and the social institutions they create.
(15) **2006:** Sean B. Carroll publishes *Endless Forms Most Beautiful* [15] describing the new science of Evolutionary Developmental Biology (evo-devo) for popular audiences (Chapter 13).

A variety of proposed modifications and additions to classical Darwinian evolutionary theory, the "Modern Synthesis", and neo-Darwinism have been published in both peer-reviewed scientific papers as well as popular accounts. They have yet to be successfully integrated into a unified and widely-known 21st century synthesis. Important new empirical discoveries and theoretical insights have a variety of names. These include, but are not limited to, **self-organization,**[x] **facilitated variation**[y] [13], **regulatory sequences**; **evolvability**, **evo-devo** [15], **evolutionary psychology** [32], and **emergent evolutionary potential** (Chapter 2), as well as interdisciplinary titles such as extended, expanded, or integrated evolutionary synthesis. *Rethinking Evolution* attempts consolidation of these major developments, and more, into a comprehensive **Updated Evolutionary Synthesis (UES)**.

Evolutionary theory expands when biological research enlarges the knowledge-base

Evolutionary theory draws upon a broad and interdisciplinary range of conceptual and empirical advances in many specialized fields of the biological sciences. The detailed explanations and our deep understanding of the ways that genomic evolution transforms genetic information, and the ways that information gives rise to complex biological organization during development have revolutionized evolutionary theory, while still

[x] https://en.wikipedia.org/wiki/Self-organization
[y] https://en.wikipedia.org/wiki/Facilitated_variation

preserving key elements of the classical Darwinian conceptual core. New opportunities are built upon previous innovations, and powerful new cellular tools facilitate new levels of complexity.

The Role of Evolving Conceptual Frameworks

Implicit framing in the natural sciences

Several factors have led to widespread confusion or otherwise prevented consolidation of an *Updated Evolutionary Synthesis* before now. These will be explored in some depth in the chapters that follow; the following list is a brief overview:

(1) Our empirical understanding and detailed knowledge of "the **gene**" have improved considerably, but the earlier understanding dominates the conceptual framework.
(2) The concept of **random variation**, an integral part of the conceptual framework of the "Modern Synthesis", assumes independence of events that are actually interdependent in a variety of ways.
(3) Darwin's classical concept that **infinitesimal variations** accumulate limits our understanding of other important types of evolutionary change, such as **emergence**.
(4) Ever since Darwin, **creationist** leaders and followers have been vocal in their opposition to the concept that human beings evolved by natural rather than supernatural forces.
(5) Creationists have unwittingly or deliberately sown confusion with false statements including, but not limited to: "evolution is just a theory", "evolution violates the second law of thermodynamics", and "how can a random process generate complexity?"
(6) Creationists have tried to circumvent constitutional separation of church and state in the US by falsely claiming to have alternative "scientific" theories such as "intelligent design", which is pseudoscience.
(7) The "Modern Synthesis" focuses on changes in the frequencies of genes in populations but fails to address how genetic information is translated into complex structure and function during development.

(8) Professional career pressures, institutional hierarchies, competition, and ego—which afflict all aspects of current institutions and human behavior—also apply to scientists. Consequently, some scientists are so eager to claim credit for disproving a theory or proposing a revolutionary new theory that they tend to accept specific specialized points of view, to the exclusion of others, rather than focusing on proven factual information.

(9) Scientific literacy among the general public in the US is quite limited, and professional scientists, communicators and educators have not deployed effective tools to remedy this problem.

(10) Science, truth, media, honesty and critical thinking are all currently under attack by powerful commercial and political special interests.

The Role of Evolving Technology and Laboratory Methods

High-level descriptions of complex phenomena vs. detailed understanding of "mechanisms"

In the mid-19th century, pioneers who began to approach the secrets of the natural world through scientific discovery were limited by the absence of available scientific knowledge about the **inner-workings of cells**. Despite that limitation, both Charles Darwin and Gregor Mendel used observations and experiments to penetrate some of these mysteries, but their progress was restricted by the few tools that were available at the time.

The tools of discovery include methods that lead to new empirical discoveries as well as advancing the conceptual framework that depends on that knowledge-base.

The impact of new methods and technology, from 1928 to the 21st century

The following list will briefly summarize how innovative laboratory methods have resulted in major advances in genetics and our understanding of the genome. Several of these topics will be discussed (and referenced) in more detail in later chapters.

(1) In the first decades of the 20th century, genetic experiments in the **fruitfly** (***Drosophila***) led to major advances in **Mendelian Genetics**.

(2) In 1928, an unknown transformation factor in bacteria was implicated to be the genetic material. In 1944, experiments that used purified enzymes to destroy specific macromolecules (such as DNA or proteins) demonstrated that DNA is the **genetic material**.

(3) In 1952, DNA was confirmed to be the genetic material by labeling the proteins and DNA of bacteriophage (viruses that prey on bacteria; Figure 2) with either radioactive sulfur or radioactive phosphorus to label proteins or DNA, respectively. This demonstrated that the genetic material from the viruses that entered the bacterial cells was DNA.

(4) In 1953, **X-ray diffraction** data of DNA, combined with published information regarding the ratios of the four nucleotides that make up DNA, led to the discovery of the double-helical structure of DNA, and suggested a copying mechanism that would allow DNA sequences to be replicated during cell division or reproduction.

(5) In 1961, radioactive labeling and **cell fractionation** techniques demonstrated that the information in DNA is transcribed into a type of RNA called **messenger RNA**, which directs the synthesis of specific **protein chains** at cellular structures called **ribosomes**. Each position in the messenger RNA sequence consists of one of four possible **nucleotides—A, C, G, or U**.

(6) Also, in 1961, **cell-free preparations** of RNA consisting of long synthetic sequences of U nucleotides were used to determine that triplets of UUU specify the addition of the amino acid **phenylalanine** to a growing protein chain. Each protein chain consists of a sequence of **amino acids**. 20 possible amino acids (Figure 3) are located at each position in the amino acid sequence.

(7) In messenger RNA, each sequential group of three nucleotides—called a **codon**—specifies the identity of each amino acid in the growing protein chain. By 1966, similar techniques were used to decipher all 64 possible triplets that specify one of 20 amino acids or the end of the amino acid sequence. This correspondence between codons and amino acids is called the **genetic code**.

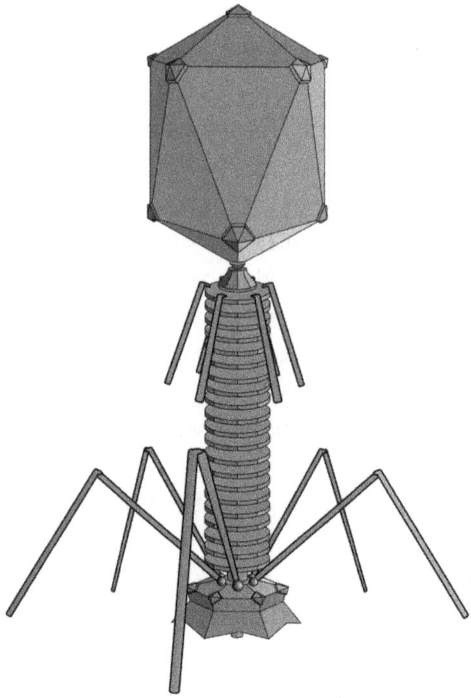

Figure 2. The outer coat of a bacteriophage, a type of virus that preys on specific bacterial cells. The outer coatings of viruses consist of protein chains and/or carbohydrates. DNA, the genetic material that subverts the bacterial cells so that they synthesize large numbers of virus particles, is contained inside. When the bacteriophage lands on the surface of a bacterial cell with its appendages resembling landing gear, the virus penetrates the bacterial cell wall and membrane and injects the DNA into the cell.

Source: Page-link: https://commons.wikimedia.org/wiki/File:PhageExterior.svg. File-link: https://upload.wikimedia.org/wikipedia/commons/4/4a/PhageExterior.svg. Attribution: By Adenosine [CC BY-SA 3.0 (https://creativecommons.org/licenses/by-sa/3.0) or GFDL (http://www.gnu.org/copyleft/fdl.html)], from Wikimedia Commons. Rendered in B&W.

(8) Bulk techniques for characterizing **heterogeneous** mixtures of DNA sequences, such as **buoyant-density centrifugation** and **nucleic acid hybridization** showed that in most multicellular organisms, only a small fraction of **genomic DNA** consists of unique sequences that "code for" proteins. Most of the genome consists of noncoding

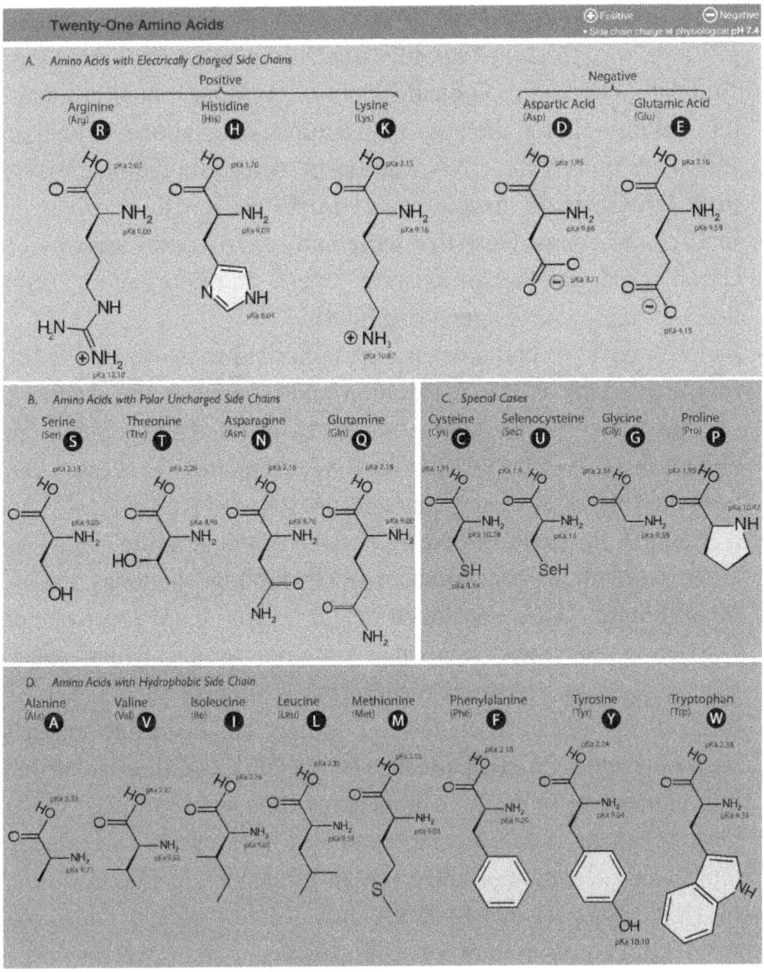

Figure 3. Structures of the 20 types of amino acids commonly found in living cells. Each amino acid has an amine group (NH_2) at one end of the molecule and an acidic carboxyl group (COOH) at the other, hence the name "amino acid". These groups are chemically bonded together during the synthesis of each protein chain. 20 possible side-chains define each type of amino acid. These side-chains determine the structure and other physical and chemical properties of protein chains.

Source: Page-link: https://commons.wikimedia.org/wiki/File:Amino_Acids.svg. File-link: https://upload.wikimedia.org/wikipedia/commons/a/a9/Amino_Acids.svg. Attribution: By Dancojocari [CC BY-SA 3.0 (https://creativecommons.org/licenses/by-sa/3.0) or GFDL (http://www.gnu.org/copyleft/fdl.html)], from Wikimedia Commons. Rendered in B&W.

DNA, including moderately and highly repetitive sequences. In 1972, it was thought that this noncoding DNA was evolutionary "junk" that had no biological functions but represented raw material for evolution. Later, techniques that allowed isolation and characterization of individual DNA sequences demonstrated that so-called **junk DNA** actually has numerous important functions.

(9) In 1970, it became clear that large numbers of genes arise by duplication and divergence of existing genes—an important evolutionary process referred to as **gene duplication**.

(10) In 1970, basic research in viral genomes led to the discovery of retroviruses (such as HIV). Isolation and cloning of the reverse transcriptase enzyme made it possible to create DNA copies (called complementary or cDNA) from RNA templates. This later made it possible to sequence and characterize RNA as well as DNA, and to distinguish the **protein-coding** portion of the **genome**.

(11) Basic research in bacterial and **bacteriophage genetics** led to the **recombinant DNA** revolution which began in 1973. Isolation of **restriction enzymes** from other bacterial strains, which cut DNA sequences at specific short **recognition sequences** such as GAATTC, made it possible to insert DNA from any organism into bacterial or bacteriophage cells, **replicate** the foreign DNA, and then isolate **molecular clones** containing large numbers of copies of a particular sequence. Once segments of DNA could be cloned, their specific sequences and molecular interactions could be characterized in great detail.

(12) In 1975, comparisons of DNA and protein sequence information from humans with that of chimpanzees demonstrated that their protein-coding sequences are so similar that it is likely that the differences are due to regulatory mutations in DNA—that is, mutations that control the expression of protein sequences.

(13) In 1977, powerful **DNA sequencing** techniques were developed.

(14) The development of the Internet protocol in 1982, along with databases for storing DNA, RNA, and protein sequence data, led to the storage of massive amounts of publicly available sequence information from a variety of organisms.

(15) The development of powerful new software methods for sequence alignment and comparison made it possible to determine the precise evolutionary lineages of related organisms.

(16) In 1987, the invention of the **polymerase chain reaction (PCR)** made it possible to rapidly **amplify** specific DNA sequences with far less effort and resources than molecular cloning.
(17) By 1987, enough sequence information had accumulated to identify the mechanism responsible for expansion of genomic DNA with repetitive sequences, known as **slipped-strand mispairing** or **replication slippage**.
(18) In the late 1980s and early 1990s, combining systematic genetic screening in ***Drosophila*** for **mutations** that affect development with new molecular techniques that allowed visualization of **gene expression** in *Drosophila* embryos led to a new understanding of the genes responsible for the formation of the entire **body plan** of the flies during development.
(19) In the 1990s, comparisons of related DNA sequences demonstrated that very similar mechanisms control the development of a broad range of animals, including both invertebrates such as *Drosophila* and vertebrates such as frogs, mice, and human beings.
(20) Beginning primarily in the 1990s, a variety of powerful new techniques in molecular, cellular, and developmental biology, major conceptual advances led to the relatively new field of evolutionary developmental biology (evo-devo).
(21) By 2001, improvements in rapid sequencing technology made it possible to sequence the entire human genome; since that time, it has become more and more practical to sequence the entire genome of a variety of organisms and use powerful software techniques (**bioinformatics**)[z] to compare them.
(22) In the 21st century, a variety of molecular, cellular, and developmental techniques are routinely combined in biological laboratories to probe the mechanisms of genomic evolution as well as the development of structure and function from genetic information. For example, we now have a detailed understanding of the complex ways that specific proteins, known as **transcription factors**, interact with specific DNA sequences to control the expression of particular genes. In general, new empirical discoveries in biology—many of which are relevant to advances in evolutionary theory—are discovered at an

[z] https://en.wikipedia.org/wiki/Bioinformatics

ever-accelerating pace, due to both conceptual advances and the development of powerful new technologies and laboratory methods.

From Darwin's observations aboard the HMS Beagle to the most recent 21st century findings made possible by powerful new methods and technology, empirical discoveries have played a major role in advancing evolutionary theory. Our conceptual frameworks have also advanced. The questions that scientists ask are as important as the answers they obtain. The greatest scientific pioneers find questions in phenomena that others take for granted, and they also challenge the answers and explanations that satisfied their intellectual forebears. The result is that theories of the causes of the deep history of life on Earth have their own fascinating history.

In recent years, evolutionary theory has dramatically improved, thanks in large part to empirical discoveries concerning the inner-workings of cells, at the level of molecular interactions, as discussed in Chapter 4.

Chapter 4

Timeless Molecular Innovations

Most anatomical and physiological traits that have evolved since the Cambrian are, we propose, the result of regulatory changes in the usage of various members of a large set of conserved core components that function in development and physiology.

—John Gerhart and Marc Kirschner, in
The Theory of Facilitated Variation[a]

The Big Picture

Evolutionary innovations lay the groundwork for subsequent innovations. They often represent tools and reusable modules that can evolve in a variety of ways—in different lineages and in different ecological contexts.

Molecular interactions involving proteins carry out virtually every function in living cells. They also mediate interactions between cells during development and are responsible for the overall patterning of the body and complex organization of individuals of every species.

The misleading concept of "random mutations" from the "Modern Synthesis" became more concrete with the discovery of DNA and its subsequent characterization. "Random mutations" were now thought of in terms of discrete changes in DNA sequences that translate into changes in specific proteins.

[a] Available online at https://www.ncbi.nlm.nih.gov/pmc/articles/PMC1876433/pdf/zpq8582.pdf

We now know that proteins are not the only type of genetic information. Evolution also acts to create, modify, and reuse regulatory sequences of DNA. A variety of regulatory DNA sequences interact in complex ways to control the expression of specific proteins.

The Updated Evolutionary Synthesis should also include six additional concepts that transcend the misleading concept that mutations are "random":

(1) Various modular and reusable tools have evolved that facilitate and accelerate genome evolution. The genome represents a metaphorical scrapyard of reusable genetic information.

(2) Platforms that bring cellular entities together create new opportunities for higher-level organization to emerge; this is a source of Emergent Evolutionary Potential (EEP).

(3) The concept of phenotype should include a distinction that emphasizes the difference between Generative and Ecological Phenotypes. The former are often tools that act during development, while the latter refers to the structures and functions of the individual that have adapted to various niches (ways of life).

(4) Random and fortuitous variation generates a metaphorical "scrapyard" of modules that can be redeployed in different ways and at different times.

(5) Fortuitous, useful phenotypic gestalts (interactive wholes) arise from the collective action of dozens or hundreds of genes. The individual genes will often be conserved, duplicated, modified, and reused in future generations.

(6) Sudden, emergent innovations, as well as infinitesimal and gradual fine-tuning of those changes, represent complementary aspects of Natural Selection.

From EEP and Natural Selection to Higher Levels of Organization

Chapter 2 added new meaning to the terms "emergence" and "potential" by bringing them together in the context of biological evolution as Emergent Evolutionary Potential (EEP). The new usage of the terms was defined in a concrete and objective way consistent with the goals of science.

Emergence describes the way that higher levels of organization arise when entities come together and interact. The new combined entities have properties that are not found in the separate entities. Water, for example, has physical and chemical properties not found in hydrogen or oxygen atoms alone.

In biology and evolution, which overlap in both empirical and conceptual ways, emergent properties are not fully described by physics and chemistry. Organic chemistry, biochemistry, cell biology, Molecular Genetics, and developmental biology are just some of the fields of natural science that provide a more adequate approximation of emergent structure and function.

Potential describes opportunities and possibilities that have not yet been manifested in actual events. Previous actual events change the probabilities of future events in a variety of ways. Future probabilities are shifted, and cannot be viewed as merely "random" possibilities. Higher-level organization—that is, new structures and functions—often lays the groundwork for actual events and interactions that generate more complex levels of organization.

When combined with Natural Selection, EEP makes it easier to understand how complex adaptations can arise by natural forces that can be described in the concrete and objective language of science. Supernatural or paranormal explanations involving teleological language such as design, intelligence, or goal-seeking behaviors of any kind are both unhelpful and misleading.

In this chapter, we'll review some concrete examples of higher levels of organization that arise via Natural Selection combined with EEP. Since these examples all represent textbook examples of biological organization, the reader would be justified in raising the following questions:

(1) How does detailed discussion of biological structures and functions that are normally covered in general or advanced biology courses add value to the UES?
(2) If more complex entities emerged from simpler entities, and were subjected to modification, modular reuse, and/or refinement by Natural Selection, then isn't a step-by-step account of evolving

entities, from simple to complex, required as necessary and/or sufficient evidence for this hypothesis?

The answer to the first question is that the close relationship between now well-established biological discoveries and the UES is a constant theme that is woven through the fabric of *Rethinking Evolution*. The goals of science demand that detailed factual account of both empirical evidence and the underlying conceptual framework and assumptions be communicated in an objective fashion, using precise and unambiguous language. This involves definitions as well as concrete examples. The UES draws upon wide-ranging aspects of the biological sciences, and is inseparable from that knowledge.

The fact that evolutionary biology is concerned with actual past events that can no longer be directly observed is an ongoing intellectual challenge that began before Darwin published *On the Origin of Species*, and continues beyond the present day.

But thanks to biological research, we obtain a better and better approximation of reality with the passing of time. Many aspects of past events can be understood by both inductive and deductive logic, based on a plethora of data drawn from numerous specialized fields—what E.O. Wilson has correctly called a consilience of evidence [33]. Sophisticated comparisons of DNA sequence data are an important source of knowledge that was wholly unavailable to Darwin, but that is only one of the many sources now available to the UES.

We do have some concrete insights into the intermediate forms that arose during stepwise increases in biological complexity that cannot be directly observed. Only a fraction of available evidence will be referenced in *Rethinking Evolution*, and there can be little doubt that increasingly powerful tools of discovery will uncover more in the near future. *Rethinking Evolution* is far from the final word on this subject, but it's a start.

The Molecular Realm of Macromolecular Interactions

At every level of complexity, the inner-workings of cells, and interactions between cells, depend on highly specific molecular interactions.

Organic molecules often contain carbon chains, which means that they can easily form chains or rings. Carbon atoms gain stability when they are bonded to four other atoms in 3D space, which can vary. Carbon atoms can readily undergo chemical reactions that bond them to other carbon atoms, or to atoms such as hydrogen, oxygen, nitrogen, sulfur, or phosphorous.

Small organic molecules can be bonded to larger ones, giving rise to four major classes of large molecules—**macromolecules**—found in all cells. These include **proteins**, **carbohydrates**, **lipids**, and **nucleic acids (DNA and RNA)**.[b]

In terms of biological evolution and reproduction of biological organization, the significance of macromolecules is usually found in the specific structures that enable individual molecules to bind and interact with other molecules (see Chapter 12). Each type of individual molecule can be uniquely described by its chemical structure and its stereochemistry—that is, its unique 3D shape. At the most fundamental levels of biological organization, EEP begins with the special properties of individual macromolecules that represent the potential to interact in specific ways.

The perceptual shift from DNA to the molecular realm of individual protein molecules

What type of molecule carries out the hereditary functions of all living cells? If you guessed DNA, you are wrong. The correct answer is proteins. Although DNA is the genetic material, in which hereditary determinants are stored and transmitted from generation to generation, DNA requires proteins to interact in various ways with other molecules to actually do the work that reproduces and maintains biological organization. The key words in that trick question were "carries out".

DNA stores two general types of genetic determinants: (1) sequences that determine amino acid sequences of specific types of protein chains, and (2) regulatory sequences control the synthesis of specific proteins at particular times and in particular types of cells.

Nonscientists are often unfamiliar with the extraordinary diversity, intricate structural beauty, and interactive capabilities of proteins. That is because we think about the nutrient value of bulk proteins in the food we eat. Since individual protein molecules (and their interactions) are too

[b] https://en.wikipedia.org/wiki/Macromolecule

small to see with the naked eye, it is helpful to think of them as metaphorical **micro-machines** that exist in the molecular realm of living cells. If we could magically shrink down to molecular size, board a tiny submarine, and embark on a sort of "Fantastic Voyage", we could visualize the variety of ways that proteins interact with other molecules to carry out the variety of functions required for life. If you like, we can set out to explore the human body, the subject of the 1966 Oscar-winning movie[c] that was later popularized in the TV series "The Magic School Bus".[d]

In 1966, of course, very little was known about the inner-workings of cells. In our modern-day 21st century voyage, we can draw upon the extraordinary breadth and depth of molecular and cellular knowledge made available by thousands of dedicated biological researchers.

Instead of entering the blood stream, let's imagine shrinking to an even smaller size, so that we can pass through the membranes of various cell types and enter the fluid-filled realm known as the cytoplasm. In the molecular realm of the cytoplasm we find a crowded, busy space that is teeming with tiny molecular micro-machines, of which many are specialized protein molecules.

In fact, several biology researchers have teamed up with animators to create a variety of informative and entertaining animated videos—such as the Virtual Cell collection—that make it much easier to visualize the hidden molecular realm. These highly-useful videos are free and available on the Internet,[e] which makes them far more accessible both to students and the general public.

Emergent Levels of Organization Involving Proteins

Protein molecules perform virtually every cellular function

When biologists think of proteins, they are thinking of the exquisite, complex and unique 3D structure of *individual protein molecules.*

---

[c] http://www.imdb.com/title/tt0060397/

[d] https://en.wikipedia.org/wiki/The_Magic_School_Bus_(TV_series)

[e] http://vcell.ndsu.nodak.edu/animations/

This requires a conceptual and perceptual shift for most nonscientists. Proteins are far, far more than what's for dinner. Bulk protein represents an essential nutrient, but the individual protein molecules are broken down into their chemical subunits, which are amino acids, when we digest them. All proteins consist of chains of amino acids, which are relatively simple organic (carbon-containing) molecules that are linked together by strong covalent bonds (called peptide bonds). Most cells have proteins that are each assembled from a subset of 20 different amino acids. The side-chains of these 20 different types define their various physical and chemical properties. For human cells, nine types of amino acids are essential nutrients—our cells cannot synthesize them from scratch. Six other types are conditionally essential (depending on specific stresses and health conditions), while five others are readily synthesized by human cells.[f] Other species vary. For example, many types of bacterial cells can synthesize a larger number of amino acids from scratch.

The important point is that the individual protein molecules that are synthesized (from amino acids) in living cells have unique and powerful capabilities that arise from their unique structure and chemical properties.

When proteins interact with other molecules in ways that are useful to living cells, they are, by definition, performing various functions. At the molecular level, evolution is responsible for the structure of individual proteins as well as the origins of useful functional interactions.

Four levels of protein structure are determined by the amino acid sequence

The structure of each specific protein molecule can be described at four levels of complexity. The first is called the **primary structure**. This is simply the amino acid sequence of a protein chain. The primary structure also defines the ways that protein chains spontaneously form higher-level structures in 3D space.

The **secondary structure** of proteins consists of the ways that specific amino acid side-chains cause the chain to form cork-screw like structures

[f]https://en.wikipedia.org/wiki/Essential_amino_acid

called **alpha helices**, or to fold up into sheets known as **beta-pleated sheets**.[g]

The third level of structure—the **tertiary structure**—describes the ways that proteins spontaneously fold into specific, reproducible 3D shapes. This folding represents an example of spontaneous **self-organization**, which is driven by the physical and chemical properties of the **side-chains** of the amino acids. Sometimes, helper proteins called **chaperonins** assist with the folding process. The 3D shapes can be determined with biochemical techniques such as X-ray crystallography combined with computer programs. Weak chemical bonding forces between amino acid side-chains and interactions with surrounding water molecules assure that protein chains will remain properly folded under **physiological conditions**.

Many proteins consist of multiple protein chains. The fourth level of structure—the **quaternary structure**—describes the ways that two or more protein chains bind together and spontaneously self-assemble into larger 3D structures. A variety of weak chemical bonding forces hold the subunits together. These bonding forces can be analyzed and predicted on the basis of biochemical principles and computer programs. Sometimes, folding and binding are supplemented by covalent bonds between the sulfur atoms on the side-chains of **cysteine** amino acids.

Prokaryotic extremophiles illustrate diversity of protein evolution

As the name implies, prokaryotic **extremophiles**[h] have evolved a variety of ways to thrive in harsh environments. In fact, many of these environments are so harsh that they would be lethal to almost every other carbon-based life form, whether **prokaryotic** or **eukaryotic**. Environmental extremes may include high temperature, low temperature, high pH, low pH, **ionizing radiation**, **UV radiation**, high **salinity**, and **desiccation** [34].

Resistance to radiation is facilitated by the evolution of special adaptations such as the presence of multiple copies of the prokaryotic genome and highly-efficient DNA repair mechanisms.

---

[g] https://en.wikipedia.org/wiki/Protein_secondary_structure

[h] https://en.wikipedia.org/wiki/Extremophile

Protein folding and assembly of subunits (tertiary and quaternary structure) are sensitive to environmental extremes such as temperature, pH, salinity, and freezing and thawing.

In extremophiles, detailed studies have shown how the amino acids sequences have evolved so that they can remain folded and assembled under extreme conditions. This is made possible by the extraordinary diversity of amino acid sequences. Protein chains of extremophiles resist unfolding or disassembly (**denaturation**) because they have evolved robust and resilient sequences of amino acid **side-chains**. These side-chain sequences have been optimized for the physical and chemical binding forces that maintain folding under extreme conditions.

Among the more striking examples of extremophiles are the modern-day thermophilic (heat-loving) archaeans that thrive in volcanic, submarine hydrothermal vents where the water temperatures range from 110°C to 121°C (230–250°F).[i]

Analysis of the proteins of these remarkable organisms show that the amino acid composition has created extraordinarily heat-stable protein structure that resists unfolding (denaturation) at temperatures that would literally cook the proteins of most other forms of life. In fact, one of the most useful types of enzymes exploited by biotechnology includes the thermostable DNA polymerases that are used in the powerful DNA amplification method known as the polymerase chain reaction (PCR) mentioned in Chapter 3.

Beyond biochemical pathways: Regulation of protein synthesis by prokaryotic genes

Prokaryotes are the masters of metabolism in the living world. The prokaryotic genome typically consists of a single circular chromosome, in which protein-coding genes are arranged in tandem along a very long double-helical molecule of DNA. DNA replication speed is at a premium among prokaryotes, because Natural Selection favors single-celled species that can reproduce more quickly than others. Prokaryotic **selective**

[i]These are distinct from the much cooler undersea alkaline hydrothermal mounds (UAHM) discussed in Chapter 5.

forces favor (1) keeping the genome as short as possible as well as (2) conserving resources such as energy and raw materials. That way, cells can replicate their DNA and grow as quickly as possible, depending on available resources.

Ideally, proteins should be synthesized only when they are needed. For **metabolic pathways** required to break down sugars, for example, the **protein chains** that make up the **enzymes** in those pathways should only be synthesized when they are required. Some bacteria, such as the ***E. coli*** that play a symbiotic role in human digestion, have evolved the ability to utilize either glucose or lactose sugars as their sources of energy and carbon skeletons. It follows that Natural Selection should favor the ability to control the production of lactose vs. glucose metabolizing enzymes. Bacteria that can turn enzyme production on and off have a **selective advantage.** This is accomplished by an innovation known as the prokaryotic **operon**.

The concept of the prokaryotic **operon** was first proposed in 1960 by a pioneering team of French scientists.[j] Operons usually consist of a regulatory sequence of DNA known as an **operator**, followed by a series of protein-coding genes required to make the enzymes of a specific metabolic pathway. When the operator is bound to another protein known as a repressor, this blocks the **RNA polymerase** from transcribing the genes, so that mRNA cannot be made. Operators are controlled by repressors and/or inducers that act as on or off switches for the production of mRNA.

In 1965, François Jacob, André Michel Lwoff and Jacques Monod won the Nobel Prize in Physiology and Medicine for this groundbreaking work.

A Brief Survey of Functions Arising from Protein Interactions

At the higher levels of organization, proteins have emergent physical and chemical properties. Those that prove useful for the survival and reproduction of cells will tend to be conserved by Natural Selection.

[j]English translation available on the web: http://www.sns.ias.edu/~tlusty//courses/landmark/JacobMonod1960.pdf

Many protein functions can be classified into three general categories as follows:

(1) The 3D shapes of protein molecules allow them to bind specifically to other proteins or other molecules, and interact with them in a variety of ways.
(2) Many proteins function as **enzymes**, which bind specifically to particular reactants and lower the activation energy required for various chemical reactions to take place.
(3) Some proteins contribute to changes in cell shape, binding to other cells, and **cell motility**.

Some proteins fall into all three categories. For example, skeletal muscles consist of proteins that use their **enzymatic activity** to split small high-energy molecules known as **ATP** (Figure 1), and use that activity to

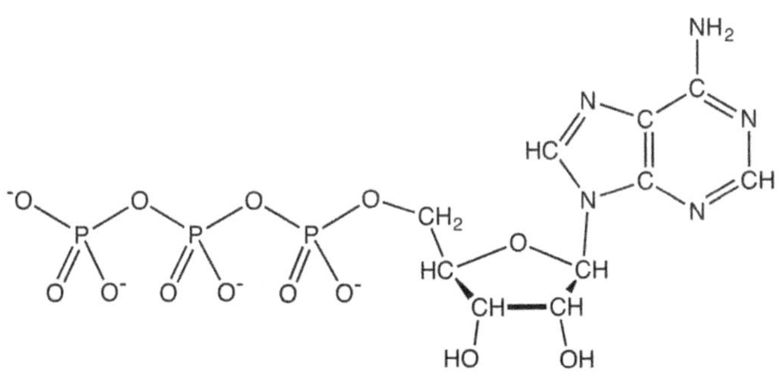

Figure 1. Structure of the ATP molecule, adenosine triphosphate, which is synthesized by mitochondria and chloroplasts. ATP represents a short-term source of energy that powers numerous cellular functions. Enzymes capture this energy by breaking chemical bonds between the second and third phosphate groups. These bonds are highly unstable due to repulsion of the electron clouds. Chemical instability often represents a source of potential energy. ATP also functions as one of the four nucleotide triphosphates that are incorporated into RNA molecules during transcription of DNA.

Source: Page-link: https://commons.wikimedia.org/wiki/File:ATPanionChemDraw.png File-link: https://upload.wikimedia.org/wikipedia/commons/f/f7/ATPanionChemDraw.png. Attribution: By Smokefoot [CC BY-SA 4.0 (https://creativecommons.org/licenses/by-sa/4.0)], from Wikimedia Commons.

change their own shapes, bind to other molecules, and cause muscle cells to contract.

Some proteins have amino acid sequences with chemical properties that cause them to interact with and become embedded into cell membranes. Some of these membrane-bound proteins act as channels that allow only certain molecules (such as water molecules) to pass through. Some use the energy of ATP to actively pump small molecules or **ions** such as sodium from one side of the membrane to the other.

Cell Signaling, Receptors, and Signal Transduction Pathways

Especially in multicellular organisms, cells frequently need to send and receive signals to and from other cells, to properly coordinate their activities. **Cell signaling**[k] takes place when cells **differentiate** and form integrated patterns during **development**. It also routinely takes place during regulation of cell division among specialized cells. Cell signaling includes a variety of short and long-range signals that bind to proteins known as **receptors**.

Signal transduction pathways are reminiscent of the "blind watchmaker"

Signal transduction pathways (Figure 2) provide a compelling example of the organizing power of EEP when combined with Natural Selection. Chapter 2 discussed the potential for emergence of higher levels of organization to arise when smaller entities are brought together and interact. When fortuitous interactions lead to cellular events that are useful in the struggle for existence, they will tend to be preserved by Natural Selection.

If Natural Selection acts as a metaphorical "blind watchmaker" [5], then selected interactions would be expected to be rather arbitrary, and

[k] https://en.wikipedia.org/wiki/Cell_signaling

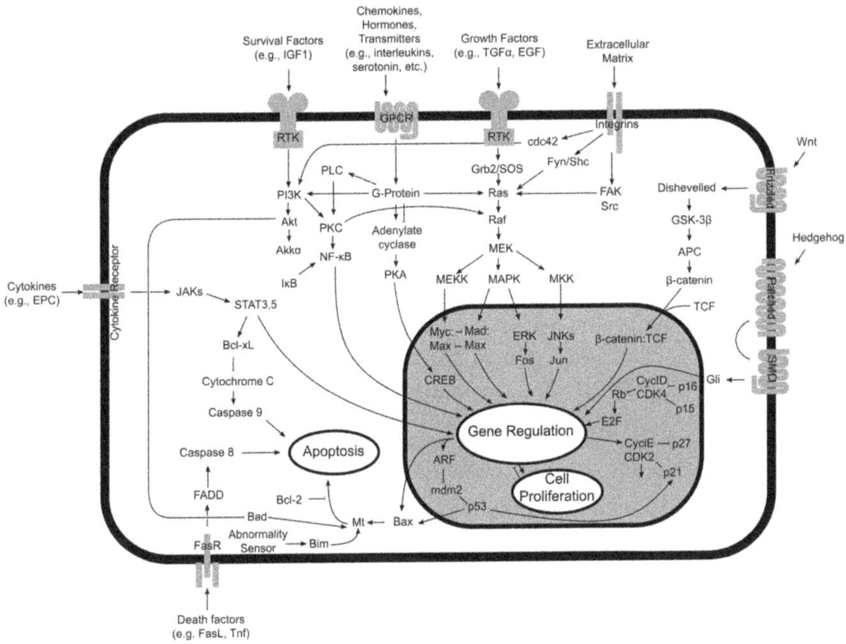

Figure 2. Diagram of major signal transduction pathways of mammalian cells. In general, proteins that act as receptors are embedded in the cell membrane. When they bind to signaling molecules, they change shape, leading to a cascade of events inside the cell that transduce the signal and trigger other cellular events. Many of these chains of events culminate in changes in gene expression and/or cell proliferation.

Source: Page-link: https://commons.wikimedia.org/wiki/File:Signal_transduction_pathways.svg. File-link: https://upload.wikimedia.org/wikipedia/commons/b/b0/Signal_transduction_pathways.svg. Attribution: By cybertory [GFDL (http://www.gnu.org/copyleft/fdl.html) or CC BY-SA 3.0 (https://creativecommons.org/licenses/by-sa/3.0)], via Wikimedia Commons. Rendered in B&W.

indirect. In contrast to any sort of conscious goal-directed intelligent design, we would expect to see cellular structures and functions that appear more like **Rube Goldberg devices** than efficient machines. Although evolution sometimes acts to increase efficiency, multicellular organisms are not subject to the same sorts of **selective pressures** of microorganisms that are constantly competing for rapid cell division. Cell division in multicellular organisms must be controlled. Extensive research has shown that most signal transduction pathways involve incredibly circuitous and complex cascades of cellular events, involving large numbers of moving parts. This is consistent with a metaphorical blind watchmaker, and inconsistent with intelligent design.

Cell-surface receptors

Cell-surface receptors are usually membrane-bound proteins that bind to molecules that act as signals. In some cases (such as **insulin**[l]) the signaling molecules themselves consist of protein chains. Some signals, such as molecules involved in the sense of smell, are a variety of small molecules. Sometimes these molecules are found in bodily fluids, and sometimes they are associated with other nearby cells. Some signaling events involve physical interactions with large proteins on the outer surface of other cells, as is the case with the mammalian immune system (see Chapter 16). And signals are not always molecular in nature: some cell-surface receptors, such as the **photoreceptor proteins**[m], respond to light.

Signal-receptor binding and interaction events trigger changes in receptor shape and/or their ability to bind to other molecules. Then, the signals are transmitted in various ways to the interior of the cell. Usually, this results in events within the cell nucleus that alter the expression of specific sets of genes.

Transmembrane receptors

Signal transduction takes place when stimulated receptors change shape and/or bind to other responsive molecules or structures. Sometimes, these responses involve changes in **enzymatic** activity.

Often, **cytoplasmic** and **nuclear** events within the interior of the cell are initiated by **transmembrane receptors**. These are protein chains that weave in and out of the cell membrane. The portions that extend outside the cell are the signal receptors, while the portions on the inside of the cell are responsible for initiating cascades of cytoplasmic and nuclear events by interacting with other molecules. These interacting molecules, in turn, initiate other

[l] https://en.wikipedia.org/wiki/Insulin

[m] https://en.wikipedia.org/wiki/Photoreceptor_protein

interactions. Ultimately, these culminate in changes in cellular activity that are often brought about by changes in the **control of gene expression**.

Some of the responsive events in the cascade involve other molecules known as **secondary messengers.**[n] Cytoplasmic events often cause enzymes to change their patterns of activity. Many signal transduction cascades involve enzymes that add or remove phosphate groups that are attached to the side-chains of particular amino acids. Such phosphorylation or dephosphorylation events change the shapes of protein chains and/or alter their ability to bind to other molecules. Sometimes, secondary messengers are small molecules such as **ions**. Ultimately, changes in the control of gene expression are often brought about by **transcription factors**[o] which interact with chromosomal DNA and alter gene **transcription** (Figure 3), an important step in the control of gene expression.

Evolution of the Genetic Apparatus

The evolution of the **genetic apparatus**[p] not only demonstrates the combined organizing power of EEP and Natural Selection, but provides a quintessential example of a complex, composite innovation that lays the groundwork for other useful innovations to arise in the future. **Protein synthesis**, storage of **genetic determinants** in the **genome, gene duplication**, control of gene expression and reproduction of complex organization during development all come under the rubric of the genetic apparatus.

Undoubtedly, as with other higher levels of complex organization, the genetic apparatus arose from simpler forebears. But even in its most primitive forms, evolution of the genetic apparatus greatly accelerated evolution of numerous other innovations. DNA has the advantage of greater stability than RNA sequences, and more easily adopts the appearance and refinement of new instances of biological organization, at various levels of complexity, under Natural Selection.

The **amino acid sequences** of protein chains, that arise via **protein synthesis**, are determined and controlled by hereditary sequences of

[n] https://en.wikipedia.org/wiki/Second_messenger_system

[o] https://en.wikipedia.org/wiki/Transcription_factor

[p] https://en.wikipedia.org/wiki/Introduction_to_genetics

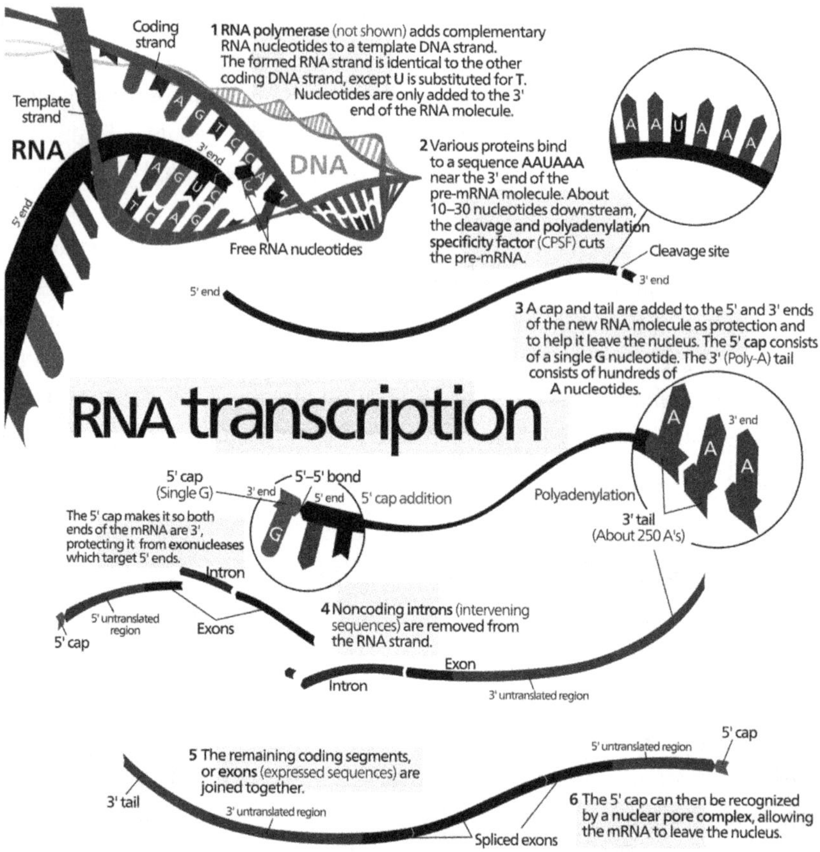

Figure 3. Overview of eukaryotic RNA transcription and processing. RNA polymerase enzymes synthesize an RNA strand from a DNA template. The sequence of the RNA strand is determined by hydrogen bonds that form between complementary DNA and RNA nucleotides during synthesis. Mature messenger RNA (mRNA) molecules are produced when introns are removed from the newly synthesized RNA by splicing, and by addition of a cap and a sequence of AAA... that are added to the ends of the processed RNA.

Source: Page-link: https://commons.wikimedia.org/wiki/File:MRNA.svg. File-link: https://upload.wikimedia.org/wikipedia/commons/9/9b/MRNA.svg. Attribution: By Kelvinsong [CC BY 3.0 (https://creativecommons.org/licenses/by/3.0)], from Wikimedia Commons. Rendered in B&W.

DNA. At the same time, the replication, expression, and evolution of those DNA sequences is dependent on a variety of different proteins, such as **DNA polymerases** and **transcription factors**. All of the inner-workings of living cells are part of an interwoven fabric of molecular interactions.

For biologists to understand various mechanisms and levels of complexity, they label them and find ways to perform controlled experiments and observations on specific units of structure and function. But although our maps of reality depend on labeling of separate units, the actual molecular realm knows no such distinctions. Statements such as "A determines B" are usually approximations of a more complex, nonlinear and interdependent reality.

Early stages of molecular evolution

We do not know very much about the steps in the origin of protein synthesis, and the steps that led to the linkage of protein synthesis to stored and heritable genetic determinants.

Several lines of evidence suggest that in the deep history of life, RNA was probably the first genetic material, rather than DNA.[q] Some RNA structures can function as enzymes. Modern-day ribosomes, for example, depend on specialized chains known as ribosomal RNA to form the peptide bonds that link amino acids together during protein synthesis. Presumably, ribosomal RNA—and ribosomes—evolved from simpler and less-refined forms of RNA that had some ability to link amino acids together. RNA molecules can be converted into DNA by two chemical modifications: the removal of a hydroxyl (OH) group of ribose, which turns it into deoxyribose, and the addition of a methyl group (CH_3) to change **uracil** bases into **thymine** bases.

Today, we find viruses that use several different forms of DNA or RNA as their genetic material. These include **single-stranded** DNA **viruses, double-stranded** DNA viruses, single-stranded RNA viruses, and double-stranded RNA viruses. Some researchers theorize that viruses may have played a role in evolutionary transition leading to replacement of RNA as the stored genomic material of cells[r] [36].

Single-stranded DNA has the advantage of greater stability than RNA sequences, and double-stranded DNA has greater capacity for repair.

---

[q] https://en.wikipedia.org/wiki/RNA_world

[r] Available online at https://biologydirect.biomedcentral.com/track/pdf/10.1186/1745-6150-7-13?site=biologydirect.biomedcentral.com

Since DNA can also base-pair with complementary RNA, the double-stranded DNA of cellular chromosomes has both the ability to replicate into two identical copies, and also to express its protein-coding information by pairing with a complementary RNA sequence, which today is known as messenger RNA.

In nucleated cells, the messenger RNA provides a way to synthesize proteins when it travels through nuclear pores into the cytoplasm and binds to ribosomes. That way, a master copy of the DNA remains protected and available in the nucleus and can serve as a template to quickly generate large numbers of a particular protein sequence by producing multiple copies of messenger RNA.

How DNA sequences can determine amino acid sequences

Transfer RNAs (**tRNA**s) are relatively short RNA chains that fold up into 3D clover-leaf shapes (Figure 4). Specialized tRNAs form covalent bonds to specific amino acids and carry them to the ribosomes.

At the ribosomes, messenger RNA **codons** (sequential three-nucleotide triplets) pair with complementary sequences (**anticodons**) on tRNA molecules (Figure 5). This involves **hydrogen bonding** between the **nitrogenous bases**.

Tables that map codons to specific amino acids are referred to as tables of the **genetic code** (Figure 6).

In this way, the inherited sequence of amino acids in each protein chain is synthesized at the ribosomes, a process known as **translation** (Figure 7). Each sequence is determined by the DNA sequence from which the messenger RNA sequence was transcribed. This is the fundamental link between the "genetic code" and protein synthesis used by all types of cells today. The complex structure of the modern genetic apparatus—that involves transcription of messenger RNA and protein synthesis at the ribosomes—must have evolved from simpler forebears. For example, the linkage of specific amino acids to tRNAs requires special enzymes called **aminoacyl tRNA synthases** (**aaRS**) that consist of protein chains. Therefore, there must have been earlier ways to couple hereditary information carried by RNA sequences to protein synthesis before tRNA synthases had evolved. The general idea of refinement and replacement of simpler molecular structures

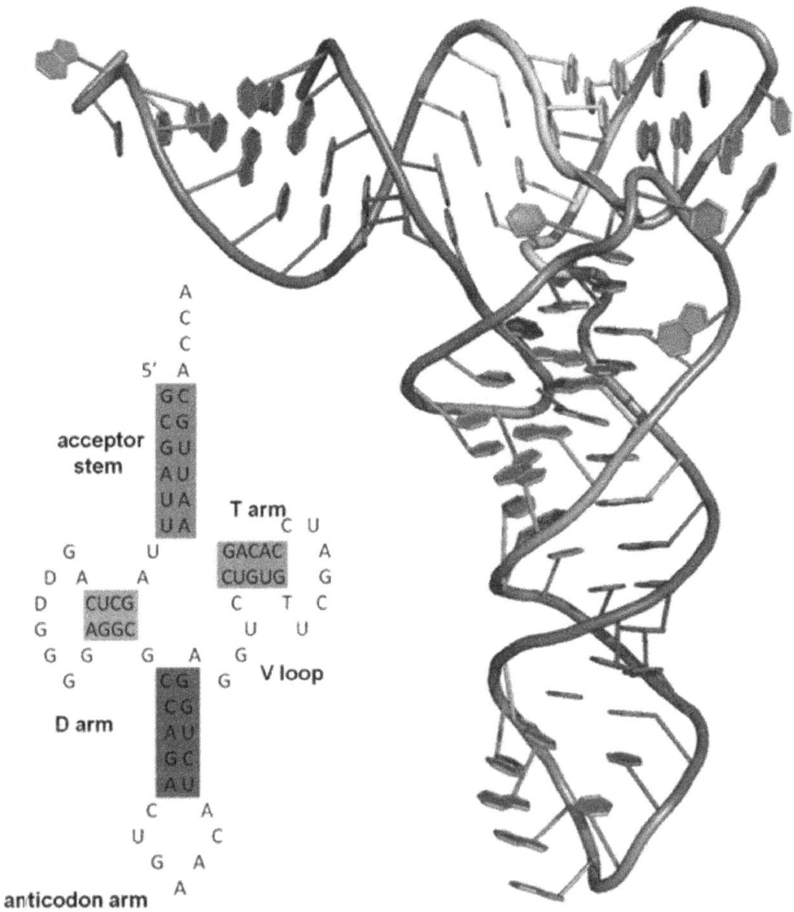

Figure 4. Diagram of folded transfer RNA (tRNA) molecule. Self-pairing of complementary RNA nucleotides causes the molecule to fold up into a functional 3D clover-leaf shape. The acceptor stem is attached to the correct amino acid by a specific enzyme. This specificity results from the precise complementary shapes and chemical properties of the enzyme, the tRNA, and the amino acid. The anticodon arm has the nucleotide triplet that is complementary to the mRNA codon. At the ribosome, enzymatic activity and base-pairing assure that the proper amino acids are added in the proper sequence to each newly synthesized protein chain.

Source: Page-link: https://commons.wikimedia.org/wiki/File:TRNA_all2.png. File-link: https://upload.wikimedia.org/wikipedia/comSmons/b/bf/TRNA_all2.png. Attribution: By Kyle Schneider (SchneiderKD) (Transferred by BQmUB2010090/Original uploaded by SchneiderKD) (Schneider KD (Original uploaded on en.wikipedia)) [Public domain], via Wikimedia Commons. Rendered in B&W.

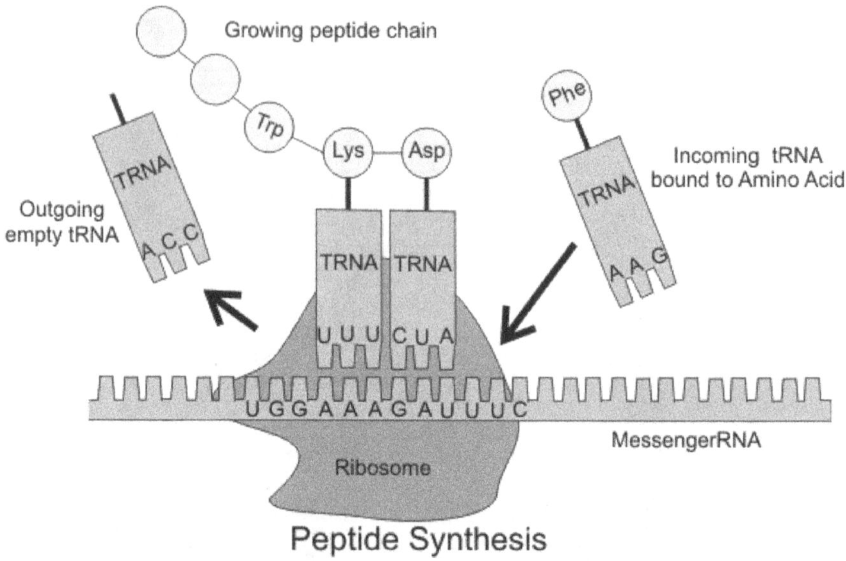

Figure 5. Peptide synthesis during production of a protein chain. Complementary base-pairing between codons of mRNA and anticodons of tRNA molecules attached to specific amino acids assure that the information transcribed from the DNA sequence determines the correct sequence of amino acids in each newly synthesized protein chain.

Source: Page-link: https://commons.wikimedia.org/wiki/File:Peptide_syn.png. File-link: https://upload.wikimedia.org/wikipedia/commons/0/0f/Peptide_syn.png. Attribution: By Boumphreyfr [CC BY-SA 3.0 (https://creativecommons.org/licenses/by-sa/3.0) or GFDL (http://www.gnu.org/copyleft/fdl.html)], from Wikimedia Commons. Rendered in B&W.

and functions by other ones is a thoroughly Darwinian concept that fits well with the ideas of the "Modern Synthesis" concerning the roles played by random mutations and Natural Selection. The discovery of DNA, protein synthesis, and the "genetic code" all took place after the "Modern Synthesis" and are generally referred to as neo-Darwinism today.

Six Updated Perspectives on Timeless Molecular Innovations

Rethinking "random mutations"

Molecular innovations serve as both platforms and tools that create new opportunities for higher levels of biological complexity to arise. All such

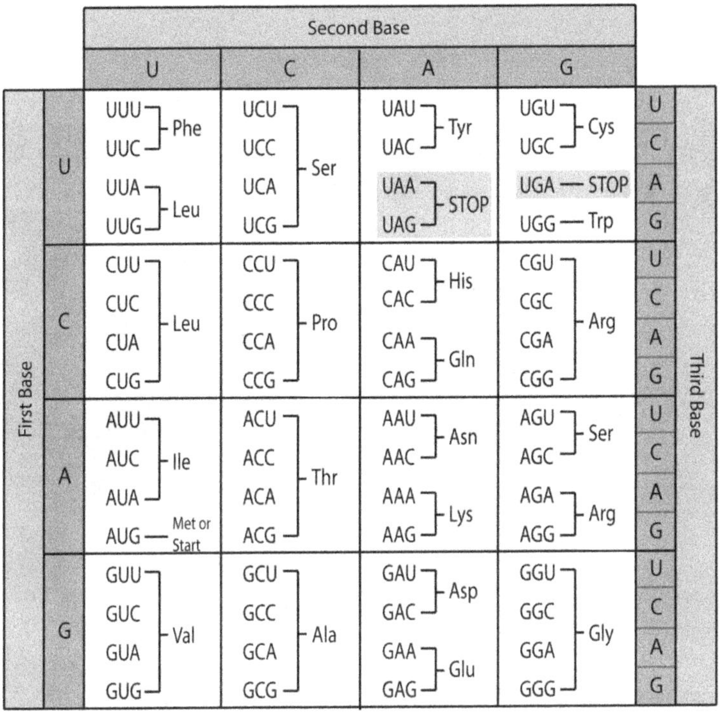

Figure 6. Table of the "genetic code". Between one and six nucleotide triplets, called codons, are associated with a particular amino acid. For example, UUU and UUC are associated with the amino acid phenylalanine, abbreviated Phe, while UUA and UUG are associated with the amino acid leucine, abbreviated Leu. In addition, UAG, UGA, and UAA serve as STOP signals that terminate the newly synthesized protein chain.

Source: Page URL: https://commons.wikimedia.org/wiki/File:Genetic_Code.png. File URL: https://upload.wikimedia.org/wikipedia/commons/8/8a/Genetic_Code.png. Attribution: By Sarah Greenwood [CC BY-SA 4.0 (https://creativecommons.org/licenses/by-sa/4.0)], from Wikimedia Commons.

innovations arise via blind natural forces yet have tremendous organizing power. Changes in DNA sequences—often referred to as **mutations**—represent an important way that changes can be stored and transmitted from generation to generation. These mutations were originally thought to be entirely random events, but that was an oversimplification.

The concept of random mutations became a key element in a revised version of Darwin's classic theory, which came to be known as the "Modern Synthesis", from the title of Julian Huxley's popular book of 1942 titled *Evolution: The "Modern Synthesis"* [23].

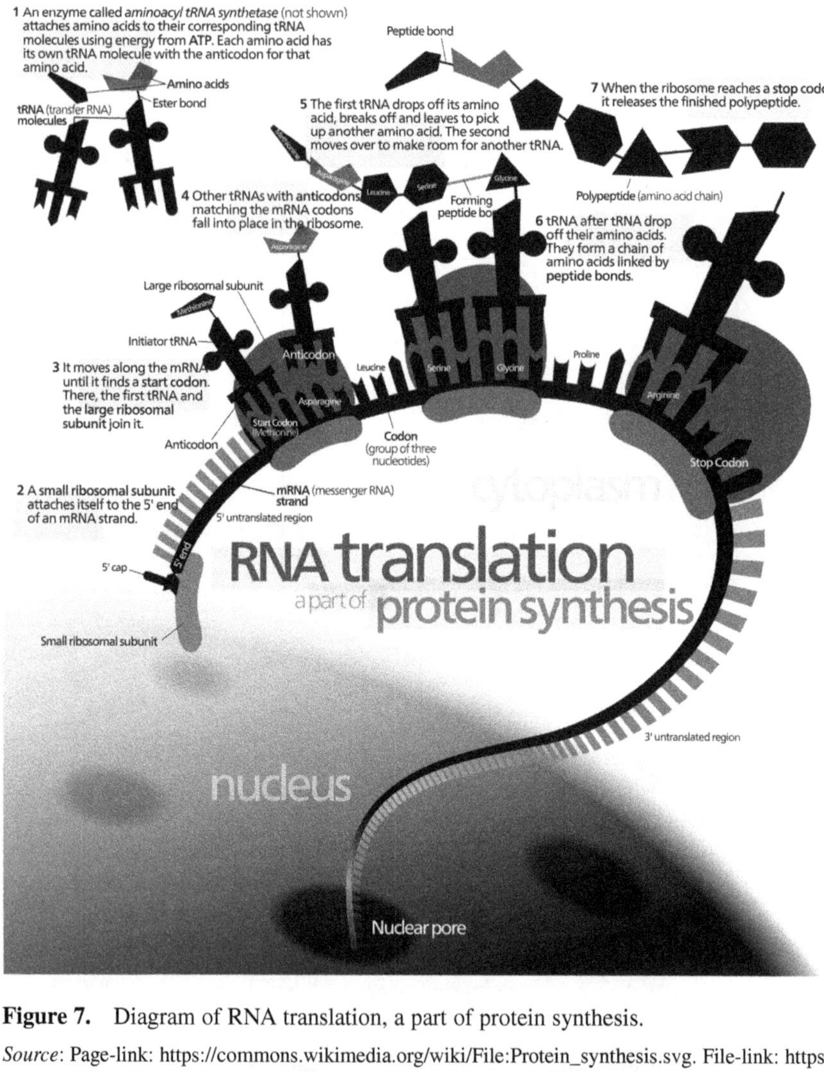

Figure 7. Diagram of RNA translation, a part of protein synthesis.

Source: Page-link: https://commons.wikimedia.org/wiki/File:Protein_synthesis.svg. File-link: https://upload.wikimedia.org/wikipedia/commons/4/44/Protein_synthesis.svg. Attribution: By Kelvinsong [CC BY 3.0 (https://creativecommons.org/licenses/by/3.0)], from Wikimedia Commons. Rendered in B&W.

The view of genetic variation from the "Modern Synthesis" asserts that random mutations serve as raw material for the evolutionary process, and that selection is the creative force. This reflects an attempt to update and defend classical Darwinian theory against attacks by creationists, and also reflects early limitations of our understanding of genome evolution.

In a popular account of the "Modern Synthesis" published in 1953, Huxley [37] neatly summarized the semantic implications of the concept of random mutations as follows:

> Mutation merely provides the raw material of evolution…it is a random affair, and takes place in all directions… [The effects of mutations] are not related to the needs of the organism, or the conditions in which it is placed.

To this day, the concept that variation arises from "random mutations", completely unrelated to the needs of the organism, remains a dominant theme in textbook accounts of Natural Selection.

The concept of "random mutations" championed by Julian Huxley requires rethinking for several general reasons. Six of those reasons arise from the ways that molecular innovations are constantly creating new opportunities and changing the potential for future innovations. Changes in DNA should not be viewed solely as independent events that arise by "chance" alone. Nor should they be viewed as arising from goal-seeking, design, or planning by a supernatural being. Evolution is subtler and more nuanced than either of those limited points of view—a deeper understanding of the natural forces involved is crucial and will be described in the following six sections.

(1) *Tools that facilitate and accelerate genome evolution*

In general, when speaking of evolution, we need to distinguish at least three categories of structure and function for molecular interactions:

(1) Molecular interactions may contribute directly to the survival and reproduction of the individual.
(2) Molecular interactions may serve as tools that generate structures and functions during the development of the individual.
(3) Molecular interactions may serve as tools that contribute to genome evolution.

The third category—tools that contribute to **genome evolution**—is the main focus of this section. To understand the significance of those tools, however, it is also necessary to understand the second category—the functional significance of various types of DNA sequences.

The genome is far more than a mere collection of DNA sequences, and genomic DNA sequences do not exist in isolation. The genome represents a platform containing several innovations that serve as tools for:

(1) Storing and transmitting genetic information from cell to cell and from generation to generation.
(2) Accelerating and facilitating further genomic evolution in a variety of ways.
(3) Linking evolved, heritable information from the genome to developmental processes that generate complex biological structures and functions (discussed more in *Generative Tools in Development* below).

Every day, laboratory biologists throughout the world expand the depth and breadth of our knowledge concerning the interactions and functions of specific sequences in the genome. Most of the new discoveries are focused on the complex chromosomes of **eukaryotic** cells.

How can we index, leverage, and keep track of these discoveries? The ENCODE Consortium represents a success story for international scientific cooperation. Here's a description from their website[s]:

> The ENCODE (Encyclopedia of DNA Elements) Consortium is an international collaboration of research groups funded by the National Human Genome Research Institute (NHGRI). The goal of ENCODE is to build a comprehensive parts list of functional elements in the human genome, including elements that act at the protein and RNA levels, and regulatory elements that control cells and circumstances in which a gene is active.

ENCODE clearly demonstrates that genomic DNA sequences perform a broad range of functions. Both nonrepetitive, complex sequences and simpler sequence elements with varying degrees of complexity contribute to these functions. The 1970 concept that much of the genome consists of "junk DNA" should be replaced with a more modern view of a "scrapyard".

---

[s] https://www.encodeproject.org/

Consider this evolutionary perspective on repetitive sequences in the genome, from the year 2000 [38]:

> Our view of the entire phenomenon of repetitive elements has to now be revised in the light of data on their biology and evolution… repetitive elements interact with the surrounding sequences and nearby genes... may serve as recombination hot spots or acquire specific cellular functions such as RNA transcription control or even become part of protein coding regions [and provide a] very efficient mechanism for genomic shuffling. As such, repetitive elements should be called genomic scrap yard rather than junk DNA.[t]

This is a clear statement of several general roles that repetitive sequences play in genomic evolution—not only as raw material, but also as tools that facilitate further genomic evolution.

Some of these roles summarized here will be revisited in the context of **gene duplication** and **transposable elements** (Chapter 16). Repetitive sequences that are either perfect or imperfect **tandem arrays** of repeating units become hotspots for additional events that alter the genome. As repetitive sequences expand, they increase the likelihood of new events that can either add or delete repeating units. This can be a **self-accelerating process**, because sequences with more repeat units can be more prone to expansion than shorter ones. Simple tandem repeats can also be readily transformed into more complex repeat units by additional mutations in individual nucleotides of DNA.

Repetitive sequences also become hotspots for unequal crossover events which can lead to the duplication of parts of genes, entire genes, or arrays containing multiple genes. Mechanisms such as **slipped-strand mispairing**[u] (**replication slippage**[v]) and **unequal crossovers** often change the genome more rapidly than **point mutations**[w] [39]. Other

---

[t] https://www.researchgate.net/publication/222538342_Genomic_scrap_yard_how_genomes_utilize_all_that_junk

[u] https://en.wikipedia.org/wiki/Slipped_strand_mispairing

[v] https://en.wikipedia.org/wiki/Replication_slippage

[w] https://www.researchgate.net/publication/19827130_Slipped-Strand_Mispairing_A_Major_Mechanism_for_DNA_Sequence_Evolution

mechanisms such as transposable elements—common mechanisms that insert duplicates of entire sequences in random locations in the genome—contribute to a substantial proportion of entire genomes, including the human genome.

When multiple mechanisms act on the same regions, they can have synergistic effects on genome evolution. For example, slipped-strand mispairing followed by point mutations can lead to unequal crossing over and can accelerate gene duplication events.

To summarize, genome evolution is more complex and diverse than the concept of "random mutations" from the "Modern Synthesis", which mostly focused on the role of simple mutations in protein-coding sequences of DNA.

(2) *Platforms that bring cellular entities together*

At a variety of levels of complexity, cellular structure and function arises from molecular interactions. These interactions often involve temporary binding events which change the shape and biological properties of proteins. Some interactions increase or decrease the amount of messenger RNA that is transcribed by particular genes. Some lead to **enzymatic** changes in macromolecules. A large class of proteins known as enzymes bind to one or more specific reactants and facilitate chemical changes that would otherwise be unlikely to occur. Sometimes, binding events are spontaneous, and require no additional energy source. For others, energy-rich molecules such as ATP are utilized by enzymes or pumps to drive nonspontaneous chemical reactions.

When evolution creates new genes, changes existing genes, or alters the expression of genes, this changes the molecular composition of various cellular compartments. This will lead to new opportunities for encounters between macromolecules that did not take place before. If by chance, any fortuitous encounters—which may at first result in only weak binding events—prove to be useful in some way to the cell or the individual, they may contribute to the survival and reproduction of that individual by Natural Selection. This can lead to those heritable changes appearing more frequently in the genome, which may be subject to additional random changes and selection that can fine-tune the interactions in useful ways. This is the probable explanation for the evolution of signal transduction pathways, which often resemble **Rube Goldberg devices** because they involve cascades of events involving large numbers of individual molecules. These cascades of events do not have the appearance of design or engineering, but rather appear to have arisen by random events and opportunities for new interactions when formerly separate entities are brought together in new contexts, as discussed in Chapter 2. When those new interactions prove useful, they tend to be preserved by Natural Selection.

When new encounters between formerly separate units create new, higher-level units of structure and function that are more complex, these new units have emergent properties. These new properties can lead to still more new opportunities for innovation. Although this involves variation and selection, this kind of Natural Selection provides a richer explanation for evolution than the classical Darwinian paradigm. Darwin's original concept of infinitesimal variations that gradually accumulate fails to capture the new opportunities that emerge when prior evolutionary events create opportunities for new interactions.

As we learn more about the inner workings of cells, an updated theory of Natural Selection provides a richer and more plausible explanation for the natural origins of complexity than classical Darwinian theory or the "Modern Synthesis". This updated perspective emphasizes the interactions between multiple macromolecules that lead to units of structure and function. New emergent properties, as well as new opportunities for further complexity, arise when those units come together. When, by trial and error, they prove useful to the organism, they will tend to be preserved in subsequent generations. This is followed by classical Darwinian fine-tuning and refinement by additional variation and selection events.

(3) *Generative tools in development*

In each generation, developing organisms use genomic information to generate the complex cellular organization of the individual. Recent discoveries in Evolutionary Developmental Biology reveal that these developmental tools are conserved throughout the deep history of evolution (at least since the **Cambrian Explosion** some 500 million years ago), and are often modified and reused, even in distantly related species. Since developmental tools can act in a variety of ways to generate the structures and functions—the phenotypes—of individuals of various species, the original concepts of genotype and phenotype—terms coined at the beginning of the 20th century—are an incomplete description of the dynamic nature of hereditary information.

Instead of simply referring to genotype and phenotype, the Updated Evolutionary Synthesis will make a distinction between the **Ecological Phenotype**—the structures and functions of the developed individual—and the **Generative Phenotype**—which consists of the tools that generate that organization during development. This distinction makes it easier to understand how biological organization arises during development, and also makes it easier to understand how Generative Tools can be modified and reused in various ways. The evolution of developmental toolkits—i.e. Generative Phenotypes of various species—have extraordinary organizing power. Such organizing power goes far beyond the gradual accumulation of useful changes in classical Darwinian Natural Selection. Such modern insights make it easier to understand how biological complexity arises, both at the level of evolution and at the level of development, in each generation. These topics will be further explored in Chapters 11–13.

(4) *The scrapyard of random or fortuitous variation*

The concept of "random variation" followed by selection—where selection is the creative force, whereas variation is just the raw material—fails to capture the numerous roles played by useful innovations that can be subsequently modified, duplicated, and reused. They may later be deployed to serve additional useful functions later on.

The collection of previous innovations that are stored in the genome and transmitted to subsequent generations of each evolutionary lineage can be viewed as a scrapyard of potentially useful modules. This means that selection is not the only creative force, since the previous innovations already have varying degrees of complex organization and utility. New innovations build on previous ones in complex ways. In particular, innovations that serve as platforms or tools that create new opportunities for further evolution must be viewed as more than mere raw material.

Today, Huxley's concept from 1953, that the effects of mutations "are not related to the needs of the organism", must be viewed as incomplete and therefore misleading. A more nuanced explanation recognizes that various types of innovations can be modified and/or reused in a variety of different contexts. Each time these innovations—or derivatives of those innovations—are reused, their significance and relation to the needs of the organism actually depends on the context in which this occurs. That context includes the changing molecular realm within cells, when formerly separate entities can form higher-level units of structure and function.

(5) *Beyond the Baldwin effect*

The **Baldwin Effect** was originally proposed to describe the connection between behaviors that individuals learn through real-world experience, and the subsequent evolution of instinctive behaviors. Proponents argue that fortuitous behaviors that happen to be useful to the individual will tend to ensure survival and reproduction. Consequently, they will increase the frequencies of any genes that those individuals transmit to future generations when they reproduce. Therefore, transmitted genes that fortuitously contribute to learning abilities will tend to be preserved. That means that learned behaviors may become fixed in populations at the genetic level. This general concept appears to have stood the test of time and was accepted by many proponents of the "Modern Synthesis". It helped to explain how complex phenomena, such as behavior, might arise from the collective effects of random variation in natural populations.

The Baldwin Effect offers a Darwinian explanation for instincts that arise from constellations of genes fortuitously preserved by selection for individuals that exhibit useful behaviors. Genes associated with learning can collectively generate lasting, instinctive phenotypes associated with specific useful behaviors.

If that is true, then it also seemed reasonable to hypothesize that the evolution of a variety of phenotypes—not only instincts—could be generated by the collective actions of constellations of genes.

When the "Modern Synthesis" was in its heyday in the mid-20th century, the cellular mechanisms that link constellations of genes to complex traits were largely a metaphorical "black box."

Since that time, continuous advances in our understanding of the inner-workings of cells have led to an increasingly detailed understanding of the molecular, cellular, and developmental mechanisms that give rise to complex phenotypes. Many of these are discussed in subsequent chapters of this book.

In each generation, the entire genotype of each individual generates, through various developmental processes, a phenotypic *gestalt* which, if useful, may be preserved by selection.

Phenotypic effects involving the actions of multiple genes are called polygenic effects. Sometimes, hundreds of DNA sequences interact in subtle ways that can only be resolved by sophisticated genetic techniques. Recent studies have revealed, for example that slight variations in hair color may be influenced by over a hundred different DNA sequences, including some sequences that contribute in ways that are not currently fully understood—a finding that may prove useful in forensic investigation [40].

(6) *Fine-tuning and incremental improvements, revisited*

In the UES, the classical Darwinian concept of gradual, incremental changes that accumulate, via selection, to generate larger changes, should be seen as just one aspect of a more complex and nuanced process that extends through multiple generations. Sometimes, changes arise from the actions of useful tools that modify the genome or participate in development.

Sometimes, emergent units of structure and function arise when previous innovations create new opportunities. In all of those cases, a variety of sources of variation may act sequentially, and sometimes synergistically, in the evolutionary process. Sometimes, sudden events, such as gene duplication events, may be subject to subsequent fine-tuning by means of incremental variations.

Our expanding knowledge of the inner-workings of cells demonstrates that innovations have potential that is constantly changing in the context of other actual events, and variation can have nonrandom aspects, and represents more than mere "raw material" for selection.

Chapter 5

The Origin of Life

I shall argue that the distinction between a 'living planet'—one that is geologically active—and a living cell is only a matter of definition. There is no hard and fast dividing line. Geochemistry gives rise seamlessly to biochemistry. From this point of view, the fact that we can't distinguish between geology and biology in these old rocks is fitting. Here is a living planet giving rise to life, and the two can't be separated without splitting a continuum.

—Nick Lane, from *The Vital Question* [11]

The Big Picture

Theories about the origin of life on Earth have evolved considerably since 1953, when Miller and Urey performed an experiment that produced some rudimentary organic molecules by simulating hypothetical conditions of early planet Earth. This, combined with the discovery of DNA, RNA, and the "genetic code", led to a family of hypotheses—known as the "Prebiotic Soup" (PBS)—that have dominated the field. However, recent empirical discoveries in geology and biochemistry led to a fundamentally new family of hypotheses that postulate that life began at "undersea alkaline hydrothermal mounds" (UAHM). These hypotheses generally provide a more cogent explanation than the PBS. The UAHM predict that the evolution of life forms resembling prokaryotic cells will be a frequent occurrence throughout the universe, but more complex life forms resembling eukaryotic cells will be relatively rare.

Two Schools of Thought Regarding the Origin of Life on Earth

Two major schools of thought characterize scientific discussions concerning the **origin of life** on Earth. These schools of thought do not represent a single unified hypothesis, but rather, a large number of detailed hypotheses that differ in a variety of ways that are discussed among specialists. Therefore, they are best described as families of hypotheses. We can call them the **Prebiotic Soup** family (**PBS**) and the **Undersea Alkaline Hydrothermal Mound** (**UAHM**) family.

At the current time, these theories are families of competing hypotheses. Each has a fascinating history. Each grew out of a distinct conceptual framework. Each is supported by an interdisciplinary set of scientific principles and by contrasting lines of empirical evidence.

Drawing from both scientific and popular publications, I will argue that the UAHM family represents a better approximation of the truth, regarding the origin of life on Earth, as well as the deep history of both **prokaryotic** and **eukaryotic** cells. To support this argument, I will consider several key questions:

First, the prebiotic soup family of hypotheses:

(1) What are the major features of the PBS?
(2) Why did the PBS family of hypotheses come first?
(3) What biochemical, genetic, and cellular principles are required to understand the PBS?
(4) What are some major milestones in the history of the PBS?
(5) What are the conceptual foundations of the PBS?
(6) How do the conceptual foundations of the PBS constrain the scientific questions that have been asked?

Next, the UAHM family of hypotheses:

(1) What are the major features of the UAHM?
(2) What biochemical, genetic and cellular principles are required to understand the UAHM?

(3) Why was the UAHM family of hypotheses proposed decades after the PBS?
(4) What are some major milestones in the peer-reviewed history of the UAHM?
(5) What other implications of the UAHM family of hypotheses were summarized in Nick Lane's popular 2015 book, *The Vital Question* [11]?

The Prebiotic Soup Family of Hypotheses (PBS)

What are the major features of the PBS?

The Prebiotic Soup family argues that life on Earth arose spontaneously by means of natural, physical, and chemical forces. The building blocks of life—that is, organic molecules such as amino acids and nucleotides—were created by energy sources such as ultraviolet light and lightning discharges, and some building blocks may have arrived when carbon-containing meteorites struck the Earth—generating tremendous heat that could create more complex molecules from simpler ones.

Mixtures of complex organic molecules were somehow concentrated and contained in proto-cells by some sort of self-organizing primitive membranes.

At some point, highly improbable events made it possible for these proto-cells to replicate themselves. After that, the classical Darwinian process of Natural Selection took over. Proto-cells that were better able to capture energy and raw materials and use them to replicate their own molecular structures preferentially survived, and gradually increased in complexity. Eventually, this gave rise to RNA, proteins, DNA, cell membranes, and the "genetic code" and genetic apparatus that we find in modern cells.

Why did the prebiotic soup (PBS) family of hypotheses come first?

In 1871, Charles Darwin wrote a famous letter in which he proposed that life first arose in a "warm little pond" by means of natural forces, and then evolved by means of Natural Selection.

The Prebiotic Soup term was coined by JDS Haldane in 1921 and the hypothesis was reformulated in terms consistent with known scientific principles.

The PBS burst into the public awareness in 1953, with the publication of the famous Miller–Urey experiment, a laboratory simulation in which amino acids were produced by means of electrical discharges in a simulated primitive Earth atmosphere.[a]

In the same year, Watson and Crick described the double-helical structure of DNA.

Along with the "Modern Synthesis", these publications provided the conceptual framework for the PBS.

What are the biochemical, genetic, and cellular principles that are required to understand the PBS?

Modern versions of the PBS are all based on our knowledge of the conserved features shared by living cells:

(1) the presence of a **cell membrane**
(2) the ability to capture and utilize external **energy** and **nutrients** and synthesize **ATP**
(3) **enzymes** and **cellular metabolism** that build up and break down **organic molecules**
(4) four types of macromolecules made up of organic subunits such as amino acids, fatty acids, nucleotides, and carbohydrates
(5) DNA and RNA synthesis and their replication
(6) Protein synthesis, DNA/RNA templates, and the "genetic code"

In addition, neo-Darwinian theory is invoked to explain how simple self-replicating organic molecules and proto-cells could become more complex by means of random variation and Natural Selection, giving rise to prokaryotic and eukaryotic cells.

---

[a]https://en.wikipedia.org/wiki/Miller%E2%80%93Urey_experiment

What are some major milestones in the history of the PBS?

Darwin's "warm little pond"

Over a decade after the publication of his *On the Origin of Species*, Darwin's thoughts turned to the origin of life itself. In his now-famous letter to J.D. Hooker in 1871, Charles Darwin included speculations that were far ahead of his time:

> It is often said that all the conditions for the first production of a living organism are now present, which could ever have been present.—But if (& oh what a big if) we could conceive in some warm little pond with all sorts of ammonia & phosphoric salts,—light, heat, electricity &c present, that a protein compound was chemically formed, ready to undergo still more complex changes, at the present day such matter [would] be instantly devoured, or absorbed, which would not have been the case before living creatures were formed.—[b]

This was a remarkable flash of insight, considering how little was known in 1871 about the critical importance of **amino groups** (related to ammonia) in **proteins** and the **phosphate backbones** (related to phosphoric salts) in DNA and RNA. Darwin's focus on energy sources such as light, heat, and electricity—which can and do contribute to synthesis of complex **organic compounds**—was also brilliant. In essence, Darwin's "warm little pond" closely resembled the "prebiotic soup" idea that dominated research many decades later, as described in the following sections.

The prebiotic soup

In the 1920s, over 50 years after Darwin's letter to Hooker, Alexander Oparin and J.B.S. Haldane separately published similar hypotheses: namely, that naturally occurring inanimate chemicals were transformed into more complex organic molecules by natural energy sources such as lightning, UV light, or heat. Then eventually these organic molecules became sufficiently complex and concentrated to spontaneously take on

[b] https://www.darwinproject.ac.uk/letter/DCP-LETT-7471.xmlr

the attributes we associate with living organisms—such as the ability to replicate themselves and to gradually change and evolve in a Darwinian fashion.

Haldane coined the term "prebiotic soup" to describe this hypothesis. This hypothesis—that life arose spontaneously from nonliving materials—is called **abiogenesis**. The philosophical and theoretical significance of the abiogenesis hypothesis—not only for evolutionary theory but also for our collective vision of our own origins—can hardly be overstated. Abiogenesis complemented classical Darwinian theory and provided a more comprehensive and durable naturalistic 20th century theory for the origin and evolution of life on Earth that lasted for many decades.

The Miller–Urey hypothesis

The Prebiotic Soup hypothesis did not take root until 1953, when a paper in the journal *Science* described what is now widely known as the famous Miller–Urey experiment. This laboratory simulation was intended to test the hypothesis that life on Earth arose spontaneously, by natural forces, and from simple gases thought—in the 1950s—to be present on the primitive Earth.

As described by Miller in his *Science* paper [26]:

> The idea that the organic compounds that serve as the basis of life were formed when the Earth had an atmosphere of methane, ammonia, water, and hydrogen instead of carbon dioxide, nitrogen, oxygen, and water was suggested by Oparin and has been given emphasis recently by Urey and Bernal…
>
> In order to test this hypothesis, an apparatus was built to circulate CH_4, NH_3, H_2O, and H_2 past an electric discharge. The resulting mixture has been tested for amino acids by paper chromatography…

Popular accounts of the results of this simulation caused quite a stir—and captured the imagination of the general public—because amino acids are the subunits of proteins, a major class of macromolecules found in all living cells. This idea—that life could spontaneously arise from the reaction of simple inorganic chemicals—was so intriguing that it has dominated textbook accounts ever since.

Popularization of the "replicator" concept

Richard Dawkins' popular 1975 book, *The Selfish Gene* [41], was another significant milestone in the popular understanding of evolutionary theory. Dawkins provided a clear explanation of the conceptual link between abiogenesis and classical Darwinian theory in his **replicator** idea:

> ...At some point a particularly remarkable molecule was formed by accident. We will call it the *Replicator*...
>
> ... Think of the replicator as a mould or template. Imagine it as a large molecule consisting of a complex chain of various sorts of building block molecules. The small building blocks were abundantly available in the soup surrounding the replicator...
>
> ...But now we must mention an important property of any copying process: it is not perfect. Mistakes will happen...
>
> ...As mis-copyings were made and propagated, the primeval soup became filled by a population not of identical replicas, but of several varieties of replicating molecules, all 'descended' from the same ancestor...
>
> ... Replicators of high longevity would therefore tend to become more numerous...

Dawkins emphasized the implicit assumption that the formation of such a "particularly remarkable molecule" that could replicate itself seems extremely improbable:

> ...This may seem a very unlikely sort of accident to happen. So it was. It was exceedingly improbable... But in our human estimates of what is probable and what is not, we are not used to dealing in hundreds of millions of years...

Similar arguments also point out that such an exceedingly rare event only needs to happen once.

What are the conceptual foundations of the PBS?

Six implicit assumptions

Up until recently, six implicit assumptions have characterized most theories of abiogenesis:

(1) The most important distinction between living and nonliving entities is self-replication.
(2) The first "living" carbon-based life form on Earth was a discrete, self-replicating entity.
(3) Self-replication was an exceedingly improbable event.
(4) All life on Earth descended from a singular, rare event.
(5) The early evolution of life was classically Darwinian—in the sense of variation, competition and selection.
(6) Complexity in early life forms arose by Natural Selection.

Influence of the "central dogma" of Molecular Genetics

Equally important, however, was the revolution in Molecular Genetics that led to the discovery of the "genetic code", and the link between nucleic acid sequences (DNA and RNA) and the amino acid sequences of proteins, discussed in Chapter 9.

The outsized 21st century influence of what Francis Crick playfully called the "central dogma" of Molecular Genetics—that hereditary information flows from DNA to RNA to proteins—is illustrated by a 2007 web page, maintained on the NASA web site for historical purposes, titled "Life's Working Definition: Does It Work?"[c]

The Astrobiology Magazine staff interviewed Carol Cleland, who described herself as a scientist "interested in formulating a strategy for searching for extraterrestrial life that allows one to push the boundaries of our Earth-centric concepts of life". Cleland argued that what we really need is a "general theory of living systems, as opposed to a definition of life".

That was an intriguing idea. But unfortunately, the astrobiology article never got off the ground. Instead of discussing a general theory of

---

[c]https://www.nasa.gov/vision/universe/starsgalaxies/life's_working_definition.html (30 July 2018).

living systems, the article fell back to Earth-bound theories, stating that the biggest hurdle to explaining the origin of life *on Earth* was "explaining the origin of the complex cooperative schema worked out between proteins and nucleic acids".

Considering that the article was supposed to be about general principles of extraterrestrial life, not life on Earth, inclusion of this perspective seemed out of place. While it is true that life forms on Earth do depend on a cooperative schema, where the "genetic code" of DNA and RNA determine the sequence of amino acids in proteins, that should not constrain a general theory of living systems.

Or should it? This is an important question. Does the origin of life depend on interactions between DNA, RNA and proteins, not only elsewhere in the Universe, but on Earth as well? This begs the following question: did the earliest origins of life on Earth depend on complex interactions between complex molecules such as DNA, RNA and proteins? Was the earliest life form a highly-improbable lucky accident? Or rather, did DNA, RNA, proteins and the "genetic code" arise later?

How do the conceptual foundations of the PBS constrain the scientific questions that have been asked?

The conceptual synergy between the Miller–Urey experiment and the emerging revolution in Molecular Genetics was described by Nick Lane in 2015 in his fascinating popular book, *The Vital Question* [11]:

> As an experimental discipline, the origin-of-life field dates back to 1953 and the famous Miller–Urey experiment, published in the same year as Watson and Crick's double helix paper. Both papers have hung over the field ever since, casting a shadow like the wings of two giant bats…

The Undersea Alkaline Hydrothermal Mound Family of Hypotheses (UAHM)

Alkaline hydrothermal vents are known undersea geological phenomena that were predicted in 1988 [31] and first empirically observed in the year 2000 [42]. They generate new mineral formations on the seafloor, known as **undersea alkaline hydrothermal mounds (UAHM).**

What biochemical, genetic and cellular principles are required to understand the UAHM?

Ions and ionization

Atoms (or molecules) that have a net negative or positive charge—due to addition or removal of electrons—are called **ions**. The most common isotope of hydrogen atoms has a single negatively charged electron and a single positively charged proton. Remove two electrons from a molecule of hydrogen gas (H_2) which contains two hydrogen atoms (H_2) and you break the chemical bond. The result is that you are left with two positive ions (H^+), equivalent to two free protons.

A variety of atoms are routinely **ionized** by natural chemical processes. **Metals** such as iron, and **nonmetals** such as sulfur—which play a major role in the chemistry of UAHM—are commonly found in ionized states. Iron atoms that have lost two electrons become Fe^{2+} ions, and if they lose a third electron, they become Fe^{3+} ions. Nonmetals such as sulfur often gain electrons rather than losing them, so sulfur ions (S^{2-}) are also quite common.

What are protons?

Protons are **hydrogen atoms** that have been stripped of their **electrons**. Hydrogen atoms are neutral in electrical charge because each atom consists of one **positively charged** proton and one **negatively charged** electron. These electrical charges add up to neutrality (zero). When an electron is removed from a hydrogen atom, this process is called ionization, and the remaining proton has a positive charge. This positive ion can also be represented as H^+.

Acidity, alkalinity, and pH

The concentration of protons that are dissolved in water determines the acidity or alkalinity (basicity) of the resulting solution. Water-based solutions, such as ocean water, hydrothermal fluids, and the contents of living cells, are known as **aqueous solutions**.

pH is the quantitative measure of proton concentrations. Each pH unit represents a 10-fold decrease in proton concentration, because pH is defined as the negative logarithm (base 10) of the proton concentration.

Equilibrium, the pH of pure water, and hydroxide ions

The pH of pure water actually represents an **equilibrium** in which a small fraction of **water molecules** spontaneously split apart, resulting in a small concentration of dissolved protons, and an equal concentration of dissolved **hydroxide** ions (**OH^-**). The pH of pure water is 7 because the **molarity** of the excess protons (**H^+ ions**) is 10^{-7} **moles** per **liter**, and the **negative logarithm** of that proton concentration, which defines pH, is 7.

Solubility of ions in water, ionic compounds, and precipitation reactions

Many free positive and negative ions are highly soluble in water—such as sodium ions and chloride ions that separate into Na^+ and Cl^- when table salt is dissolved. However, certain combinations of ions—such as iron and sulfur ions—($Fe^{2+}$and S^{2-}) are more strongly attracted to each other than they are to the **dipoles** in water. Therefore, they come out of the **solution** as solid ionic compounds. This process is known as **precipitation**, and the resulting solids are known as **precipitates**.

The effect of pH on solubility

Since protons and hydroxide ions can combine with a variety of ionic substances, changes in pH will often result in changes in the solubility of those ions.

Neutralization reactions at undersea alkaline hydrothermal mounds

When alkaline hydrothermal fluids meet with acidic ocean water, a neutralization reaction takes place, resulting in the formation of more water molecules and a shift in pH that is closer to the neutral pH of 7. In UAHM, this neutralization reaction, combined with ion–ion attractive forces, causes certain ionic compounds to come out of the solution and form precipitates.

The catalytic capabilities of iron monosulfide and other precipitates

Iron monosulfide (FeS) precipitates are of special interest, because they can function as catalysts—that is, they can participate in chemical reactions and lower their activation energy, increasing the likelihood that those reactions will take place, as well as making them occur at a faster rate.

Catalysts are essential for many organic chemical reactions

Catalysts such as FeS are essential for the formation of many of the key organic compounds required for the earliest life forms to arise. Unlike the PBS, UAHM can readily explain the availability of mineral catalysts long before organic catalysts such as proteins or RNA molecules would have had a chance to evolve.

Porous mineral containers and semi-permeable membranes

Another key aspect of the precipitates that form UAHM mounds is that the resulting rock-like structures routinely form hollow tubes that are highly **porous**—that is, they have large numbers of tiny openings that allow certain substances to pass through. If these openings selectively allow certain dissolved substances to pass through, while retaining others, they are referred to as **semi-permeable**. This also provides an essential requirement for the earliest life forms to arise: namely, the presence of **semi-permeable membrane-**like structures. This would permit containment of solutions of specific organic molecules, long before lipids—the organic components of membranes of modern cells—had a chance to form.

Prebiotic and biological membranes

Tubes that are **porous** can function as **semi-permeable membranes** that retain specific **inorganic** or **organic** materials. They can also function as partitions to separate different components. Although the concept of a **biological membrane** is associated with **lipids**—such as the **fatty-acid** based lipids found both within and at the cell wall of bacteria, in organelles, and surrounding the **cytoplasm** of **eukaryotes**—the earliest forms of life would benefit from naturally-occurring mineral membranes that precipitate when hydrothermal fluids and ocean water are mixed. These precipitates contain high concentrations of key **catalytic** ionic compounds such as **iron monosulfide**.

Three abundant and continuous sources of potential energy at the undersea mounds

In addition to the heat generated at alkaline hydrothermal mounds by distant volcanic activity, two continuous energy sources are as follows:

(1) The chemical energy of **oxidation** and **reduction** reactions.
(2) The **potential energy** of gradients of ions such as protons, or electrons, or both.

The potential energy of gradients of ions or electrons

This potential energy becomes highly significant when higher concentrations of **ions** or **electrons** are separated by semi-permeable, catalytic barriers, and when selected ions can flow through these semi-permeable membranes to dissipate that potential energy. These energy flows can perform potentially useful tasks, and in the context of the origin of life, useful tasks would include the large-scale production of complex organic chemicals of all sorts.

Proton gradients within the alkaline hydrothermal mounds

When alkaline hydrothermal fluids and acidic ocean water are separated by semi-permeable membranes resulting from precipitation, proton gradients can form across those mineral membranes. The concentration of **protons** (hydrogen ions) will be lower on the alkaline side of the membrane and higher on the acidic side of the membrane; this is called a **proton gradient**, which represents potential energy.

Proton motive force and proton flow

Pores in the membrane provide a path for protons to flow from one side of the membrane to the other. The protons will naturally flow from the side with the higher proton concentration to the side with the lower concentration, driven by their own kinetic energy. This energy flow can drive chemical reactions and can result in the production of more complex organic molecules from simpler ones.

Reacting carbon dioxide with hydrogen gas in the hydrothermal mounds

The special conditions of proton gradients, embedded mineral catalysts, and other factors such as heat and pressure, are conducive to the dissolved carbon dioxide in ocean water reacting with the hydrogen gas from hydrothermal flows. Other compounds containing nitrogen, phosphorous, and other substances can also react. As a result, complex organic compounds can be produced, driven by the abundant flows of raw materials and energy, and catalyzed by iron monosulfide (FeS) and other mineral catalysts in the hydrothermal mounds.

Electrons, oxidation–reduction reactions, and reduction potential

Atoms can gain or lose electrons. When atoms lose electrons, they become oxidized, and this process is called oxidation. When atoms gain electrons, they become reduced, and that process is called reduction. Often, these reactions occur simultaneously, when one atom donates one or more electrons to another atom. Since oxidation and reduction reactions tend to occur together, they are usually referred to as redox reactions.

Since electrons carry energy, energy tends to flow from oxidized atoms or compounds to reduced ones.

The relative tendency of atoms to get reduced—and gain energy—is known as their **reduction potential**. In general, differences in reduction potential represent a source of potential energy, and an important driving force of chemical reactions.

Strong chemical bonds and organic molecules

Strong chemical bonds arise when a pair of atoms shares one or more pairs of electrons. These are called **covalent bonds**, and the resulting combinations of two or more atoms are referred to as **molecules**.

Organic molecules—that is, molecules that contain carbon but are more complex than carbon atoms or carbon dioxide—are usually held together by covalent bonds.

Activation energy, spontaneous and nonspontaneous reactions

Chemical reactions involve making and breaking of chemical bonds. Usually, this requires a threshold of energy, known as the activation energy, for a chemical reaction. Sometimes, this activation energy will be supplied by sources of heat, such as a spark. Other times, the activation energy will be lowered by a catalyst, so that less energy is required to get the reaction started. Chemists also distinguish between reactions that, once activated, require an input of energy from an external source. These are called nonspontaneous reactions. Alternatively, reactions that, once activated, do not require any external energy source—and often will release energy to the environment—are called **spontaneous reactions**.

Often, chemical reactions that break down covalent compounds into smaller units tend to be spontaneous, whereas those that build up larger, more complex molecules tend to be nonspontaneous.

For that reason, the synthesis of more complex organic molecules from simpler ones generally requires an external source of energy. In the protected environments of the alkaline hydrothermal mounds, external sources of sustained energy are found in abundance in the form of heat, redox potential, and proton gradients.

The concept of energy dissipation and equilibrium

In any closed system, energy will tend to become uniformly distributed. A proton gradient maintained by a membrane is in a state of **disequilibrium**, because there is more energy on the side of the membrane that has more protons. The gradient represents potential energy, and protons will tend to flow from higher to lower proton concentrations through any opening in that membrane. When the protons are uniformly distributed, the system has reached equilibrium. The process of reaching **equilibrium**—which depends on the proton flow from one side of the membrane to the other—represents one example of **energy dissipation**.

Covalent chemical bonds also represent potential energy. A mixture of different types of molecules that have higher and lower amounts of potential energy in their covalent bonds is in a state of disequilibrium.

Differences in reduction potential are one example of differences in chemical energy. A mixture of molecules that have different reduction potentials is inherently unstable—it is in disequilibrium. Given the proper conditions—namely, when the proper amount of activation energy is supplied, the bonds will become unstable, electrons will flow, and the molecules will tend to undergo chemical reactions that distribute the chemical bond energy—and electrons—more evenly. As with the proton gradients, chemical bond energy—called enthalpy—will tend to flow from where the energy is higher to where it is lower. In general, chemical reactions will result in the flow of energy from compounds with higher-energy bonds into compounds with lower-energy bonds. Excess energy will tend to be released as **heat**. When activated, such chemical reactions will tend to occur spontaneously.

In order to get chemical bond energy to flow in the opposite direction—from low to high—among activated molecules, an external source of energy must be provided. Such reactions are called **nonspontaneous** chemical reactions.

This is an important consideration in origin of life theories, because in general, life depends on larger organic molecules that are formed by nonspontaneous chemical reactions. Therefore, the continuous production of larger organic molecules depends on a continuous external energy source. Such energy is abundant in UAHM.

In the presence of an abundant source of external energy, and especially, when activation energy has been lowered by catalysts, energy can flow from low to high, and this can result in production of molecular products of chemical reactions that have more energy than they started with.

Organization, entropy, and information

If you drop an egg on the floor, it will break and splatter. It is highly unlikely, however, that you will ever see a broken egg spontaneously rearrange itself into an intact egg once it has been broken.

If you take a cable or wire, such as a charging cable or a wired set of earbuds for a mobile phone, and put it in your pocket, it will tend to get tangled up, simply because there are more random ways for the wire to get tangled than for it to be nicely arranged in an ordered coil. This is an example of the fact that things tend to move from **order** to **disorder**.

If you take a tangled cable out of your pocket and arrange it in a coil, you will need to expend energy in order to do that. In general, transforming a disordered state into an ordered one requires an energy input.

In general, transforming small inorganic molecules such as carbon dioxide, water, hydrogen and ammonia into larger organic molecules such as **amino acids**, **fatty acids**, **sugars**, or **nitrogenous bases**, will require external energy. Additionally, arranging those molecules into macromolecules such as protein chains, lipids, carbohydrates, or nucleic acids will require external energy. This is partly because of the potential energy contained within the chemical bonds of the larger organic molecules, and it is also because specific, ordered arrangements of atoms into molecules—and small molecules into macromolecules—represents a specific ordered state rather than a random state.

The concept of **entropy** can be used to quantitatively describe the amount of **organization** (order) in a particular object or arrangement of objects. The greater the degree of organization, the lower the entropy.

Organized molecules, held together by covalent bonds, can also be thought of as containing information. In order to replicate this organization, the specific arrangement needs to be reproduced, and this specific arrangement represents a kind of information. The higher the information content of a molecule, the lower the entropy. More complex molecules have lower entropy, and more implicit informational content.

DNA sequences consist of specific arrangements of four different types of **nucleotides**, which have four different kinds of **nitrogenous bases**, designated **A**, **C**, **G**, **and T**. Because of our prior knowledge regarding DNA and the "genetic code", we tend to think of these

nucleotide sequences in DNA as representing **information**, but we are less likely to think of the arrangement of amino acids in protein chains as representing information. But this distinction is completely subjective on our part. In terms of **information content**, complex macromolecules such as proteins *do* contain **implicit information content**. Additionally, specific arrangements of enzymes that are embedded into biological membranes, and any complex organic molecules—or complex interactions between molecules—also contain implicit informational content.

This shift in perspective, in which organic molecules of all kinds—not just DNA or RNA—contain implicit information—may seem unfamiliar, but it's an important one, because it helps us to get past the limiting assumption that life depends on information that can only be found in DNA or RNA. This will be important when considering the origin of life in the UAHM. The earliest life forms must have depended on interactions between specific arrangements of complex organic molecules and catalysts. These must have evolved prior to the origins of DNA, RNA, proteins, or the "genetic code". Presumably, those specific arrangements of complex organic molecules already contained implicit information before DNA, RNA, or proteins could appear.

What are the major features of the UAHM?

UAHM provide continuous sources of geochemical energy, in the form of proton gradients, heat, and redox potential. The mounds provide abundant sources of energy and carbon as well as catalysts and containment. Both in the present day and on the primitive Earth, UAHM mounds provide continuous flows of inorganic raw materials containing carbon, hydrogen, oxygen, and nitrogen. Additionally, particular mounds can be active for hundreds of thousands and perhaps many millions of years. The vents consist of hollow tubes that provide containment of large quantities of newly synthesized organic materials arising from natural chemical reactions. This is a means of venting waste products and mineral catalysts, such as compounds of iron, nickel, sulfur, and other metals that represent natural catalysts which lower the energy barriers so that—starting with carbon dioxide and hydrogen gas—complex organic materials can be spontaneously produced in abundance.

The UAHM family of hypotheses focuses on energy and metabolism rather than nucleic acids, proteins, replication or heredity, and assumes that all of these things came later, within the protected natural incubators of the undersea mounds. Unlike the rare events postulated by PBS, sound geological and cosmological principles and empirical observations predict that UAHM will be common natural phenomena and are expected to arise by means of natural forces on planets throughout the universe [11].

Modern forms of the UAHM family of hypotheses include discussions that attempt to explain the deep history of the known cellular forms of life, including the early origins of free-living **archaeal** and **bacterial** (eubacterial) cells as well as the later evolution of **eukaryotic** cells [11]. Detailed **peer-reviewed** discussions are thoroughly grounded in modern biochemical principles, as well as modern DNA-based comparisons. DNA-based comparisons use sophisticated computer programs to systematically compare differences in DNA sequences, to determine ancestry and evolutionary distances between ancient lineages of cells.

For prokaryotes known as **acetogens** and **methanogens**, peer-reviewed papers provide plausible and highly-detailed scientific discussions regarding the early origins of highly conserved **core metabolic pathways**. Generally accepted biochemical principles guide discussions regarding the physical and chemical constraints on generation of **high-energy compounds**, as well as **macromolecules**, **template-based replication**, and the **genetic code**.

Why did the UAHM family of hypotheses come after the PBS?

We have already seen that the PBS family of hypotheses have early roots in the history of biology, driven by the early suggestion of Darwin in his letter to Hooker in 1871, and the subsequent publication of the Miller–Urey experiment and the structure of DNA in 1953. Discovery of the "genetic code", combined with Darwinian principles such as the replicator concept popularized by Dawkins, also played an important role in framing discussions concerning the origin of life in terms of templates, replication, and Natural Selection ever since.

The UAHM hypotheses did not begin to take shape in the scientific literature until 1988, however, long after discussions of abiogenesis had already been framed in terms of the PBS, replication, the "genetic code", and Natural Selection.

In terms of the geology of UAHM, rocks of volcanic origin that have especially high content of metals such as iron and magnesium—called **ultramafic** rocks—become hydrated to become rocks known as serpentinite, in a process known as **serpentinization**. "Serpentinization—A Review" by Judith Moody appeared in the peer-reviewed literature in 1976 [43] and UAHM were not predicted until the 1980s. This later discovery, combined with the QWERTY effect described above, all but guaranteed that UAHM hypotheses would have to compete with a firmly entrenched PBS school.

Consequently, even though the peer-reviewed arguments are now more favorable for UAHM, these are still largely absent from textbook discussions concerning the origin of life. If it were not for Nick Lane's 2015 publication of *The Vital Question* [11], this topic would be largely off the radar for nonscientists, despite the fact that some popular articles have appeared over the years and could easily be found with a Google search. Searches, after all, do require a searcher who has a good reason to take the time to enter particular key words.

What are some major milestones in the peer-reviewed history of the UAHM?

In contrast to modern PBS hypotheses, modern UAHM hypotheses focus primarily on energy and the origins of core metabolic pathways, rather than DNA, RNA, proteins, and the "genetic code".

This important difference was driven by the prediction of both the existence of special geological features of UAHM, beginning in the 1980s, which led to the discovery of the first currently active UAHM mound in 2000, which was later characterized in some detail [42].

Here, we will consider a sample of some of the peer-reviewed papers that represent milestones in the evolution of the modern UAHM family of hypotheses. Only a small selection of the extensive scientific literature on this subject can be presented here. Among the many contributors to this fascinating scientific literature are Mike Russell, A.J. Hall, Bill Martin, and Nick Lane.

But in 1961, long before the UAHM hypothesis came of age, Peter Mitchell proposed his extraordinary theory of cellular proton pumps, known as chemiosmosis[d], as lucidly summarized in an earlier popular book by Lane about mitochondria [125]. Briefly, Mitchell's chemiosmotic hypothesis proposed that energy could be stored across membranes in living cells—and in particular, in mitochondria—by pumping protons across a biological membrane. Mitchell went on to explain how that stored energy could be recovered in the form of high-energy ATP molecules, by using the proton motive force from the protein reservoir to drive the remarkable micro-machines known as ATP synthases. This hypothesis led to testable predictions that turned out to be true, and the chemiosmotic theory is now widely accepted throughout the scientific community.

Our UAHM story begins decades later, in 1983 when geochemist Mike Russell had a brilliant flash of insight. As beautifully retold by journalist Tim Requarth in his popular account published in *Aeon*:

> Mike Russell found his moment of inspiration on a warm spring evening in Glasgow in 1983, when his 11-year-old son broke a new toy. The toy in question was a chemical garden, a small plastic tank in which stalagmite-like tendrils grew out of seed crystals placed in a mineral solution. Although the tendrils appeared solid from the outside, when shattered they revealed their true nature: each one was actually a network of hollow tubes, like bundles of tiny cocktail straws.[e]

At the time, Russell, a geologist, was struggling to understand an unusual rock he had recently found. It too was solid on the outside but inside was full of hollow tubes, their thin walls riddled with microscopic compartments. It dawned on him then that this rock—like the formations in his son's toy—must have formed in some unusual kind of aqueous solution. Russell posited a whole new geological phenomenon to explain it: undersea hydrothermal hotspots where mineral-rich water spewed from the Earth's interior and then precipitated in the cool surrounding water, creating chemical gardens of towering, hollow rocks growing up from the ocean floor.

---

[d] https://en.wikipedia.org/wiki/Chemiosmosis

[e] https://aeon.co/essays/why-life-is-not-a-thing-but-a-restless-manner-of-being

Russell's early insight led to a remarkable series of peer-reviewed, collaborative scientific papers. In a short letter to Nature in 1988 [31], Russell and co-authors formally predicted the existence of a special kind of undersea vent that had not yet been discovered—called an alkaline hydrothermal vent—that should form towering, porous structures at the interface where alkaline, mineral rich warm fluids contact the cooler, acidic seawater.

The Russell/Hall synthesis of 1997

A peer-reviewed paper by Russell and A.J. Hall, published in 1997 [44] proposed several key aspects of the UAHM family of hypotheses. These include the following arguments:

(1) Life emerged about 4.2 billion years ago at the interface between alkaline hydrothermal fluids and acidic sea water.
(2) Thin, semi-permeable membranes consisting of minerals precipitated at the interface.
(3) The semi-permeable membranes included natural catalytic materials such as iron monosulfide (FeS) and nickel (Ni), which facilitated the production of organic materials.
(4) A natural pH gradient, and electron transfer across the membranes, provided the driving force for chemical oxidation and reduction reactions.
(5) The authors provide detailed examples of the ways that primitive chemical reactions could have led to the core, conserved biochemical reactions that resemble key metabolic pathways of living cells.

Discovery of the Lost City Hydrothermal Field

In the year 2000, 12 years after the Nature publication predicted their existence, the Lost City Hydrothermal Field (LCHF) was discovered at depths ranging from 750 to 900 m below the surface, in the Mid-Atlantic Ocean (Figure 1). Later explorations revealed mounds consisting of numerous alkaline hydrothermal vents that are home to a variety of

Figure 1. A modern-day undersea alkaline hydrothermal mound (UAHM). A 5-foot-wide flange, or ledge, on the side of a chimney in the LCHF is topped with dendritic carbonate growths that form when mineral-rich vent fluids seep through the flange and come into contact with the cold seawater.

Source: Page-link: https://commons.wikimedia.org/wiki/File:Lost_City_(hydrothermal_field)02.jpg. File-link: https://upload.wikimedia.org/wikipedia/commons/4/4d/Lost_City_%28hydrothermal_field%2902.jpg. Attribution: By National Science Foundation (University of Washington/Woods Hole Oceanographic Institution) (http://www.nsf.gov/od/lpa/news/press/01/pr0156.htm) [Public domain], via Wikimedia Commons. Rendered in B&W.

species that utilize the energy and raw materials provided by the vents today [42].

Early bioenergetic evolution

Fast-forward to 2013, and we find an extensive and detailed biochemical analysis [45] that summarizes several facets of a modern version of the UAHM family of hypotheses, and reviews numerous earlier peer-reviewed journal articles. Based on sound biochemical principles and empirical evidence, the review outlines a hypothetical path from the protected, energy-rich environment of alkaline hydrothermal mounds to the origin of free-living cells. The review begins with an elegant definition:

> Life is the harnessing of chemical energy in such a way that the energy-harnessing device makes a copy of itself.

It focuses on several aspects concerning the origin of life. Some of these are summarized as follows:

(1) Alkaline hydrothermal vents (which form UAHM) provide rich and continuous sources of potential energy and raw materials that are not at equilibrium, and where energy tends to dissipate.
(2) A specific biochemical pathway, known as the **Wood-Ljungdahl** (**acetyl-CoA**) pathway, provides a plausible route by which carbon could be fixed by processes driven by UAHM energy flows.
(3) The hydrothermal mounds provided natural mineral-based compartments containing simple precellular replicating organic entities.
(4) **Acetogenesis** and **methanogenesis** were probably the ancestral forms of energy metabolism that arose in the UAHM.
(5) The early evolution of life, formation of high-energy compounds such as **ATP**, and evolution of macromolecules including carbohydrates, lipids, proteins and nucleic acids, and the "genetic code" itself, all evolved within the protected environments of the UAHM mounds. This process was driven by abundant energy and raw materials and facilitated by the inorganic catalysts and containment that were all provided by the UAHM.
(6) The authors also summarize highly detailed and technical discussions that lead to detailed hypotheses concerning the origins of eukaryotic cells.

Additional technical reviews on this subject were later provided by Martin, Garg and Zomorski [46] and by Archibald [47] in 2015. Detailed technical discussions of specific hypothetical steps in the evolutionary process that presumably gave rise to both prokaryotic and eukaryotic cells can be found in those reviews.

The deep history of prokaryotic and eukaryotic cells

Nick Lane's popular 2015 book, *The Vital Question* [11] summarizes key aspects of the peer-reviewed literature in an engaging and accessible way. It is a great resource for further reading on the subject.

In addition to the points summarized above, Lane summarizes additional ramifications of the UAHM family of hypotheses, including the fact that UAHM are likely to arise, with high probability, on other planets throughout our Milky Way galaxy and throughout the universe. He predicts that cells resembling prokaryotes, which rely on proton gradients, are likely to have arisen independently by natural processes quite frequently on other planets. This point of view—that the origin of life is probable, rather than improbable—represents an important departure from the textbook idea of the PBS family of hypotheses, which argue that life first arose as an improbable event, and that the improbability of life is due to the improbability of the generation of a self-replicating set of organic chemicals from scratch in the prebiotic soup. The UAHM family of hypotheses argue that on the basis of fundamental principles, the abundant energy, raw materials, and other key conditions in alkaline hydrothermal mounds are powerful drivers of change. They are expected to generate highly complex organic chemicals and primitive metabolic pathways that dissipate the energy flows. These changes will be entirely consistent with known laws of chemistry, physics, and biochemistry.

One other point emphasized by Lane should be included in this chapter. Lane briefly summarizes the now-abundant empirical evidence for the common ancestry of two ancient lines of prokaryotic cells—the **archaea** and the **bacteria** (eubacteria). He argues these cell types arose relatively early in the deep history of life, in the protected environments of the hydrothermal mounds.

The origins of **eukaryotic** cells came much later, however. Lane summarizes a modern form of the **endosymbiotic** family of hypotheses to explain how eukaryotic cells first arose from a combination of archaeal and bacterial ancestral prokaryotic cells. Additionally, the modernized endosymbiotic hypotheses further explain why organelles such as **chloroplasts** (Figure 2) and **mitochondria** (Figure 3) resemble bacterial cells, are powered by proton gradients, and have their own DNA which is separate from

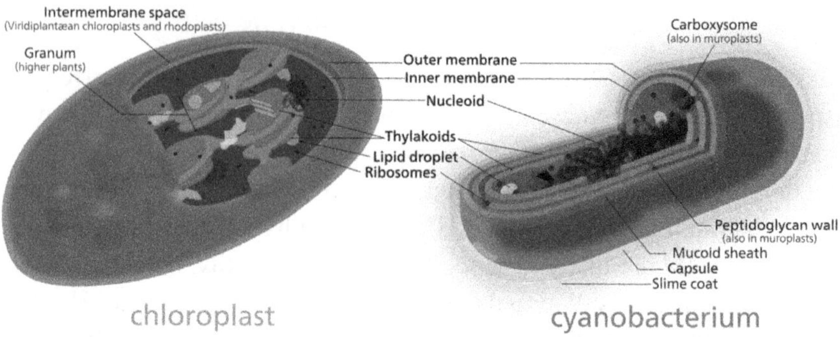

Figure 2. Diagram of a chloroplast—an organelle commonly found in plants—and a cyanobacterium. There is fossilized evidence—in stromatolites—that ancient cyanobacteria-like cells lived on the ancient earth some 3.5 billion years ago. The endosymbiotic theory—that ancient cyanobacteria-like organisms were engulfed by eukaryotic cells and evolved into chloroplasts of plants—is widely supported. Both have their own circular chromosomes. Both cyanobacteria and chloroplasts capture rays of sunlight to generate ATP and other high-energy molecules. Both generate proton gradients that depend on internal membranes, and those proton gradients generate more ATP. Both rely on the high-energy molecules to synthesize glucose molecules from carbon dioxide and water.

Source: Page-link: https://commons.wikimedia.org/wiki/File:Chloroplast-cyanobacterium_comparison.svg. File-link: https://upload.wikimedia.org/wikipedia/commons/9/92/Chloroplast-cyanobacterium_comparison.svg. Attribution: By Kelvinsong [CC BY-SA 3.0 (https://creativecommons.org/licenses/by-sa/3.0)], from Wikimedia Commons. Rendered in B&W.

the nuclear DNA of eukaryotes. **Chloroplasts** are the organelles frequently found in plant cells that capture the energy of sunlight and use it to synthesize glucose and other sugars. Mitochondria are found in both plant and animal cells, and transform the energy stored in glucose into ATP.

Lane argues that the evolutionary success of eukaryotic cells was a complex process requiring a great deal of trial-and-error, and that is the reason that eukaryotic cells did not appear until more than a billion years later, despite the fact that free-living bacterial and archaeal lineages of prokaryotic cells had already appeared on Earth well over 3.5 billion years ago. In fact, fossilized marine structures known as stromatolites that are about 3.5 billion years old closely resemble stromatolites produced by modern-day cyanobacteria (Figure 4).

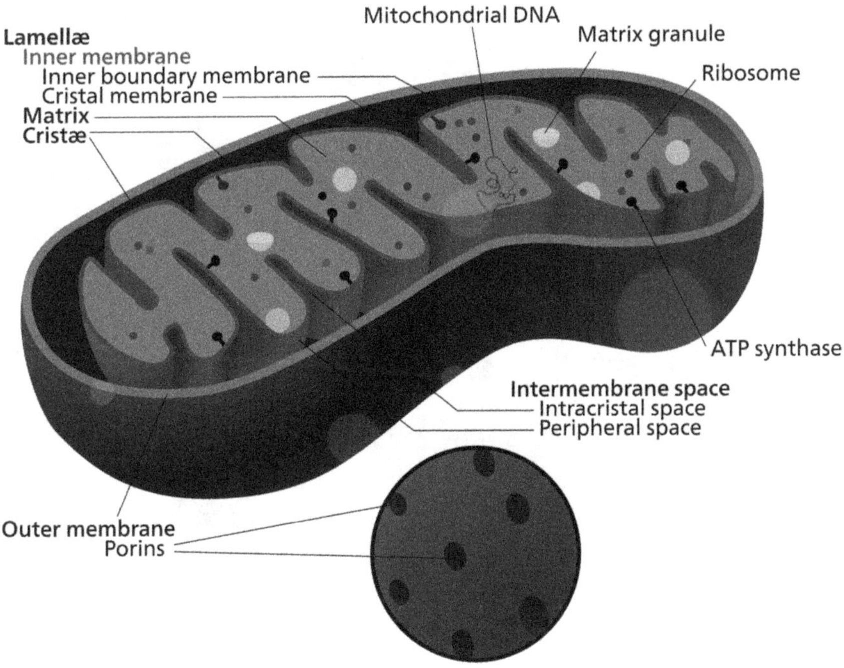

Figure 3. Diagram of a mitochondrion. Mitochondria bear several similarities to bacterial cells. Today, several lines of evidence support the hypothesis that eukaryotic cells evolved from a combination of archaeal and bacterial prokaryotic cells. Both mitochondria and bacteria have their own circular chromosomes. Mitochondria, archaea, and bacteria all use energy from sugars such as glucose to generate proton gradients and make ATP, an important universal energy currency of all living cells and one of the four nucleotides incorporated into RNA.

Source: Page-link: https://commons.wikimedia.org/wiki/File:Chloroplast-cyanobacterium_comparison.svg. File-link: https://upload.wikimedia.org/wikipedia/commons/9/92/Chloroplast-cyanobacterium_comparison.svg. Attribution: By Kelvinsong [CC BY-SA 3.0 (https://creativecommons.org/licenses/by-sa/3.0)], from Wikimedia Commons Rendered in B&W.

Complex Proteins and Organelles Involved in Cellular Energy Production and Conversion

All known cell types have evolved their own ways of generating and utilizing self-contained proton gradients. Early carbon-based life forms were presumably dependent on the naturally-occurring proton gradients of the UAHM. Cells evolved internal proton gradients, which freed them

Figure 4. Modern day stromatolites in Shark Bay, Australia. Stromatolites are calcified structures built from cyanobacteria. Ancient fossilized stromatolites indicate that cyanobacteria-like cells had already evolved on the ancient Earth as long as 3.5 billion years ago.

Source: Page-link: https://commons.wikimedia.org/wiki/File:Stromatolites_in_Sharkbay.jpg. File-link: https://upload.wikimedia.org/wikipedia/commons/1/1b/Stromatolites_in_Sharkbay.jpg. Attribution: By Paul Harrison [GFDL (http://www.gnu.org/copyleft/fdl.html) or CC-BY-SA-3.0 (http://creativecommons.org/licenses/by-sa/3.0/)], via Wikimedia Commons. Rendered in B&W.

from the UAHM and allowed them to seek other sources of energy and colonize the far corners of the land and seas. A hypothesis advanced by Lane and Martin [48] suggests that escape from the UAHM required the evolution of a membrane-bound protein that can pump two types of ions (including protons) in opposite directions, called an **antiporter**.[f]

Several complex innovations make it possible for cells to capture energy from the environment and convert it into useful forms such as ATP. In every case, these complex structures depend on protein innovations. In

[f]https://en.wikipedia.org/wiki/Antiporter

addition, membrane-bound enzymes that perform sequential steps must be properly inserted into the membranes. Today, **cell biologists** understand a great deal about the physical and chemical properties of proteins and **lipids** that cause them to routinely self-organize and self-assemble within living cells.[g] For example, membrane-bound proteins have evolved amino acid sequences called **signal peptides**[h] that lead to their transport and proper insertion. When combined with Natural Selection, this sort of detailed biological understanding adds depth and plausibility to the UES.

The endosymbiotic origins of chloroplasts and mitochondria transcend classical Natural Selection theory because the way that they arise cannot be described as gradual accumulation of variation. Nevertheless, Natural Selection clearly did transform both host and endosymbiont in each case, and the organelles have been modified and refined in an incremental fashion.

The **ATP synthase** enzymes that produce most of the ATP in chloroplasts and mitochondria is one of the most amazing examples of micromachinery in living cells. They convert the potential energy of proton gradients into ATP by means of a channel and a literal rotor that turns, effectively transforming the potential energy of the proton flow into mechanical energy and then into the potential energy of the high-energy bond of ATP[i] (Figure 5).

Debunking "irreducible complexity"

ATP synthase enzymes are examples of the organizing power of evolution. The extraordinary complexity and efficacy of complex molecules like ATP synthases has led certain creationists such as Michael Behe—a biochemistry professor—to construct pseudoscientific arguments intended to persuade nonscientists to reject evolutionary theory, and instead adopt the religious argument of **Intelligent Design**.

Behe argues [18] that such examples of complex structure and function represent examples of **Irreducible Complexity**. The claim assumes that since proteins like ATP synthases have so many complex subunits

---

[g] https://en.wikipedia.org/wiki/Cell_biology

[h] https://en.wikipedia.org/wiki/Signal_peptide

[i] https://en.wikipedia.org/wiki/ATP_synthase

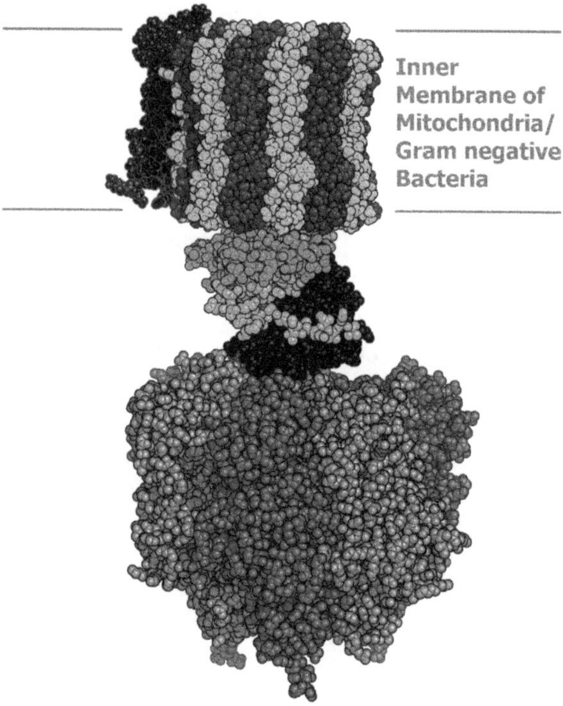

Figure 5. Computer-generated model of ATP synthase. This enzyme, consisting of multiple protein chains (different shades of gray), functions as a complex micro-machine. A rotor converts the potential energy of a proton gradient into mechanical energy, which is subsequently converted into chemical bond energy, during the synthesis of ATP. ATP is a major energy currency of all living cells.

Source: Page URL: https://commons.wikimedia.org/wiki/File:Atp_synthase.PNG. File URL: https://upload.wikimedia.org/wikipedia/commons/0/00/Atp_synthase.PNGAttribution: By Alex.X (enWiki (PDB.org for coordinate)) [GFDL (http://www.gnu.org/copyleft/fdl.html) or CC-BY-SA-3.0(http://creativecommons.org/licenses/by-sa/3.0/)]. Rendered in B&W.

(protein chains) that must all work together in order to function, complex structures could not arise via infinitesimal variations that are subject to Natural Selection, because too many different structures must change in a concerted fashion.

However, these arguments have been thoroughly discredited by biologists who have shown that complex structures such as ATP synthases, consisting of multiple protein chains, have evolved from simpler structures, with fewer subunits, that performed *different* functions that also contributed to survival and reproduction, and therefore arose at an earlier time by Natural Selection.

In other words, the earlier, simpler structures represented opportunities for Emergent Evolutionary Potential (EEP) to modify and reuse components in different ways. Those earlier structures already had some of the organized features that laid the groundwork for other potentially useful interactions to take place, and for various subunits to be reassembled in various ways—including instances where new structures are modified, and functions are improved and refined by the addition of new interacting subunits.

Proteins represent examples of the timeless aspect of EEP. Each sequential protein innovation is an opportunity. Each innovation lays the groundwork for further innovations to arise at later times—in other contexts involving different species and different environmental constraints. Innovations reflect a logical aspect of reality that transcends particular events—where the potential is constantly changing. New opportunities with increased organizing power are continually emerging in ways that transcend time and space.

Three Concluding Thoughts

The first conclusion is that our 21st century UES replaces the PBS family of hypotheses with the UAHM because they are better supported by relatively recent empirical discoveries and well-established principles of biochemistry.

The second conclusion is that we also need to dispense with the idea that DNA, RNA, proteins, and the "genetic code" provided the driving force for the earliest carbon-based life-forms on Earth. Instead, the prerequisites for the origin of life are found in the energy-rich protected environments of the UAHM, which also provided raw materials, venting of wastes, catalysts, and containment.

Third, although Darwinian Natural Selection clearly played an important role that was greatly accelerated when DNA, RNA, proteins, and the "genetic code" evolved, it is not the only principle that explains how complex organization arises among living things. These topics and more will be further explored in Chapter 6.

Chapter 6

Combining the Origin of Life with the Origins of Complexity

All life is organized as cells—compartments separated from their surroundings that spontaneously multiply with energy gleaned through self-contained, thermodynamically favourable redox reactions. From our viewpoint, physical compartmentalization from the environment and self-organization of self-contained redox reactions are the most conserved attributes of living things, hence inorganic matter with such attributes would be life's most likely forebear.

—William Martin and Michael J. Russell [49]

The Big Picture

This chapter focuses on the implications of an updated evolutionary perspective that bridges the gap between the origin of life (Chapter 5) and its subsequent evolution. Before Natural Selection could begin to take place, the protected, energy-rich environment of undersea alkaline hydrothermal mounds (UAHM) were sufficient to serve as an incubator for the earliest forms of biological complexity to arise. Energy flux, combined with the intrinsic Emergent Evolutionary Potential (EEP) of organic compounds, drove the origins of complexity before RNA, DNA, proteins, or the "genetic code" had a chance to evolve.

Later, once macromolecules and the genetic apparatus appeared (presumably within the UAHM), Darwinian Natural Selection provided a powerful tool that accelerated evolvability, dramatically increasing the organizing power of the early life forms. Subsequent innovations provided the ability to capture energy and nutrients, maintain internal proton gradients, and ultimately escape from the UAHM and reproduce their own kinds.

That critical role for Natural Selection—providing cells (and organisms) with the organization required to capture energy and nutrients in novel ways, has continued throughout the entire history of life on Earth.

Introduction

From Darwin's warm little pond to UAHM

William Martin and Michael Russell presented their developing hypothesis—based on the protected environment of the **undersea alkaline hydrothermal mounds** (**UAHM**; see Chapter 5)—to the Royal Society of London. In so doing, they continued a tradition extending back to the 19th century. Both Charles Darwin and Alfred Russel Wallace presented their evolutionary theories to the Royal Society in 1858[a] [50]. Martin and Russell's presentation was published in 2003 [49]. The title of that milestone publication was "On the origins of cells: a hypothesis for the evolutionary transitions from abiotic geo-chemistry to **chemoautotrophic** prokaryotes, and from prokaryotes to **nucleated** cells".

Clearly, theories concerning the origin of life had come a long way from Darwin's "warm little pond" of 1871.[b] It is time to bridge the conceptual gap between the origin of life and Darwinian theory. In this chapter, I hope to provide a new perspective that stimulates discussion as well as new experimental approaches.

Perhaps the best way to present this combined theory is to list its essential features in a modular way, followed by a corresponding list of predictions for each item on the list. Together, they offer a deeper perspective on the forces driving the origin of life, and the role of Natural Selection in its subsequent evolution.

---

[a]Available online at https://www.uv.mx/personal/tcarmona/files/2010/08/Kutchera-2003.pdf (30 July 2018).

[b]https://www.darwinproject.ac.uk/letter/DCP-LETT-7471.xml

Twelve Assertions to Help Bridge the Gap Between the Origin of Life and Darwinian Evolutionary Theory

Organic compounds

Carbon-containing molecules—organic compounds—offer a uniquely diverse range of possibilities that can be predicted on the basis of well-established chemical, biochemical, and physical principles. The structure and function of organic molecules is essential to the origin and evolution of life on Earth.

Natural organic synthesis

Natural synthesis of organic compounds requires raw materials, energy, catalysts, and permissive physical and chemical conditions such as pH, pressure, aqueous solutions, containment, and release of waste products.

Undersea alkaline hydrothermal mounds

Undersea alkaline hydrothermal mounds (UAHM), arising from **serpentinization** (a geological chemical reaction), provide both the **necessary and sufficient** conditions for the production of organic compounds.

Terrestrial and extraterrestrial UAHM

Stable UAHM probably appeared frequently on both the ancient and modern-day Earth. Similar conditions promoting undersea serpentinization are thought to be widespread on planets throughout the universe [11].

Driving force for organic synthesis and interactions

Facilitation of capture and utilization of raw materials and **energy** are universal drivers for both the origin and continuing evolution of carbon-based life forms, which all depend on organic synthesis and interactions.

Diverse and general ways to naturally facilitate organic synthesis and interactions at UAHM

Here is a partial list of possible ways to naturally facilitate organic synthesis and interactions at UAHM, that is meant to stimulate further thought, discussion, and predictive experimental designs:

(1) Channel natural energy flows.
(2) Improve efficiency and diversity of catalysts.
(3) **Tether** or embed **reactants** or catalysts in proximity to others to create sequential multi-step **metabolic pathways**.
(4) Concentrate reactants and/or useful **products**.
(5) Partition metabolic reactants, pathways, and/or products.
(6) Improve venting of waste products.
(7) Adjust **porosity** of **membranes**.
(8) Improve containment.
(9) Adjust chemical or physical conditions conducive to specific organic syntheses.
(10) Couple **exergonic** (energy-releasing) and **endergonic** (energy absorbing) reactions.
(11) Mobilize or release containers or tethered combinations of reactants, catalysts, and/or products.
(12) Mobilize or release functional sets for organic synthesis to seed reactions at other locations.
(13) Occupy "real estate" (i.e. strategic locations within UAHM) with optimal flows of energy and raw materials.

Expanding metabolic pathways and range of organic products at UAHM

Synthesis and **stabilization** of a variety of classes **of organic compounds** contributes to a pool of potentially useful **innovations** and **modules** and expands the **metabolic scrapyard**.

Bringing organic products into proximity in potentially useful ways at UAHM

Emergent Evolutionary Potential (**EEP**; Chapter 2) plays an important role in the origin and subsequent evolution of carbon-based life forms. Bringing organic products into physical proximity, so that they have the **potential** to interact in various ways, opens the door to increased **complexity** and new innovations.

Metabolic innovations preceded sequence-based information

In the protected environments of the UAHM, innovations in metabolic pathways led to increased complexity before the advent of subsequent innovations such as RNA or proteins. These innovations, along with increases in molecular diversity and emergent, higher-level interactions, could have been driven by several plausible mechanisms. Sequence information, storage, transmission, variation, templating, and replication were neither necessary nor sufficient for early increases in organic complexity.

Rethinking the "RNA world" family of hypotheses

RNA sequences could play several diverse roles in molecular evolution because of their catalytic capabilities, activated amino acid transport capabilities, and informational capabilities—storage, transmission, replication, variation, and templating. Once any sort of primitive "genetic code" was in place, Natural Selection and Molecular Genetics could begin in earnest (see the 1989 review by Gerald Joyce [51]).

Separating the early evolution of complexity from the RNA world

Although many questions remain regarding the sequence of events leading to genomic storage of amino acid sequence information and protein synthesis, it is clearly time to free ourselves from the conceptual framework of presuppositions that argue for the origins of life via the prebiotic soup,

RNA world, and Natural Selection. In the 21st century, it is more reasonable to consider the RNA world, the "genetic code", protein synthesis and Natural Selection as powerful emergent innovations that were facilitated by increases in metabolic complexity that presumably took place in the UAHM. Increases in the complexity of primitive metabolic pathways came first and laid the groundwork for all of these subsequent innovations. This perspective provides evolutionary theory with a more comprehensive explanation for the origins of biological organization than the persistent gene-centered conceptual frameworks that have dominated evolutionary theory before now.

The extraordinary added organizing power of Natural Selection

Once RNA, proteins, the genetic apparatus and Natural Selection *did* arise—presumably within the protected environment of the UAHM—these innovations accelerated further refinement and innovation tremendously. Such acceleration represents a general increase in **evolvability** and is just one reason why the concept of "random variation" fails to acknowledge the historical component of increased complexity.

Predictions

Organic compounds

Biochemists are currently able to analyze, predict, validate, and simulate synthesis of increasingly complex organic compounds and biosynthetic pathways in the lab. This will continue to provide new and more detailed insights into actual or plausible primitive compounds and pathways and fill in the gaps in our current understanding of the early evolution of life.

As empirical knowledge of modern biosynthetic pathways continues to expand (e.g. acetogens, methanogens, prokaryotic **metabolomics**, **electron bifurcation**, and more—see Chapter 5), so does our ability to find supporting evidence for plausible advances in the complexity and efficacy of primitive metabolic pathways. Examples appear on a regular basis in the specialized scientific literature.

Natural organic synthesis

New laboratory simulations of hypothetical metabolic evolutionary advances will draw on the expanding empirical knowledge pool, using various raw materials, **intermediate products**, energy, catalysts, and permissive physical and specific chemical conditions. Controlled variables will include pH, pressure, concentrations, containment, and release of waste products. The ability to analyze the results will continue to improve with modern analytical techniques.

Undersea alkaline hydrothermal mounds

Active UAHM will be discovered and characterized, as was done for the Lost City Hydrothermal Mound.

New metabolic pathways will be found when newly-described microbes from various UAHM are analyzed. Some may hold important clues concerning early metabolic evolution.

Terrestrial and extraterrestrial UAHM

Empirical evidence for serpentinization and UAHM will be found on other planets, either through astronomy, space exploration or both.

According to Lane [11], if carbon-based life is found on other planets, it is likely to resemble terrestrial prokaryotic cell types in some ways but will likely differ in others. Such life forms are likely to make use of self-contained proton gradients as energy sources.

Lane also argues [11] that life forms resembling eukaryotes will be rarer than life forms resembling prokaryotes, due at least in part to the increased difficulties of providing sufficient energy for expanded nuclear genomes and repertoires of proteins, while maintaining local control over the efficiency of proton pumps in multiple mitochondria. This may also limit the evolution of multicellular life forms. This, in turn, would limit the evolution of central nervous systems. Assuming that brains are required for intelligent life to exist, that would mean that highly-intelligent life forms would not be frequently found throughout the universe.

Driving force for organic synthesis and interactions

Biochemists will find additional evidence supporting the hypothesis that stepwise increases in organic and metabolic complexity took place before RNA or proteins had a chance to evolve.

Diverse and general ways to naturally facilitate organic synthesis and interactions at UAHM

New UAHM discoveries as well as evidence of primitive metabolic activity at ancient UAHM will uncover a variety of ways that organic synthesis was facilitated, before Natural Selection was possible.

Expansion of metabolic pathways and range of organic products at UAHM

Biochemists will continue to learn more about the most plausible ways that core metabolism may have expanded to include other types of biosynthesis such as high-energy compounds, RNA, proteins, and DNA. This should provide new insights that are not constrained by the conceptual framework that assumes that a highly unlikely combination of fortuitous, complex interactions between ATP, RNA, DNA, and the "genetic code" were required for the origin of life.

Bringing organic products in proximity in potentially useful ways at UAHM

Additional evidence for the independent evolution of various metabolic pathways, and emergent interactions at ancient or modern-day UAHM, followed by consolidation into more complex, unified pathways, will be found from various lines of evidence ranging from characterization of newly-discovered UAHM, new prokaryotic life forms, new insights into biochemical pathways, new laboratory simulations, and more.

Metabolic innovations preceded sequence-based information

Biochemists will gain a more detailed theoretical understanding of the timing of the appearance of various metabolic innovations and will find additional empirical evidence to support those claims. This will fill in gaps in our existing theory concerning the evolution of both core and peripheral biochemical pathways.

Rethinking the "RNA world" family of hypotheses

Evidence will be found for the dramatic acceleration of innovations and complexity as a consequence of Natural Selection. The advent of Natural Selection presumably emerged as a consequence of earlier metabolic innovations—prerequisite modules and tools—such as RNA, proteins, heredity, the "genetic code", and DNA.

Separating the early evolution of complexity from the RNA world

The complementary nature of energy-driven metabolic innovations and Natural Selection will become increasingly clear as we further explore the inner-workings of a broad range of modern prokaryotic and eukaryotic cells.

Extraordinary organizing power of Natural Selection, made possible by the RNA/DNA genome and the genetic apparatus for protein synthesis and evolution

A new appreciation of the organizing power of Natural Selection will capture the imagination of the general public and will strengthen evolutionary theory overall.

PART 2

From Genes to Complex Organization

Thus, from the war of nature, from famine and death, the most exalted object which we are capable of conceiving, namely, the production of the higher animals, directly follows. There is grandeur in this view of life, with its several powers, having been originally breathed into a few forms or into one; and that, whilst this planet has gone cycling on according to the fixed law of gravity, from so simple a beginning endless forms most beautiful and most wonderful have been, and are being, evolved.

—Charles Darwin, from *On the Origin of Species*, 1859 [20]

Chapter 7

Levels of Organization That Transcend Species

If you really want to study evolution, you've got to go outside sometime, because you'll see symbiosis everywhere!

—Lynn Margulis[a]

The Big Picture

Emergent Evolutionary Potential (EEP) is a principle that applies to every imaginable level of complexity among living things. When separate species come into contact with one another, each may represent opportunities for the other to benefit in some way. This often causes the evolution of the interacting species to become intertwined. Darwin described several examples of coevolution in his *On the Origin of Species*, and field biologists and ecologists have found numerous remarkable examples ever since.

Even more remarkable, however, are the numerous examples in which formerly separate lineages of cells come together to form new, emergent forms of life which *transcend* individual species. These include composite cells, composite organisms, and composite genomes. Both EEP and classical Darwinian mechanisms play a synergistic role in the evolutionary history of these transcendent forms of biological complexity.

[a] http://www.wiseoldsayings.com/evolution-quotes/

Introduction

How do natural forces give rise to complex, organized innovations that may benefit multitudes of interacting species?

How can a blind, natural process account for complex interactions between multiple species? Explanations based solely on incremental accumulation of random variation may seem implausible. But organization and complexity do not arise from Natural Selection alone—even though that is an essential ingredient. The combination of natural forces that give rise to biological complexity and organization is far more nuanced, and far more interesting, than 19th or 20th century classical or neo-Darwinian explanations. Quite often, previous innovations create reusable modules that come together to generate new innovations. The results are even higher levels of complexity. And quite often, those higher levels of complexity drive symbiotic interactions and coevolution between two or more species.

Often, the result in question is some incredibly complex symbiotic interaction—that is, an interaction that benefits two or more cooperating species. Or it may be some elaborate and seemingly clever and devious deception in a defensive or offensive predator–prey relationship, such as a caterpillar that resembles a poisonous viper, or a praying mantis that resembles an orchid that offers nectar to an unwitting victim, respectively.

Darwin, of course, realized that it is no mere accident that most of these results can be traced to the universal struggle for existence in natural history, in one way or another. Both symbiotic and predator–prey interactions often involve obtaining energy, water, minerals and other nutrients, shelter from the elements, protection from predators or capture of prey, or obtaining mates and successfully rearing the young.

The human mind appears to be wired to see design and intelligence and purpose and goal-seeking in the wonders of natural history. But we are also endowed (by evolution) with curiosity, rational thought, and the search for logical answers.

But here is one major problem for the public awareness and acceptance of the evolutionary explanation: the supernatural explanation tends to occur in the fast mode of human cognition, which requires less effort, less study, less time, and less careful consideration than the evolutionary explanation. The supernatural explanation is deeply embedded in the human psyche, and is akin to a knee-jerk reflex, an instinct.

The evolutionary explanation requires more study, effort, critical thinking, and a detailed working knowledge of biological principles and vocabulary. It is virtually impossible to understand and appreciate evolution without a knowledge of biology, which requires study, time, and effort.

But the classical 19th century theory of Natural Selection is not enough. There are other critical conceptual components in the **Updated Evolutionary Synthesis (UES)**, including new empirically supported facts and perspectives (such as **Emergent Evolutionary Potential (EEP)**, Chapter 2), **gene duplication** (Chapter 16), and others that are discussed throughout this book.

Many nonscientists do not currently realize that a full understanding and appreciation of conceptual beauty of modern evolutionary theory requires an understanding of modern biological principles. Hopefully, this book will provide a watershed moment that will change this trend.

The importance of the public understanding of evolution goes far beyond the life of the mind. Improved communication that facilitates the public understanding of evolution is not only worth the effort: it is crucial to the continued post-industrial survival of our species, given our great power to destroy the planet by unleashing CO_2 and toxic byproducts that accelerate global warming and other aspects of climate change,[b] and mass extinctions of entire ecosystems on which our very lives depend.[c]

In the following sections, we consider several extraordinary examples of evolutionary symbiosis and coevolution. In every case, previous innovations by each formerly separate species laid the groundwork for subsequent innovations. They all demonstrate the combined, complementary effects of Natural Selection and other natural forces.

---

[b] https://www.weforum.org/agenda/archive/climate-change

[c] https://www.theguardian.com/commentisfree/2019/may/06/biodiversity-climate-change-mass-extinctions

How the UES Explains Complex Symbiotic/ Coevolutionary Innovations

How dogs stole our hearts

The heading of this section is taken from the title of a recent popular explanation of just one aspect of the remarkable coevolutionary history of dogs and humans. The story of human–canine cooperation and coevolution began thousands of years ago. An engaging account is provided in the documentary film, *Man's First Friend*:

> The dog…has helped us hunt, and find food. He helped us travel, and transport our goods, looked after our animals, protected us from our enemies, saved us from the cold and from drowning, guided us through even the harshest terrain…He has found a place for himself in our homes, he cherishes us, comforts us, helps us endure loneliness and sickness…He has become our unwavering ally.[d]

This section describes the findings of a new research[e] [52] which shows that:

> When our canine pals stare into our eyes, they activate the same hormonal response that bonds us to human infants. The study—the first to show that this hormonal bonding effect between humans and another species—may help explain how dogs became our companions thousands of years ago.[f]
>
> "It's an incredible finding that suggests that dogs have hijacked the human bonding system," says Brian Hare, an expert on canine cognition at Duke University…who was not involved in the work.

The hormone is oxytocin,[g] and the response occurs when dogs and humans gaze into each other's eyes. Remarkably, the behavior can be linked to genetic changes (mutations) that are present in the dog genome,

---

[d] https://curiositystream.com/video/2131

[e] https://www.pnas.org/content/early/2019/06/11/1820653116

[f] https://www.sciencemag.org/news/2015/04/how-dogs-stole-our-hearts

[g] https://en.wikipedia.org/wiki/Oxytocin

but absent in the genome of wolves. A popular *New York Times* account of the scientific research tells the story as follows[h]:

> You know that face your dog makes, the one that's a little bit quizzical, maybe a bit sad, a bit anticipatory, with the eyebrows slanted? Sometimes you think it says, "Don't be sad. I can help." Other times it quite clearly asks, "No salami for me?"
>
> Scientists have not yet been able to translate the look, but they have given it a very serious label: "AU101: inner eyebrow raise." And a team of evolutionary psychologists and anatomists reported…in the *Proceedings of the National Academy of Sciences*…[000]…that dogs make this face more often and way more intensely than wolves. In fact dogs, but not wolves, have a specific muscle that helps raise those brows.

So, cooption of oxytocin-mediated bonding, and its facilitation by changes in the musculature of the dogs' eyes, illustrate some of the extraordinary ways that evolution actually takes place. Previous innovations (such as oxytocin-mediated infant bonding) pave the way for those organized functional modules to be reused—or coopted—for new innovations that lead to higher-level organization (dog–human bonding and cooperation). Such reuse of organized modules can have major impacts on the subsequent evolution of the *niches* (ways of life) of cooperating species.

Robert Sapolsky's popular book, *Behave* [53], provides an engaging and accessible discussion of the behavioral effects of mammalian hormones such as oxytocin, and the biology and evolution of human behavior in general.

Coevolution in the deep history of symbiotic fungi and plants

One should not overlook the power of streaming media to beautifully convey an accessible summary of recent discoveries on important topics in evolution. Although the visuals add a great deal to the content, consider

[h] https://www.nytimes.com/2019/06/17/science/dogs-eyebrows-evolution.html

these brief excerpts from the closed captions in *The Kingdom: How Fungi Made Our World.*[i]

To begin with, the documentary points out that the "body" of soil fungi is hidden from view, but represents a vast network of connected cells:

> The mycelium is the body of the fungus. It spreads through the soil eating everything in its path, and even penetrates solid wood… They can get there by physical penetrative force, but also by producing enzymes... These enzymes are the teeth and claws of fungi… The fungi are unusual because they are a single, interconnected network…

Fungi are intimately connected with opportunities to recycle or harvest food from their immediate environment:

> The whole of their growth and development depends on what's going on in their environment… Unlike animals, the body of the fungus is constantly changing shape in a relentless search for food… It can keep spreading, and it can recycle material that's not useful, and use all of that material to grow somewhere else…

Fungi played a major role in the evolution of land plants (and therefore, the animals that depend on them):

> During a billion years of evolution, fungi became the masters of survival...About 500 million years ago, a group of freshwater algae started to move from the ocean to freshwater ponds on land...But to take a foothold on ground, they'd need to make a deal with fungi.

The potential advantages of symbiosis between plants and fungi arise from their distinct capabilities and requirements:

> The arrival of land plants offers fungi an easier way to access food. By exploiting a living organism to get sugar out of them, rather than having

[i] https://curiositystream.com/video/1853/the-kingdom-how-fungi-made-our-world

> to excrete their own digestive enzymes and then assimilate all the nutrients. That's energetically a lot less expensive for the fungus, so it makes economical sense. When algae emerge from lakes, they're ready for a trade.

The algae offer the fungi sugars, and in return, they receive minerals. Algae have the photosynthetic capacity to convert the energy of sunlight into a form that fungi can use to power their mitochondria, while the enzymatic "teeth and claws" of the fungi extract essential minerals from rocks and soil, and make them available to the algae.

> This mutually beneficial relationship is a form of symbiosis, and it will become one of the most powerful forces of evolution.... And so when that first algal cell hits the terrestrial ground, it was already ready to say 'I'm here, let's form a relationship.'

The documentary film mentions published work by Erik Hom and Andrew Murray[j] [54], in which obligate mutualism arose spontaneously in the laboratory, when *Chlamydomonas* algal cells were brought together with various fungal species ranging from yeast to filamentous forms. These cells aggregated into emergent physical structures consisting of species from two entirely separate branches of the phylogenetic tree, which diverged over half a billion years ago. Under stringent selection for carbon and nitrogen exchange, the evolution of new, higher-level symbiotic structures and functions was engineered, under real-time laboratory conditions.

The film goes on to describe the global influence of obligate symbiotic relationships between fungi and plants that have had a profound effect on the deep history and evolution of entire terrestrial ecosystems. These include numerous symbiotic and coevolutionary relationships that depend on fungi, giving rise to diverse and interdependent food webs consisting of land-based producers and consumers from the plant and animal kingdoms, respectively.

---

[j] https://www.ncbi.nlm.nih.gov/pmc/articles/PMC4409001/

The entire subsequent natural history of the vast terrestrial ecosystems of planet Earth depends on several key innovations, including, but not limited to:

(1) the ability of fungi to form vast underground webs of interconnected cells,
(2) the ability of fungi to break down wastes and extract minerals from the ground by secreting enzymes,
(3) the ability of plants to synthesize sugars via photosynthesis, and
(4) the ability of plants to transport water, minerals, nutrients, and sugars, back and forth between roots, stems, and leaves by means of vascular tissues.

Previous events create reusable modules, new opportunities, and change the probabilities of potential future interactions and consequences of those actions. When fungi and algae reached land, they both carried frameworks of previous innovations that made their subsequent symbiosis—in the context of the environmental constraints of the land—all but inevitable. Each was poised to benefit from a close association with the other, and the structures of both—particularly the vast network and rock penetration of fungi—were ideal for forming intimate associations with the roots of land plants of all kinds.

The association of fungi with the roots of land plants—usually symbiotic, occasionally pathogenic—is called mycorrhizae,[k] and they form vast symbiotic networks with local plant life.[l]

Without fungal associations, the rich evolutionary history of plants on land would have been severely limited—just as the rich evolutionary history of consumers and predator–prey food webs, complex behaviors, etc.— and depend on the presence of producers. In each case, essential nutrients and energy sources are shared.

Dr. Hom demonstrated how even lineages of fungi and algae that have been separate for hundreds of millions of years carry toolkits that allow

[k] https://en.wikipedia.org/wiki/Mycorrhiza

[l] https://en.wikipedia.org/wiki/Mycorrhizal_network

them to quickly form intimate associations and share nutrients and energy in experimental laboratory encounters.

There is little that is neither gradual nor random about the way that previous innovations change future probabilities. Previous innovations and complex organized entities create opportunities for higher levels of organization to arise. The rich molecular architecture of macromolecules, as well as cellular structures and developmental toolkits, all make it much easier to understand why, under the aegis of Natural Selection, new, higher-level organizations and associations would be expected to evolve. There is no sentients or goal-direction involved in this process, but in most senses of the word, these associations are not strictly random—they build on the modular, reusable framework of the past. When a previous innovation is explicitly redeployed in a new environmental context for a different purpose, we call that exaptation.[m] But there are innumerable other evolutionary sequences that involve multiple, modular innovations that are reused in different ways. Those are more complex than exaptation, but carry the essential concept of reuse of previously evolved complex modules.

Three kinds of composite biological organization

Virtually every biological species is actually derived from several distant lineages of cells and from distant genomes. This applies to both domains of prokaryotic cells (archaeal and bacteria) as well as all eukaryotic cells and multicellular organisms, including our own species.

In this chapter, we will focus on three general types of high-level **composite biological organization** as follows:

(1) **Composite genomes** (involving frequent **lateral gene transfer** between prokaryotic lineages);
(2) **Composite cells** (**eukaryotic cells,** which arose via **endosymbiosis**);
(3) **Composite organisms** (**lichens**, which are composites of diverse **fungal** and **algal** lineages).

[m] https://en.wikipedia.org/wiki/Exaptation

What are the broad implications of composite biological organization for the updated evolutionary synthesis?

Strictly speaking, when composite biological organization first arises, it does not fall under the rubric of classical Darwinian Natural Selection. Darwin's classical theory focused on the creative power of accumulated, incremental variation, within a lineage of organisms. When two different organisms suddenly fuse and give rise to a new, emergent life-form, this is *not* a classical Darwinian event. When the genetic material from one species is laterally transferred to another, this is also different than classical Natural Selection. However, once a new, viable composite entity has emerged, classical Natural Selection *does* continue to act as a driving force for evolutionary change. Variation, selection, and accumulation of incremental, useful changes will lead to competition between the descendants of that composite life form. Also, it is likely that the struggle for existence will become more acute, because the composite life form will need to make internal adjustments and refinements that minimize the organizational problems that can arise when two or more genomes must find ways to coexist as one.

This is a simple explanation that embraces classical Darwinian theory while recognizing the complementary role played by other factors that are clearly supported by empirical evidence.

Coevolution and mutualism

In addition to composite organisms, we'll consider the widespread evolution of a fourth kind of high-level organization that transcends individual species, but has already been widely recognized and described, ever since Darwin:

(4) **Coevolution** and **mutualism** (where the evolutionary history of two or more lineages is closely intertwined, yet the interacting lineages can still be distinguished throughout their life cycles).

The opportunities and potential for two or more completely distinct species to come together, interact, and combine to form a new, more complex, emergent entity is ubiquitous in nature.

The general term that comes closest to describing these sorts of phenomena is **symbiosis**, which refers to "any type of a close and long-term biological interaction between two different biological organisms, be it **mutualistic**, **commensalistic**, or **parasitic**".[n] I would argue that this definition not only applies to item (4), but the term should now be expanded to include composite cells, genomes, and organisms, as mentioned in items (1)–(3).

Emergent Evolutionary Potential (**EEP**) was explored in Chapter 2, primarily in the context of molecular interactions and the inner-workings of cells. The same general principle extends throughout the biosphere and is largely responsible for the evolution of diverse **food webs** and **ecological relationships** and **ecosystems**.

Lateral Gene Transfer Between Bacterial Genomes

The deep history of prokaryotic evolution as well as endosymbiotic theory have been confounded by the fact that exchanges of genomic material, particularly between prokaryotes, have been so frequent (on an evolutionary timescale) that it is very difficult to construct evolutionary trees by comparing genomic data.[o] This process of exchanging genetic information between species is an example of the well-documented phenomenon of **lateral gene transfer**, also known as **horizontal gene transfer**.[p]

In fact, although we do still refer to both **bacteria** and **archaea** as individual species—using the binomial nomenclature of Genus and species, such as *Escherichia coli* or *E. coli*—the reality among prokaryotes is that each species *does not* represent a relatively isolated pool of genetic information—which is required for the strict definition of a biological species to apply. In reality, prokaryotes actually share a large gene pool—a "metagenome" that is probably more than four times as large as the genomes of individual species [11].

---

[n] https://en.wikipedia.org/wiki/Symbiosis

[o] https://en.wikipedia.org/wiki/Horizontal_gene_transfer_in_evolution

[p] https://en.wikipedia.org/wiki/Horizontal_gene_transfer

The Endosymbiotic Origins of Eukaryotic Cells

As with most areas of evolutionary theory in particular and biological sciences in general, our detailed understanding continues to improve as empirical discoveries expand our knowledge-base. Additionally, the conceptual framework improves, but there is generally a significant lag between new empirical discoveries and widespread revisions at the conceptual level. There is also a wide gap between what is known to the general public and ongoing discussions—in scientific meetings and published research and review papers—among scientific specialists.

The endosymbiotic theory concerning the origins and deep history of eukaryotic cells follows this pattern. Here we will briefly review some of the history of ideas and cross-reference this discussion with some selected popular discussions and scientific milestones or reviews.

On a personal note, I recall that my own dream and determination to become a biological research scientist—and my early interest in evolution in particular—began with a popular *Scientific American* article[q] by Lynn Margulis, titled *Symbiosis and Evolution*. The article was largely based on her own pioneering work in the field. It included several high-magnification, high-resolution images of the **ultrastructure** of cells, made possible by the **transmission electron microscope**. Invention of the **scanning electron microscope** also made it possible to observe the surfaces of tiny 3D objects, ranging from pollen grains (Figure 1) to proteins sandwiched between the two layers of biological membranes. Those images provided one of the more dramatic lines of evidence supporting the endosymbiotic theory of eukaryotic cell evolution, but certainly not the only one. They showed that there is a close resemblance between mitochondria and chloroplasts—the **organelles** of eukaryotic cells that utilize proton gradients (Chapters 5 and 6)—and the internal structure of bacteria.

Another line of evidence is the fact that mitochondria and chloroplasts have their own DNA. Today, several additional lines of molecular and biochemical evidence are also available, and there can be little doubt that mitochondria and chloroplasts are endosymbionts. However, as

[q] I am grateful to my parents for making accessible and intellectually stimulating publications such as *Scientific American* available to me throughout my childhood.

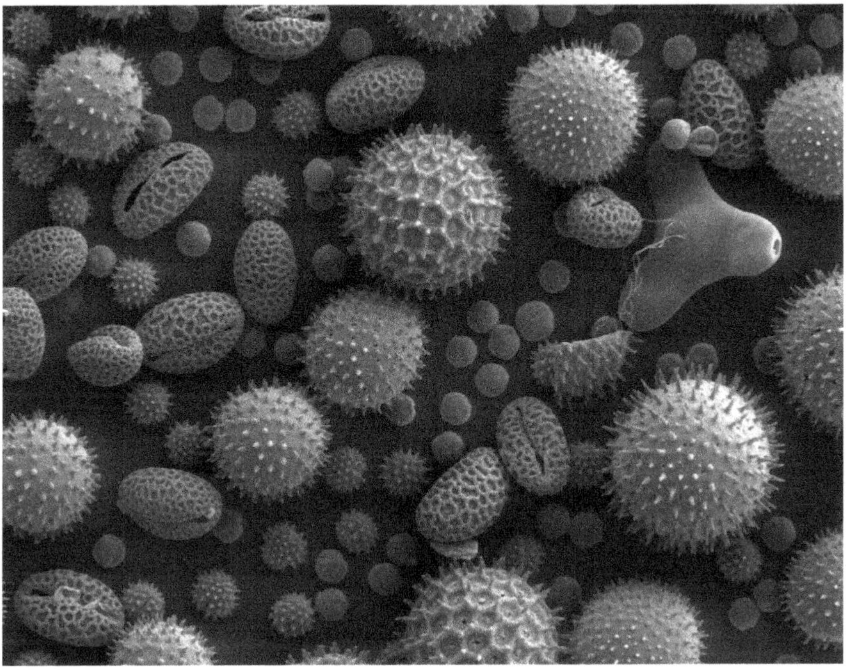

Figure 1. Scanning electron micrograph of pollen grains.

Source: Page-link: https://commons.wikimedia.org/wiki/File:Misc_pollen.jpg. File-link: https://upload.wikimedia.org/wikipedia/commons/a/a4/Misc_pollen.jpg. Attribution: By Dartmouth College Electron Microscope Facility [Public domain], via Wikimedia Commons.

mentioned in Chapter 5, those same lines of evidence call for revision of several aspects of Margulis' original theory.

A eulogy to Lynn Margulis, in the form of a *Scientific American* blog by John Horgan, had this to say:

> Lynn Margulis was among the most creative challengers of mainstream Darwinian thinking of the late 20th century. She challenged what she called "ultra-Darwinian orthodoxy" with several ideas. The first, and most successful, is the concept of symbiosis. Darwin and his heirs had always emphasized the role that competition between individuals and species played in evolution. In the 1960's, however, Margulis began arguing that symbiosis had been an equally important factor—and perhaps more important—in the evolution of life…

> Margulis proposed that eukaryotes may have emerged when one prokaryote absorbed another…
>
> …she argued that conventional Darwinian mechanisms could not account for the stops and starts observed in the fossil record. Symbiosis, she suggested, could explain why species appear so suddenly and why they persist so long without changing.[r]

In addition to scientific discussions concerning her pioneering work, Margulis was a female scientific pioneer who also held strong political and philosophical beliefs—and she refused to make a clear distinction between the scientific ideas that were supported by empirical evidence, and more far-reaching speculations and opinions concerning society, which she was not afraid to express. It is therefore not difficult to see why the scientific community had a professional love-hate relationship with her.

Personally, even though I do not agree with all of her ideas, I do agree with some of them, and I do admire her courage and willingness to challenge authority and conventional beliefs. While some of her ideas—notably her support for the Gaia hypothesis[s]—are highly speculative, certain key aspects of her endosymbiotic theory—notably, the endosymbiotic origins of **mitochondria** and **chloroplasts**—continue to be well-supported by several lines of solid evidence.

Today, the endosymbiotic theory has been revised and refined, thanks to new lines of evidence that have only recently become available. There can be little doubt that eukaryotic cells did evolve when prokaryotic host cells captured other prokaryotic cells, but failed to digest those captured, internalized cells, and instead, coevolved with them.

Margulis was not the first to propose a theory of endosymbiosis, but she did more to advance the theory—both among the scientific community and among the general public—than any other scientist.

---

[r] https://blogs.scientificamerican.com/cross-check/r-i-p-lynn-margulis-biological-rebel/
[s] https://en.wikipedia.org/wiki/Gaia_hypothesis

A brief history and update of endosymbiotic theory

Let's briefly look at the history of empirical discoveries that have led to the current status of the updated endosymbiotic theory. Much of the list that follows was nicely summarized in a 2015 review by John Archibald [47]:

(1) It is firmly established that **mitochondria** evolved from an **alpha-proteobacterial** ancestor by endosymbiosis.
(2) The precise nature of the host cell that first partnered with the mitochondrial endosymbiont remains an open question.
(3) The host cell that was first colonized by cyanobacteria (leading to the chloroplast and perhaps other plastids) was already eukaryotic.
(4) The movement of plastids during algal diversification remains an open question.
(5) DNA sequencing and sophisticated bioinformatic approaches to genomic comparisons have played a transformative role in endosymbiotic evolutionary theory.
(6) Constantin Mereschkowsky was a Russian botanist who played an important role as one of the early contributors to endosymbiotic theory, which included discussions of the composite nature of lichens as well as analysis of eukaryotic cell origins from plastids.
(7) Mereschkowsky's landmark 1905 paper firmly rejected the now widely-accepted fact that mitochondria evolved by endosymbiosis (see the 1999 annotated English translation by William Martin and Klaus Kowallik) [55].
(8) Lynn Margulis, who introduced endosymbiosis to the English-speaking world for both scientists and nonscientists, claimed to have been heavily influenced by her mentors Hans Riis and Walter Plaut, who were both familiar with the earlier research in Germany and Russia.
(9) Margulis' early landmark scientific publications included a 1967 theoretical paper (published as Lynn Sagan) [56] as well as a widely-cited 1970 book on the subject [57].
(10) Ford Doolittle, Carl Woese and colleagues demonstrated a strong evolutionary link between **algal plastids** and **cyanobacteria** by

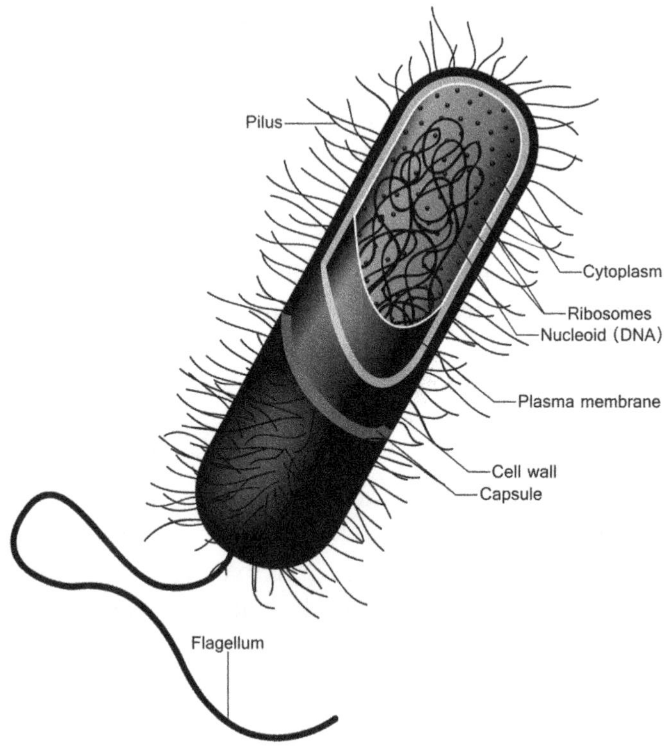

Figure 2. Diagram of a prokaryotic cell: in this case, a motile bacterium.

Source: Page-link: https://commons.wikimedia.org/wiki/File:Prokaryote_cell.svg. File-link: https://upload.wikimedia.org/wikipedia/commons/c/c5/Prokaryote_cell.svg. Attribution: By This vector image is completely made by Ali Zifan [CC BY-SA 4.0 (https://creativecommons.org/licenses/by-sa/4.0)], from Wikimedia Commons. Rendered in B&W.

comparing snippets of ribosomal RNA sequences in the mid-1970s [58]. Woese is most famous for using a similar technique in 1977, to discover that prokaryotic cells (Figure 2) actually consist of two domains—not only bacteria, but also archaea [59]. Today, state-of-the-art, high-throughput sequencing and powerful software have made it possible to construct a detailed phylogenetic tree that illustrates the deep ancestry of the three domains of life: bacteria, archaea, and eukaryotes (Figure 3).

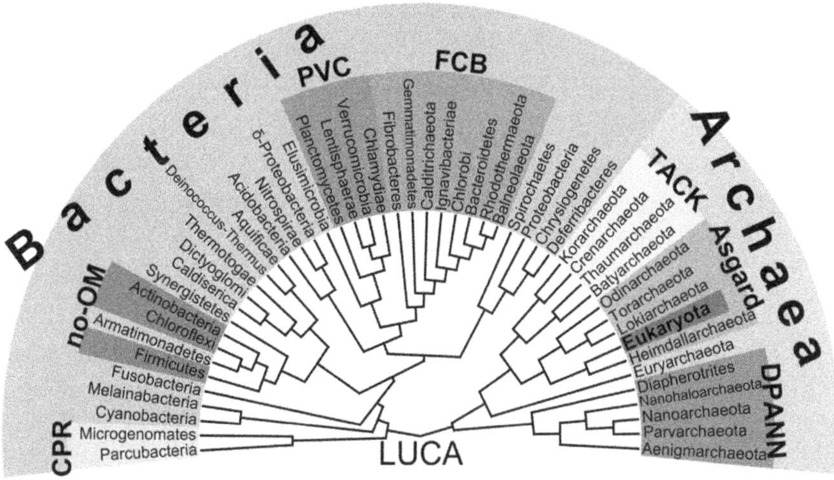

Figure 3. A phylogenetic tree illustrating the ancestral relationships of the major groups of bacteria and archaea, assuming a common ancestral cell referred to as the Last Universal Common Ancestor (LUCA). Notice that the third domain of eukaryotic cells is most closely related to archaea.

Source: Page-link: https://commons.wikimedia.org/wiki/File:Phylogenetic_Tree_of_Prokaryota.png. File-link: https://upload.wikimedia.org/wikipedia/commons/5/5b/Phylogenetic_Tree_of_Prokaryota.png. Attribution: By 投稿者本人 [GFDL (http://www.gnu.org/copyleft/fdl.html), CC BY-SA 3.0 (https://creativecommons.org/licenses/by-sa/3.0), GFDL (http://www.gnu.org/copyleft/fdl.html) or CC BY-SA 4.0. (https://creativecommons.org/licenses/by-sa/4.0)], via Wikimedia Commons. Rendered in B&W.

(11) By the mid-1980s, a **consilience** of empirical data on the biochemistry and molecular biology of organelles led to a clear consensus that endosymbiosis was the only reasonable explanation for both mitochondrial and plastid evolution.

(12) Archibald summarizes two competing evolutionary scenarios for the deep origins of eukaryotic cells and the mitochondria they now contain. The traditional view envisioned a step-wise process of evolution preceding the later uptake (by a process of engulfment known as phagocytosis) of the alpha-proteobacterium that evolved into the mitochondrion. This is designated as the "mitochondrion late" point of view.

(13) The alternative view, known as the hydrogen hypothesis of Martin and Müller [35], proposed an early metabolic symbiosis between methane-producing archaea and **alpha-proteobacteria**, followed by the later evolution of eukaryotic complexity. This is the "mitochondrion early" point of view.

(14) The "mitochondrion early" hypothesis is consistent with the idea that the earliest origins of metabolic pathways leading to both prokaryotic and eukaryotic cells took place in the protected environments of **undersea alkaline hydrothermal mounds (UAHM)**. Metabolic pathways were first driven by naturally-occurring geological proton gradients, which later gave rise to prokaryotic cells that could escape from the mounds by generating their own membrane-bound, internal proton gradients.

(15) In 2010, Nick Lane and William Martin hypothesized [60] that genome expansion in the eukaryotic nucleus, along with multiple smaller mitochondrial genomes, were strictly dependent on mitochondrial power derived from ATP production via mitochondrial proton gradients.

(16) State-of-the-art phylogenetic comparisons led to publications such as that of Cox *et al.* in 2008, demonstrating that it was an archaeal host, rather than a bacterial host, that acquired the alpha-proteobacterium that evolved into the mitochondrion [61].

The above list demonstrates not only the empirical discoveries driving changes in evolutionary theory, but also helps to identify the areas where further work is needed. Today's endosymbiotic theory supports a view in which an archaeal host acquired an alpha-proteobacterial endosymbiont that gave rise to mitochondria. It also supports ancient cyanobacterial origins for photosynthetic organelles such as the chloroplast. Research specialists will undoubtedly continue to expand and refine our future understanding of endosymbiosis and eukaryotic cell origins with both empirical facts and advances in our conceptual framework.

Scientific theories evolve over decades as new empirical discoveries are made. It took about 50 years for the early endosymbiotic theory, first championed by Lynn Margulis in the late 1960s and early 1970s, to mature into our present-day understanding. We know far more today than we did then, thanks to the hard work and dedication of specialists from a broad range of scientific disciplines, ranging from the geochemistry of undersea alkaline hydrothermal mounds to the sophisticated phylogenetic analysis of genomic DNA. In addition to the critical contributions of research scientists who provided a consilience of evidence from empirical facts, we can also thank the **generalists** and theorists who brought those ideas together in a new synthesis. Although the conceptual frameworks do tend to lag behind empirical discoveries for a time, they do eventually lead to a consensus that reflects a better and better approximation of reality. The scientific method really does work, and today's endosymbiotic theory of eukaryotic cells is quite sound.

Composite Organisms: Lichens

Lichens transcend mutualism between species

Most people who have spent any time outdoors have seen examples of lichens. As with most subjects, a variety of fascinating discussions and beautiful images[t] can be found on the web.[u] Lichens display[v] a dazzling range of shapes, textures, and colors,[w] and grow on surfaces ranging from bare rock to roofs, soil, tree branches, as well as extreme environments that are too hot, dry, cold, or toxic for other organisms.

Lichens are classified in the special category of **composite organisms**. They have arisen at numerous times and from a broad range of different combinations of algal and fungal species as well as cyanobacteria. The sheer diversity of different combinations of species that make up these composite organisms demonstrates evolutionary opportunism as its finest.

---

[t]Photo of hummingbird in nest constructed from lichens, by Brendan Lally: license at https://creativecommons.org/licenses/by/2.0/legalcode

[u] http://lichen.com/

[v] https://www.fs.fed.us/wildflowers/beauty/lichens/biology/images/cladonia_fimbriata_lg.jpg

[w] http://www.michwildflowers.com/

The remarkable history of lichen research

In 1851, Charles Darwin published a note on lichens in *An Enumeration of the plants of the Galapagos Archipelago* [62] which reads:

> "[*Usnea plicata*] Hab. James Island, 'hanging from the boughs of the trees in the upper damp region, where it forms a considerable proportion of the food of the large tortoise.'" (Charles Darwin, Esq.)

Darwin was interested in this particular lichen because it was a favorite food source of the now-famous Galapagos tortoises.

But Darwin was not the only revolutionary 19th century naturalist to notice lichens. Less than a decade after publication of the first edition of *On the Origin of Species*, Simon Schwendener hypothesized in 1867 that lichens are formed by two different organisms: a fungus and an alga.[x] As is often the case with ground-breaking hypotheses, this symbiotic association was not widely accepted until many years later.

Beatrix Potter, who was most famous for her timeless classics in children's literature—notably *The Tale of Peter Rabbit*[y]—also had strong interests in the art and science of fungi. As it turns out, she too was opposed to Schwendener's symbiotic hypothesis.[z]

Eventually, however, Schwendener's hypothesis was accepted, and for 140 years, the symbiotic hypothesis held that lichens consisted of two species: an algal species and a fungal species.

Quite recently, however, a 2016 research paper by Toby Spribille *et al.* in *Science*[aa] demonstrated that lichens often consist of three (or more) different species, not just two, and the third member of this symbiotic *ménage a' trois*[ab] is often a type of yeast. In fact, these and other authors have showed that a variety of different three-species lichen symbioses can be found all over the world.

---

[x] https://en.wikipedia.org/wiki/Simon_Schwendener

[y] https://www.gutenberg.org/files/14838/14838-h/14838-h.htm

[z] http://www.bbc.com/Earth/story/20160215-beatrix-potter-pioneering-scientist-or-passionate-amateur https://en.wikipedia.org/wiki/Beatrix_Potter

[aa] http://science.sciencemag.org/content/sci/early/2016/07/20/science.aaf8287.full.pdf

[ab] https://www.youtube.com/watch?v=d167NrioW7c

Throughout the history of biology, but especially in recent decades, new insights have quickly followed when powerful new techniques could be applied to ongoing questions. Even the most brilliant pioneers are limited by available empirical techniques. Truth be told, in the hands of scientists who know how to ask the right questions, modern laboratory methods have led to more major discoveries during the past three decades than in the 2,000 years that preceded them. This includes both basic theoretical discoveries and important breakthroughs in biomedical applications.

It's worth noting that anyone who still believes that "all of the good stuff" in biology has already been discovered should reconsider recent findings such as these. The presence of three species in the symbiosis was revealed by recently developed state-of-the-art empirical techniques in genomics and bioinformatics.

In addition to containing two types of fungi, some lichens[ac] also contain both cyanobacteria and green algae as their photosynthetic symbionts.[ad]

Returning to our focus on *emergent symbiosis*, it is not difficult to imagine how and why a close association between fungi and photosynthetic organisms evolved.

The first requirement is *proximity*, which provides the opportunity for new—and potentially useful interactions to take place. Lichens are generally found on otherwise inhospitable surfaces, such as rocks. Let's look at this from the metaphorical point of view of a photosynthetic organism. For producers that depend on sunlight to make a living, rocks provide ideal substrates to "set up shop". Not only are they flat surfaces that are often exposed to direct sunlight, there is also very little competition from other producers, because bare rocks are not covered with soil, which most plant species require. They do, however, receive moisture via direct rainfall or condensation from the morning dew. What are in short supply, however, are the small molecules and minerals that all producers require, such as nitrates, phosphates, and salts.

[ac] http://people.oregonstate.edu/~mccuneb/FunctionalRoles.htm

[ad] https://en.wikipedia.org/wiki/Cyanolichen

Now, let's consider the metaphorical "point of view" of an enterprising fungus. Fungi are consummate decomposers and are often capable of secreting caustic substances that can break down surrounding materials which can then be absorbed as nutrients.

Now consider the possibilities if these two diverse forms of life—photosynthesizers and decomposers—were to form a close association, to divide up the labor and to share resources. Photosynthetic organisms produce sugars that provide an excellent all-purpose source of energy, and decomposers can break down rock and free the minerals. Both types of organisms can produce cell walls to reduce water loss from evaporation, and the photosynthesizers can produce **accessory pigments** that limit DNA damage from direct sunlight.

Given all of these potentially useful features, there is every reason to expect that a close symbiotic association—one that can provide both protective and nutrient services for the symbiotic partners—might evolve. In order for this association to be maximally effective, there must be a very close association of the photosynthetic and fungal cells. The resulting products of evolution are so closely associated that it took several years for other scientists to accept Schwendener's hypothesis. Schwendener postulated two organisms. It took another 140 years before high-tech genomic studies and bioinformatics revealed that there are often three partners in these associations, rather than two.

Lichens exemplify the broader phenomenon of EEP

So, what do lichens tell us about **EEP**? First, they clearly demonstrate that when two species that can provide useful materials or services for each other come into close contact, this represents an extraordinary opportunity for close associations to evolve.

With the possible exception of some **primary producers**, virtually every cellular species on Earth—including consumers, ranging from apex predators to decomposers—depends on other species for their survival. Those other species provide needed energy, nutrients, and other useful features such as shelter or substrates to which to attach. In coral reefs, for example, we find complex tropical ecosystems and elaborate food webs. These depend not only on sunlight, but also on the surfaces and hiding places provided by the highly-branched structures produced by the corals.

From this point of view, virtually every kind of ecological relationship between species represent specific examples of evolutionary emergence. Emergence is almost always an essential aspect of evolution of virtually every kind of biological adaptation involving biotic resources of virtually every ecosystem.

Since capture of energy and nutrients are essential aspects of the struggle for existence—for *all* living things—there are innumerable cases where pairs of individuals (and sometimes even more than two) of distinct species that coexist, have opportunities to *interact* in various ways that prove useful to one, or both, of those individuals. New ways to transfer energy and nutrients can arise—that is, emerge—from those interactions. Natural Selection—ever the tinkerer and opportunist—finds extraordinary ways to exploit those opportunities for interaction.

Those interactions have been empirically described in a scientific way with terms such as **mutualism**, **coevolution**, **symbiosis**, and **endo-symbiosis**, and are frequently part of the core curriculum in high school biology classes. All of these phenomena involve emergence, because the interacting organisms form a new entity that has new interactions that depend on the properties and capabilities of a whole. Often, emergent phenomena can be described between two individuals. Although scientists often use the term "emergence" to describe macroscopic phenomena—often called *systems*—involving very large numbers of particles—such as the thermodynamics of heat transfer among trillions of atoms in a gas or liquid—in biology, most examples of emergence can be found in the interactions between much smaller numbers of interacting parts. Enzymes and the reactants that they will transform, for example, usually form enzyme-substrate complexes that require interaction of only two or three separate molecules at the active site of the enzyme. Predator–prey interactions take place between pairs of individuals. In biology, emergence often does *not* usually require large numbers of interacting components—or the dynamics of systems. Quite often, emergence involves the binding and complementary shapes of small numbers of subunits. There is nothing mysterious about the evolution of these wonders throughout the deep history of life.

Coevolution, Mutualism, and Other Ecological Symbionts

As denizens of planet Earth, we are truly blessed with Darwin's "endless forms most beautiful and most wonderful".[ae] Perhaps the most remarkable of these—which provide compelling support for Natural Selection—are examples in which species have coevolved structures, functions, and behaviors that are deeply and mutually interdependent. Clearly in such cases, each of the interacting species provides a crucial, biotic component of the environment that directly shaped the evolutionary history of the other. In such cases, both species usually have their own gene pools. Quite often, we can observe a metaphorical "arms race" between parasites and hosts, predators and prey, and consumers and producers.

Alongside the more violent aspects of the struggle for existence—the "nature red in tooth and claw"[af]—cooperative relationships also abound. Any aspect of the requirements of life—formation of a *niche*, protection from the elements, acquisition of energy and raw materials—including metabolic steps that cannot be completed by one species but are routinely completed by others—and dispersal of pollen and seeds—will be subject to the principle of EEP. In general, a rule of thumb is that any opportunity that brings two or more species into proximity and would serve to increase the abundance of one or more of those species, will probably be "discovered" by trial-and-error, and quite possibly may lead to a coevolutionary relationship between those interacting species.

Sometimes those relationships will benefit one species and not harm the other (e.g. squirrels making their nest in the treetops). Sometimes one will benefit at the expense of the other (predator–prey relationships and parasitism), and sometimes both will benefit (normal strains of E. coli benefiting from the food and warmth of the human gut while providing essential nutrients to the host).

Fascination with these natural wonders often leads curious, adventurous people to professional careers in ecology, marine biology, or field

[ae] https://www.amnh.org/exhibitions/darwin/endless-forms-most-beautiful/from-so-simple-a-beginning/

[af] https://www.phrases.org.uk/meanings/red-in-tooth-and-claw.html

work in remote locations. Thanks to books [63–65], modern media,[ag] and the Internet,[ah] millions of nonscientists can experience these wonders of nature up close and in person in their own daily lives.

On an evolutionary time frame spanning tens of millions of years, the appearance of major branches on the **tree of life** is often driven by powerful coevolutionary forces. For example, the appearance of insects around the same time as flowering plants is no coincidence.[ai] There is no shortage of examples where flowers provide nectar for insects while the insects serve as **pollination vectors** on which the plants depend for sexual reproduction and for extending their range. The opportunistic nature of EEP would also predict that numerous other species—such as birds and mammals—would also play similar roles. Fruits, for example, provide essential nutrients for animals, and the animals in turn serve as **dispersal vectors** for the seeds. Still other seeds hitch a ride on passing animals for free by evolving structures that enable them to stick to the appendages and fur of four-legged mammals or to the feathers of birds.

The opportunistic nature of EEP also predicts that any and all environmental mechanisms that do not involve other species—such as wind—will be exploited. For example, the helicopter-like action of spiraling maple seeds increases their range when they are caught by the wind. Similarly, the fluffy and intricate structures attached to dandelion seeds that allow them to catch the wind and blow far and wide.

Prokaryotic cells have evolved a variety of ways to make use of diverse sources of energy and nutrients

In the struggle for existence, finding new ways to utilize available sources of energy and nutrients—or escape from potential environmental threats—can mean the difference between surviving and leaving descendants or going extinct.

In recent years, scientists have found that prokaryotes have evolved a variety of ways to cooperate with other cells in mutually beneficial ways.

[ag] https://en.wikipedia.org/wiki/David_Attenborough

[ah] https://video.nationalgeographic.com/

[ai] https://en.wikipedia.org/wiki/Coevolution

In biology, mutualism is generally defined as relationships between different species in which each individual can benefit from the activity of the other.[aj] However, we need a broader definition to describe mutually beneficial interactions between individual prokaryotic cells, because the cells that benefit often do so by blurring the distinctions between species.

Among these mutually beneficial interactions are **horizontal gene transfer**,[ak] **quorum sensing** [66], **biofilms**,[al] **microbial consortia**,[am] and even extracellular exchanges of electrons.

Horizontal gene transfer has taken place extensively in the deep history of life. Prokaryotic cells frequently incorporate foreign DNA (or RNA) sequences into their genomes, and many exchange small, self-replicating circular DNA molecules known as **plasmids**, which can carry genes from one prokaryotic individual or species to another. This also has implications for human medicine, because plasmids sometimes bear antibiotic resistance gene. This is one of the ways that bacterial pathogens actively acquire **antibiotic resistance** which makes them far more dangerous to human health.[an]

Microbial consortia, which can include a variety of archaean, bacterial, or eukaryotic cells allow communities to benefit one another by sharing various nutrients and metabolic intermediates.

Biofilms are biological systems that facilitate binding to surfaces as well as a variety of community interactions between multitudes of single cells, either of the same species or of multiple species. Biofilms facilitate quorum sensing as well as microbial consortia that can benefit each other by sharing various nutrients or metabolic intermediates.

According to Wikipedia,[ao] quorum sensing (QS):

> enables bacteria to restrict the expression of specific genes to the high cell densities at which the resulting phenotypes will be most beneficial. Many species of bacteria use quorum sensing to coordinate gene expression according to the density of their local population.

[aj] https://en.wikipedia.org/wiki/Mutualism_(biology)

[ak] https://en.wikipedia.org/wiki/Horizontal_gene_transfer

[al] https://en.wikipedia.org/wiki/Biofilm

[am] https://en.wikipedia.org/wiki/Microbial_consortium

[an] https://en.wikipedia.org/wiki/Plasmid-mediated_resistance

[ao] https://en.wikipedia.org/wiki/Quorum_sensing

Extracellular exchanges of electrons between prokaryotic cells [67] can facilitate metabolism by serving as a means by which some cells can donate high-energy electrons (act as **reducing agents**) (and prevent an excess from building up) while other cells act as oxidizing agents and receive those electrons. This can facilitate complex metabolic pathways involving the enzymes that evolved in two or more separate lineages of cells.

All of the mutually beneficial prokaryotic interactions described above are consistent with the extraordinary synergistic organizing power of Natural Selection combined with **EEP** (see Chapter 2). Numerous opportunities to facilitate energy capture and metabolism have driven the evolution of a variety of innovations involving mutually beneficial prokaryotic interactions. Natural Selection alone cannot account for this, because these collaborative cross-domain interactions do not arise in an incremental fashion. Rather, the general hypothesis presented in *Rethinking Evolution* argues that new, higher levels of organization first tend to arise by EEP, and are subsequently refined by Natural Selection.

Predictions after the fact

Notice that I have used the word *predict* to describe these examples of evolution. Strictly speaking, these are not the best sorts of scientific predictions, because they are, of course, explanations that must be made after the fact. This limitation of evolution—that literally tens of thousands of empirical examples of evolutionary products must be analyzed retrospectively—has been a favorite theme among two groups of people with quite different objectives:

(1) Creationists who wish to discredit evolutionary theory by pointing out that retrospective "predictions" are not predictions at all.
(2) Philosophers such as Karl Popper, who was entirely comfortable with pointing out the problems with falsifiability in evolutionary theory, yet weighed on the side of evolutionary theory as a legitimate **epistemological** approach.[ap]

[ap] https://en.wikipedia.org/wiki/Falsifiability (30 July 2018).

Stephen J. Gould, a prolific contributor to both the popular and scientific literature on evolution, also challenged the tendency of theorists to wax poetic with subjective "just so stories"[aq] when describing the presumptive Darwinian evolution of biological structures and functions as adaptations[ar] [68].

The importance of scientific rigor is a genuine concern. The tendency to attribute one particular hypothetical mechanism as "the cause" of an observed phenomenon can be misleading. This is especially problematic with the retrospective observations of evolution. It is better to describe them as correlations. This problem has been the bane of existence for scientists who try to attribute the Cambrian Explosion—the relatively sudden appearance of numerous new and diverse phyla about 550 million years ago—to particular causes [69]. Life involves complex adaptive systems that are often better described as both cause and effect of particular phenomena, and where causality is, at best, multifactorial.

Nevertheless, there is an overwhelming abundance of examples that *are* consistent with a scientific explanation involving trial-and-error of fortuitous events and variation. Sometimes, these events or variation prove useful, and tend to be retained. They can also enhance evolvability *per se*, so that they tend to influence EEP in ways that transcend time and space (see Chapter 2).

Complementary driving forces in evolution

The fact that certain aspects of evolutionary theory are retrospective does *not* disqualify it as rigorous science. There is, as E.O. Wilson put it, a consilience of evidence [33] supporting a naturalistic approach to evolutionary theory. But we need to incorporate a broader perspective than classical Darwinian theory alone as our best approximation of reality. I feel sure that, were he alive today, Darwin himself would be among the first to agree.

We are led to the general conclusion—well supported by both empirical evidence and theoretical rigor—that Darwinian mechanisms involving

---

[aq] https://en.wikipedia.org/wiki/Just_So_Stories (30 July 2018).

[ar] Available online at https://faculty.washington.edu/lynnhank/GouldLewontin.pdf

vertical inheritance and incremental accumulation of random events, is certainly one of the mechanisms that contribute to evolution. But it is not the only one. We have focused on at least one other major driver for evolutionary change, which arises when two or more entities come together, which represent opportunities for new entities to emerge. These opportunities can be found at every level of complexity ranging from simple molecular interactions to the inner-workings of cells, organelles, tissues and organ systems, as well as interactions between species that inevitably come together at various times and in various contexts. These new opportunities can lead to complex interactions that are sometimes external to each species—called **ectosymbiotic** relationships—or that actually result in fusion of two or more entities—**endosymbiotic** relationships. This distinction between outside and inside is, of course, an arbitrary one when it comes to biological structures. Lichens, for example, are composite organisms that do not fit neatly into either category.

EEP and classical Darwinian mechanisms are long-term, complementary natural partners in the deep history of life. Yet, although I would argue that this is an important update for a 21st century evolutionary synthesis, it is by no means the final word on the subject. Rather, such concepts should inevitably lead to further discussions, new conceptual frameworks, new empirical discoveries, and the emergence of new and better ideas concerning the driving forces of evolutionary change. To cling to any particular set of ideas, and to resist changing one's mind, are natural, conservative human tendencies. Skepticism is useful in rejecting pseudoscience and distinguishing speculation or belief from well-supported theories and facts. The key, in my view, is to strike a reasonable balance, and to use the skepticism not as a means of mere rejection of particular ideas, but rather to stimulate new experiments, observations, and thought, driven by two other capabilities that lead to further progress: the desire to discover and understand the truth, the whole truth, and nothing but the truth, as well as an unfailing curiosity and sense of wonder. This provides comfort and courage when facing the reality of science—that we never have a final answer to any questions, and each answer brings forth new questions. But science in general, and evolutionary science in particular, certainly do provide us with better and better approximations of the truth.

Chapter 8

From 19th Century Natural Selection to 20th Century Mendelian Genetics

Neo-Darwinism is a gene-centred theory of evolution. Yet, its central notion, the "gene", is an unstable concept. Surprising as it may seem, there is no single agreed definition of "gene". Even more seriously, the different definitions have incompatible consequence for the theory. [a]

—Denis Noble [70]

The Big Picture

As is true for all scientific pioneers, Charles Darwin drew upon the theories and discoveries of those who came before him and expanded and modified those theories using his own discoveries. Darwin was not the first to propose that species change over time. Based in part on his voyage of discovery aboard the HMS Beagle, and in part on other theories such as those of Gradualism and Malthusian competition, Darwin proposed that new species arise in the struggle for existence by natural forces. His major contribution was publication of *On the Origin of Species* in 1859 which provided strong evidence to support his new theory of Natural Selection.

In 1866, Gregor Mendel published some early insights into the nature and transmission of hereditary factors. Mendel's Laws of Heredity, as

[a]Noble D. (2015) "Evolution beyond neo-Darwinism: a new conceptual framework". Originally published in Ref. [70]. Corrected web version available as a pdf: http://jeb.biologists.org/content/jexbio/218/1/7.full.pdf, 14 March 2018.

they are known today, resulted from his brilliant experimental design and his diligent accounts of his detailed observations.

Both Darwin's and Mendel's theories represent the foundations for evolutionary theory, but they were limited by the knowledge-base and empirical methodology of their times.

Mendelian Genetics truly came of age in the first decades of the 20th century when, armed with the new Chromosomal Theory of Inheritance, Thomas Hunt Morgan and colleagues utilized powerful new experiments in fruitflies to link Mendel's Laws to the behavior of genes carried by chromosomes, during sexual reproduction.

The Origins of Classical Darwinian Theory

Darwin's voyage of discovery

Try to put yourself in Charles Darwin's shoes for a moment during his five years as a gentleman naturalist aboard the HMS Beagle on its second voyage of discovery, from the 27 December 1831 to the 2 October 1836.[b] His original plan was to see the tropics before becoming a **parson**. During the voyage, Charles Lyell's *Principles of Geology* captured his imagination.

Three years of the voyage were spent exploring the land, and most of his careful and extensive notes concerned geology, although he did also take extensive and carefully organized zoological notes. For the latter, he collected specimens, and his direct observations were supplemented with dissections and microscopic observations. Comparisons of the structures in the fossil record, of living species on remote islands, as well as diverse species on various continents, offered unprecedented scientific insights for this meticulous and thoughtful observer of natural history.

When he published his findings from the voyage,[c] Darwin's reputation as a geologist and fossil collector soared. Perhaps the most significant personal consequence of the voyage, however, was that his keen and

[b] https://en.wikipedia.org/wiki/Second_voyage_of_HMS_Beagle

[c] http://darwin-online.org.uk/contents.html

extensive empirical observations shook his belief—widely shared by his contemporaries—that biological complexity and species diversity had divine origins. Darwin's direct observations, combined with other lines of evidence such as the effects of artificial selection by humans in domesticated species, raised a number of questions for Darwin that ultimately led to his classical theory of Natural Selection.

If species are not fixed, then how did they diversify?

Darwin was methodical in the way that he painstakingly observed and documented his findings, thanks in part to his excellent training as a naturalist. He relied on empirical evidence—that is, the knowledge he received through observation and experimentation.[d]

Consider, for example, the conclusions that he drew regarding the now famous marine iguanas of the Galapagos Islands. Today, thanks to National Geographic, and thanks to the fruits of modern technology, such as diving equipment, underwater video cameras, the Internet, and mobile devices, millions of viewers throughout the world can observe the behavior of this remarkable species from virtually any location in the world.[e] I urge readers to view this video, and also to make full use of the powerful search tools and multimedia that are freely available on the web, as these are bound to provide a deeper understanding and appreciation to any discussion of natural history and evolution.

In contrast to what we have available today, Darwin's 19th century empirical tools were decidedly "low-tech". Consider, for example, the way that he learned about the diet of this unique species[f]:

> I opened the stomachs of several and found them largely distended with minced sea-weed (Ulvae), which grows in thin foliaceous expansions of a bright green or a dull red colour. I do not recollect having observed this sea-weed in any quantity on the tidal rocks; and I have reason to believe

[d] https://en.wikipedia.org/wiki/Empirical_evidence

[e] https://video.nationalgeographic.com/video/160509-fbtos-galapagos-marine-iguana

[f] https://charles-darwin.classic-literature.co.uk/the-voyage-of-the-beagle/ebook-page-188.asp

> it grows at the bottom of the sea, at some little distance from the coast. If such be the case, the object of these animals occasionally going out to sea is explained. The stomach contained nothing but the sea-weed. Mr. Baynoe, however, found a piece of crab in one; but this might have got in accidentally, in the same manner as I have seen a caterpillar, in the midst of some lichen, in the paunch of a tortoise... The intestines were large, as in other herbivorous animals. The nature of this lizard's food, as well as the structure of its tail and feet, and the fact of its having been seen voluntarily swimming out at sea, absolutely prove its aquatic habits...

Darwin was an avid reader, so he knew of the geological theory of gradualism, and the extraordinary power of gradual, infinitesimal changes to transform the face of the Earth, put forward by Charles Lyell, as evidenced in this passage[g] from Darwin's *On the Origin of Species* [20]:

> He who can read Sir Charles Lyell's grand work on the Principles of Geology, which the future historian will recognise as having produced a revolution in natural science, and yet does not admit how vast have been the past periods of time, may at once close this volume.

Darwin also knew of the geometrical ratio of increase, known as the doctrine of Malthus, which he also discussed[h] in *On the Origin of Species* [20]:

> A struggle for existence inevitably follows from the high rate at which all organic beings tend to increase... Every being, which during its natural lifetime produces several eggs or seeds, must suffer destruction during some period of its life, and during some season or occasional year, otherwise, on the principle of geometrical increase, its numbers would quickly become so inordinately great that no country could support the product. Hence, as more individuals are produced than can possibly survive, there must in every case be a struggle for existence, either one individual with another of the same species, or with the individuals of distinct species, or with the physical conditions of life. It is the doctrine of Malthus applied with manifold force to the whole animal

---

[g] http://www.literaturepage.com/read/darwin-origin-of-species-309.html
[h] https://www.bartleby.com/11/3003.html

> and vegetable kingdoms; for in this case there can be no artificial increase of food, and no prudential restraint from marriage. Although some species may be now increasing, more or less rapidly, in numbers, all cannot do so, for the world would not hold them.

Darwin thought more deeply than the average gentleman of his time about more commonplace observations[i] concerning domesticated plants and animals [20]:

> ...domestic races of the same species differ from each other in the same manner as do the closely allied species of the same genus in a state of nature, but the differences in most cases are less in degree. This must be admitted as true, for the domestic races of many animals and plants have been ranked by some competent judges as the descendants of aboriginally distinct species, and by other competent judges as mere varieties.

If human beings can cause such dramatic changes in the shape and form of animals and plants over the course of several generations, Darwin could only imagine how much change could occur, by means of natural forces, over tens of thousands or millions of generations.

The argument from design[j]

Darwin was also aware of the changing views of other naturalists regarding how species change over time, and he was not the first to conceive of the idea of evolution. The classical theory of Natural Selection, however, was the product of Darwin's own creative genius and determination to understand the deep history of life and the origin of species as a product of natural forces[k] [20]:

> At the present day almost all naturalists admit evolution under some form. Mr. Mivart believes that species change through "an internal force or tendency," about which it is not pretended that anything is known.

---

[i] http://www.literaturepage.com/read/darwin-origin-of-species-23.html

[j] https://en.wikipedia.org/wiki/Teleological_argument

[k] http://www.literaturepage.com/read.php?titleid=darwin-origin-of-species&abspage=236

> That species have a capacity for change will be admitted by all evolutionists; but there is no need, as it seems to me, to invoke any internal force beyond the tendency to ordinary variability, which through the aid of selection, by man has given rise to many well-adapted domestic races, and which, through the aid of Natural Selection, would equally well give rise by graduated steps to natural races or species.

It therefore came to pass that Charles Darwin—following the empirical traditions of Natural Theology—came to understand that species arise by means of natural forces, rather than by supernatural acts of creation. For a 19th century gentleman naturalist who intended to join the parsonage, this was a dramatic transformation.

Consider the contrast between the conclusions drawn by Darwin and those of William Paley, who is most famous for his watchmaker analogy[1]:

> In crossing a heath, suppose I pitched my foot against a stone, and were asked how the stone came to be there; I might possibly answer, that, for anything I knew to the contrary, it had lain there forever: nor would it perhaps be very easy to show the absurdity of this answer. But suppose I had found a watch upon the ground, and it should be inquired how the watch happened to be in that place; I should hardly think of the answer I had before given, that for anything I knew, the watch might have always been there.... There must have existed, at some time, and at some place or other, an artificer or artificers, who formed [the watch] for the purpose which we find it actually to answer; who comprehended its construction, and designed its use.... Every indication of contrivance, every manifestation of design, which existed in the watch, exists in the works of nature; with the difference, on the side of nature, of being greater or more, and that in a degree which exceeds all computation.

The title of Paley's book, *Natural Theology, Or, Evidences of the Existence and Attributes of the Deity. Collected from the Appearances of Nature* [71], reflects the deeply held beliefs that were widespread in 19th century Europe. We have already seen that similar views were held by

---

[1] https://en.wikipedia.org/wiki/Watchmaker_analogy

other Natural Theologists such as William Buckland,[m] another of Darwin's contemporaries, in Chapter 3.

The tendency of human beings to interpret empirical observations in terms of what they already expected to find—which is largely determined by the intrinsic bias resulting from their underlying core beliefs—is deeply embedded in the human psyche. Scientists who are genuine pioneers strive to put aside personal bias. They consequently tend to develop more revolutionary general theories that are at odds with those held by their peers. Such individuals have always been, and continue to be, quite rare in the history of science.

Darwin's theory was a simple one that grew out of his careful observations and his determination to find honest answers to insightful questions. If the first key question concerned the natural forces responsible for changes in the deep history of life and the origin of species, then a second key question for Darwin concerned the extraordinary matches that he found between the diverse shapes and forms of organisms and their numerous ways of life.

Adaptations arise by Natural Selection

In addition to the origin of species, Darwin's **Theory of Natural Selection** addressed several critical questions regarding the origins of complex biological organization and the close match between the shape and form of particular species and their various ways of life:

(1) Why and how do species acquire their specialized structures and functions?
(2) Why do particular structures such as wings, the heart or the brain appear as if they are designed?
   a. Why do inherited structures and functions so effectively equip each individual?
   b. What is responsible for the close match between the shape and form of various species and the tools that would be required to survive in a particular environment and reproduce their own kind?

[m] https://en.wikisource.org/wiki/Geology_and_Mineralogy_considered_with_reference_to_Natural_Theology

Darwin's brilliant answer was that natural forces were responsible, both for the origins of species and for the **efficacy** of biological structure and function. Here is a brief summary of the key elements of his Theory of Natural Selection:

(1) Since far more individuals are born than can possibly survive, they must struggle for their existence.
(2) Slight, infinitesimal variations in the structure of function of individuals will, by chance, equip some of them with advantages that will enable them to survive longer than others and to reproduce more offspring that are like them.
(3) The variations that are inherited by each individual are passed on to their offspring.
(4) Over long periods of time, advantageous variations will gradually accumulate, leading to the remarkable diversity of structure and function we observe in the deep history of life.
(5) These structures and functions appear as if they are designed to match the requirements of life in various environments yet arise by purely natural forces.
(6) The origins of species, and their diverse adaptations that are so well suited to aiding them in their struggle for existence, are the product of evolution by means of Natural Selection.

Despite the fact that Charles Darwin's *Theory of Natural Selection* was brilliant and ground-breaking—and despite the fact that the solid intellectual core of classical Darwinian theory has stood the test of time—it is important to recognize that Darwin's vision was limited by the empirical tools and the background knowledge that were available to him at the time.

In particular, Darwin could not possibly know about the inner-workings of living cells, and for this reason, his theory of Natural Selection could not penetrate the mysteries of biological *variation*. It should therefore come as no surprise that, by today's rigorous scientific standards of detailed empirical evidence and proof, Darwin's classical theory, as described in his *On the Origin of Species* of 1859, was incomplete. Some of the questions that could *not* be answered by Darwin were:

(1) What are the units of hereditary variation in living cells?
(2) How are hereditary units transmitted from parents to offspring?
(3) What is the source of the variation that is carried by hereditary factors?

Answers to those questions were ultimately provided by the 20th century work of Thomas Hunt Morgan and colleagues (see below). However, the earliest significant empirical data concerning the nature and transmission of hereditary factors was provided by Gregor Mendel. Mendel is widely known as the 19th century scientific pioneer, Augustinian friar and abbot who lived in obscurity in Brno, Moravia. He published his now-famous Laws of Heredity in 1866.[n]

From Gregor Mendel to the Chromosomal Theory of Inheritance

Mendel's Laws of Heredity

The field known as **Mendelian Genetics** refers to the behavior of the hereditary factors that had first been discovered by Gregor Mendel in the 19th century but were not well-understood until the 20th century. Although Mendel was Darwin's contemporary, Mendel's work with pea plants was obscure at the time. It has often been said that if Darwin had met Gregor Mendel personally, it would have transformed the history of science. However, a careful and thorough analysis by Pablo Lorenzano published in 2011, which examines several relevant historical artifacts, shows that this is unlikely.[o]

Mendel was one of the earliest pioneers in biology who performed systematic, controlled experiments in order to discover the hidden laws of biology. By performing crosses with true-breeding strains of pea

[n] https://en.wikipedia.org/wiki/Gregor_Mendel

[o] http://www.academia.edu/download/37687693/Lorenzano-History_and_Philosophy_of_the_Life_Sciences_33_2011.pdf

plants, he was able to deduce what are today commonly referred to as Mendel's Laws:

(1) **First Law:** The Law of Segregation of Genes.
(2) **Second Law:** The Law of Independent Assortment.
(3) **Third Law:** The Law of Dominance.

Mendel focused on the passage of seven specific visible characteristics from parents to offspring, including:

(1) **Pea shape:** Round or wrinkled.
(2) **Pea shape:** Constricted or inflated.
(3) **Pea color:** Green or yellow.
(4) **Pod color:** Green or yellow.
(5) **Flower color:** Purple or white.
(6) **Plant height:** Tall or dwarf.
(7) **Position of flowers:** Axial or terminal.

Mendel observed that the hereditary factors seem to occur in pairs. In true-breeding strains of pea plants, the factors are both the same. When they are crossed, however, the offspring will randomly inherit one member of each pair from the male plant, and one from the female.

By counting the ratios of the visible characters associated with each member of the pair of factors, Mendel was able to draw conclusions about the nature of heredity:

(1) It consists of paired factors that determine specific characters.
(2) One factor from the male parent and one from the female parent are randomly passed to the next generation.
(3) One factor can be **dominant** over the other in determining a characteristic in the offspring.

The power and limitations of 19th century biology

Darwin and Mendel asked deep questions, but neither could perform experiments that provided detailed information concerning the inner-workings of living cells because of the limited conceptual framework and empirical knowledge-base in the mid to late 19th century.

Darwin was a skilled naturalist, and he relied on direct observations in largely natural settings. He could not discover the nature or causes of the variation on which Natural Selection depends, and he clearly stated those limitations[p] in *On the Origin of Species* [20]:

> Our ignorance of the laws of variation is profound. Not in one case out of a hundred can we pretend to assign any reason why this or that part has varied.

Mendel's experimental findings—which were presented in 1865 and published in 1866—were rediscovered in 1900,[q] and remained controversial until they could be integrated with new information about the inner-workings of living cells.

Mendel's understanding of hereditary factors was limited to the three laws that he cleverly deduced with his systematic experimental approach. But new empirical tools were required before the fledgling field that we now call Mendelian Genetics could advance to the next level.

That next critical step was to establish a link between visible characteristics of the organism and the cellular structures that carry the hereditary factors—the chromosomes.

The chromosomal theory of inheritance

Theodor Boveri dedicated much of his productive experimental career to proving unequivocally that chromosomes carried the hereditary Mendelian factors. His approach to linking heredity to chromosomes was to demonstrate that specific visible characteristics of one species would develop when particular chromosomes were experimentally combined with the cytoplasm of a different species, in developing sea urchin embryos. His finding, combined with those of Walter Sutton and others, became known as the **Boveri–Sutton Chromosomal Theory of Inheritance**. This would ultimately clarify the connection between **hereditary factors**, **chromosomes**, and **sexual reproduction**.

---

[p] http://darwin-online.org.uk/Variorum/1860/1860-167-c-1866.html

[q] https://en.wikipedia.org/wiki/Mendelian_inheritance

Most sexually-reproducing species are diploid, which means that they have two sets of similar chromosomes. With the Boveri–Sutton chromosomal theory, it was now possible to establish a conceptual linkage between Mendel's pairs of segregating hereditary factors, and the pairing of the two sets of chromosomes.

Sexually reproducing organisms have two major modes of cell division: **mitosis** (Figure 1), which produces genetically identical diploid cells that make up all of the organs in the body of the individual, and **meiosis** (Figure 2), which takes place among a special set of cells—called the **germ line** cells—and results in the production of sperm cells in males and egg cells in females.

The link between chromosomes, meiosis, and Mendel's Laws of Heredity

Meiosis reduces the number of chromosomes in each sperm or egg cell to one set instead of two. This occurs by a random process in which the pairs

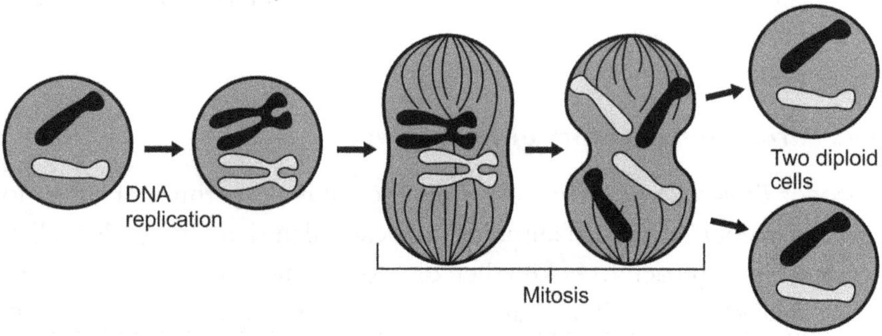

Figure 1. Diagram illustrating the way that newly replicated chromosomes are distributed during normal cell division (mitosis). Chromosomes replicate during DNA synthesis, and then, two identical sets of chromosomes are distributed equally to the two daughter cells during mitosis. Most sexually reproducing animals and plants have two sets of chromosomes during the diploid stage of their life cycle. If only one set of chromosomes is present during the haploid stage of the life cycle, then identical copies of that single set of chromosomes are equally distributed (not shown).

Source: Page-link: https://commons.wikimedia.org/wiki/File:Major_events_in_mitosis.svg. File-link: https://upload.wikimedia.org/wikipedia/commons/e/e0/Major_events_in_mitosis.svg. Attribution: By Mysid [Public domain], via Wikimedia Commons. Rendered in B&W.

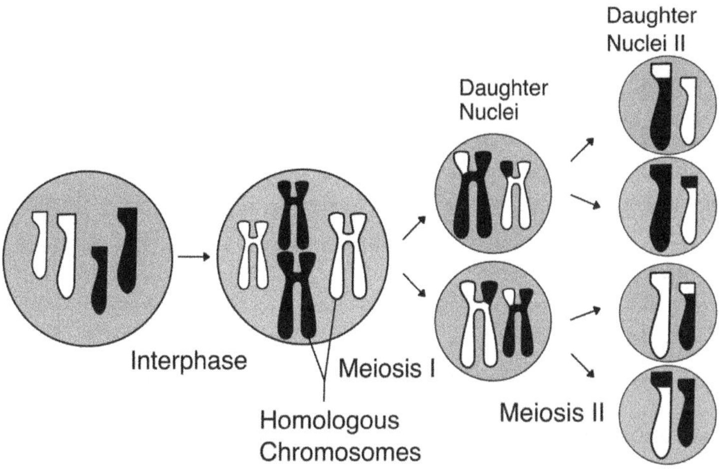

Figure 2. Illustration of DNA replication, chromosomal recombination (crossing-over) and distribution of chromosomes during production of sperm cells or egg cells, in sexually reproducing organisms. The germ line cells start out with two similar, but not identical, sets of chromosomes. One set was inherited from the mother, and one from the father. The germ line cells are therefore diploid, prior to meiosis. The similar pairs of chromosomes from each of two sets are called homologous chromosomes, and may carry the same or different alleles at each position along the chromosome. After DNA replication, random recombination events take place between homologous pairs, resulting in recombinant chromosomes. During meiosis, two cell divisions take place. In the first division, the homologous chromosomes (or recombinant chromosomes) are randomly distributed to the daughter cells, resulting in two genetically different cells. During the second division, the resulting sperm or egg cells have one set of chromosomes that is a random mixture of maternal and paternal chromosomes.

Source: Page-link: https://commons.wikimedia.org/wiki/File:Meiosis_Overview_new.svg. File-link: https://upload.wikimedia.org/wikipedia/commons/9/96/Meiosis_Overview_new.svg. Attribution: By Rdbickel [CC BY-SA 4.0 (https://creativecommons.org/licenses/by-sa/4.0)], from Wikimedia Commons. Rendered in B&W.

of chromosomes segregate into separate daughter cells. This random segregation of pairs of chromosomes represents the cellular mechanism that is responsible for Mendel's First and Second Laws.

When one sperm cell fertilizes an egg cell, the diploid chromosome number is restored in the next generation. The fertilized egg will then divide by mitosis to produce the identical somatic cells of this individual. Each cell will contain a random assortment of half of the hereditary factors of the mother and half of the hereditary factors of the father. During the process of development, a fraction of the cells in this individual will become specialized germ line cells, and these will form the sperm or egg cells of the next generation, depending on the gender of the individual.

The chromosomal theory of inheritance represented a tipping point for the wide acceptance of Mendelian Genetics in the first decades of the 20th century.

Taking Mendelian Genetics to the Next Level

Although most biology textbooks portray Mendel as the father of the field of genetics—and he was a pioneer, to be sure—Mendel's laws were taken much further by the group of intrepid laboratory scientists at Columbia University led by Thomas Hunt Morgan.

The brilliant genetic research program of Thomas Hunt Morgan and colleagues

Mendelian Genetics truly came of age in 1915,[r] when Morgan, along with his students A.H. Sturtevant, H.J. Muller, and C.B. Bridges, published a remarkable book. *The Mechanism of Mendelian Heredity* [22] was remarkable as much for its clarity as for the ways in which it explained Mendel's three laws. The book explained heredity in terms of concrete biological mechanisms. Hereditary factors are carried by chromosomes. In their own words[s]:

> Mendel's law was announced in 1865. Its fundamental principle is very simple. The units contributed by two parents separate in the germ cells of the offspring without having had any influence on each other... Mendel did not know of any mechanism by which such a process could take place... But in 1900, when Mendel's long-forgotten discovery was brought to light once more, a mechanism had been discovered that fulfils

[r] https://en.wikipedia.org/wiki/Boveri%E2%80%93Sutton_chromosome_theory
[s] https://books.google.com/books?id=lMO9goTre2UC&pg=PA263

> exactly the Mendelian requirements of pairing and segregation. The sperm of every species of animal or plant carries a definite number of bodies called chromosomes. The egg carries the same number. Consequently, when the sperm unites with the egg, the fertilized egg will contain the double number of chromosomes... Thus the behavior of the chromosomes parallels the behavior of the Mendelian units... These units will henceforth be spoken of factors; the two factors of a pair all called allelomorphs of each other. Their separation in the germ cells is called segregation.

Today, we call those **hereditary factors genes**, and we call the two hereditary factors of each pair **alleles**. Although the term "gene" (from the German or Danish word "gen") had already been coined in 1909 by Wilhelm Johannsen, a Danish botanist, Morgan and his colleagues continued to refer to them as "factors" in 1915.

In addition to coining the term "gene" (from the German or Danish word "gen") in 1909, Wilhelm Johannsen, a Danish botanist, coined two other very useful terms: **genotype** and **phenotype**. Genotype refers to specific alleles inherited by an individual, whereas phenotype refers to the specific visible characteristics that are somehow determined by those alleles.

The common fruitfly provides a powerful new empirical tool

One of the most important empirical tools used by Morgan and his colleagues, was their choice of an ideal model organism to study and experimentally manipulate in the lab—the tiny **fruitfly** known as ***Drosophila melanogaster*** (Figure 3).

With an incredibly short generation time of about two weeks, a small set of only four pairs of chromosomes, the ability to thrive in laboratory settings, and a broad range of visible phenotypes associated with specific alleles, *Drosophila* has turned out to be one of the most informative multicellular organism models for generations of biologists. This continues to be true today.

Morgan's intrepid group of laboratory explorers made an extraordinary number of key discoveries that extended Mendelian Genetics and

Figure 3. Illustration of a fruitfly, *Drosophila melanogaster* that has been ensnared by the sticky droplets produced by the sundew (*Drosera capensis*), which is a carnivorous plant. The fruitfly is arguably the most important model organism in terms of its contributions to both Mendelian and Molecular Genetics.

This photo illustrates one of the extraordinary ecological relationships that have evolved between species that function as producers and consumers, and as predators and prey. In this case, the plant is acting as both a producer and a predator, and the fruitfly is acting as both consumer and as prey.

Carnivorous plants tend to live in soils that are deficient in nutrients such as nitrogen. In addition to normal photosynthesis, carnivorous plants have evolved fascinating ways to attract and ensnare small prey, such as insects. The plant secretes enzymes that break down the macromolecules of the prey so that their nutrients can be absorbed by the plant.

Source: Page-link: https://commons.wikimedia.org/wiki/File:Drosophila_melanogaster_%E2%99%80_Melgen,_1830,_Drosera_capensis_Linnaeus,_1753_1100.1.2171.JPG File-link: https://upload.wikimedia.org/wikipedia/commons/5/58/Drosophila_melanogaster_%E2%99%80_Melgen%2C_1830%2C_Drosera_capensis_Linnaeus%2C_1753_1100.1.2171.JPG. Attribution: By Parent Géry [CC BY-SA 4.0 (https://creativecommons.org/licenses/by-sa/4.0)], from Wikimedia Commons.

linked hereditary factors to the formerly hidden inner-workings of cells. These new principles and experimental approaches included, but were not limited to the following:

(1) Mendel's hereditary factors—which are now commonly referred to as genes—are physically linked to chromosomes. Most chromosomes contain hundreds or thousands of gene factors called linkage groups.
(2) Only when factors are located on separate linkage groups—that is, when they are located on separate chromosomes—do they assort independently. Mendel's seven factors in pea plants were located on seven different linkage groups.
(3) Most pairs of chromosomes—called **homologous pairs**—contain the same set of genes in the same locations.
(4) Genes often have two or more distinct versions—or alleles—which can determine different visible characteristics, or phenotypes.
(5) The pairs of homologous chromosomes may carry two copies of the same allele—in which case the individual has a **homozygous** genotype. Alternatively, they can carry two different alleles, in which case the individual is said to have a **heterozygous** genotype.
(6) Mendel only discovered one type of dominance relationship, in which the **dominant** allele, when present, will mask the phenotype of the other allele, which is said to be **recessive**.
(7) Although some pairs of alleles obey Mendel's Law of Dominance, others behave differently. Sometimes the phenotype is intermediate—which is called **incomplete dominance**—and sometimes both alleles produce different visible characteristics—which is called **co-dominance**. Human blood types (Figure 4) provide an excellent textbook example of the relationship between co-dominant and recessive alleles.
(8) Sometimes one gene produces a visible phenotype and masks the phenotype of another gene—this is called **epistasis**.
(9) Sometimes one gene is responsible for several phenotypes and is said to be **pleiotropic**.
(10) Sometimes multiple genes contribute to the same visible characteristic in an additive fashion—this is one type of **quantitative inheritance**.
(11) Most of the chromosomes are in homologous pairs known as **autosomes** and carry the same **linkage groups** of genes.
(12) One pair of chromosomes—the **sex chromosomes** (X and Y)—determines the gender of the individual. In *Drosophila*, the gender is

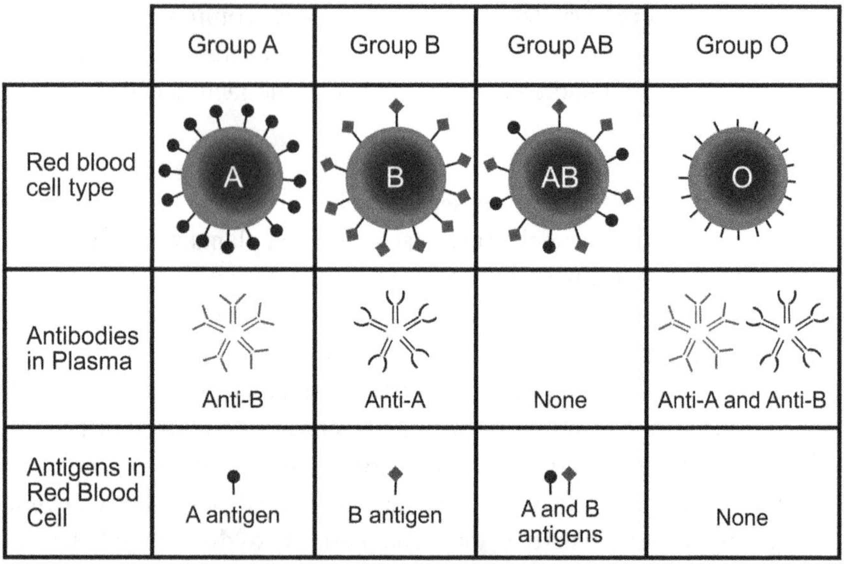

Figure 4. Relationships between alleles that determine human A, B, AB, and O blood types. Human blood types result from the expression of specific protein chains on the surface of red blood cells. Type A and Type B result from two co-dominant alleles, the A and the B allele, that determine slightly different amino acid sequences in these protein chains. Type AB blood results from expression of both the A and the B allele in the same individual. In Type O blood, no functional protein is transported to the surface. Therefore, the allele that determines Type O is recessive to both A and B alleles.

The medical consequences are that foreign alleles are recognized by the human immune system as antigens. Therefore, Type A individuals will make antibodies against the B allele antigen (which is foreign to them) and Type B individuals will instead make antibodies against the A allele antigen. Type AB individuals will not make antibodies, so they can be safely transfused with any type of blood—(Type A, B, AB, or O)—and are called universal acceptors. Type O individuals will make antibodies against both A and B alleles, so they can only be safely transfused with Type O blood. But since Type O individuals lack antigens on their own red blood cells, they can be universal donors to Type A, B, AB, or O individuals.

Source: Page-link: https://commons.wikimedia.org/wiki/File:ABO_blood_type.svg. File-link: https://upload.wikimedia.org/wikipedia/commons/3/32/ABO_blood_type.svg. Attribution: By InvictaHOG [Public domain], from Wikimedia Commons. Rendered in B&W.

determined by the ratio of X chromosomes to the autosomes; in humans, the presence or absence of the Y chromosome determines gender.

(13) Since X and Y chromosomes each carry different linkage groups, males will inherit only one set of X-linked genes. This means that alleles of certain genes carried on the X chromosome will produce visible phenotypes in males, even when they are recessive.

(14) Autosomal pairs of chromosomes undergo a random process known as **crossing over** or **recombination**, which takes place before the alleles separate during meiosis. This results in recombinant chromosomes in which combinations of alleles from both the mother and the father are carried on the same chromosome.

(15) Since genetic analysis of random crossovers shows that they are proportional to distances between genes, this provided a powerful tool for mapping specific genes to specific chromosomal locations.

(16) The fine-structure of chromosomes is revealed by consistent banding patterns when the chromosomes are stained.

(17) Special organs such as the salivary glands of *Drosophila* undergo mitosis without division of the cytoplasm, resulting in **polytene chromosomes** that exhibit "puffs" (extended fuzzy regions visible under the light microscope) when specific genes are actively expressed. Today, we attribute active expression to the production of **mRNA** (see Chapter 9).

(18) Literally thousands of mutations, including some that are found in nature—**spontaneous mutations**—as well as a larger number that are produced artificially in the lab with **X-rays** or **chemical mutagens**, have provided detailed information concerning the visible effects of alleles of thousands of *Drosophila* genes.

(19) Pairs of similar chromosomes that carry the same genes, but often carry different alleles—one from the male parent and one from the female parent—line up during sperm or egg production, respectively, and randomly separate such that each sperm or egg ends up with a random assortment of alleles attached to those chromosomes.

(20) Two of the chromosomes, known as X and Y chromosomes—carry genes that determine the sex of the offspring. These are called sex chromosomes.

(21) Germ line cells generate **sex cells**—the sperm and eggs—that transmit hereditary factors from the father and the mother, respectively, to the offspring, during sexual reproduction.

(22) By performing systematic genetic crosses, investigators can now routinely characterize newly discovered genes or alleles by following their visible effects.

(23) Today, although *Drosophila* still have many advantages as model organisms for research related to Mendelian Genetics, other organisms, including zebrafish and mice, are routinely used to study vertebrate or mammalian genes, including those that are more closely related to genes of medical importance in humans.

Conclusion

Charles Darwin's theory of Natural Selection provided a brilliant 19th century insight into the natural forces that are responsible for the origins of diverse species, with their remarkably effective and complex structures and functions. These adaptations equip each species to survive and reproduce their own kind in almost every imaginable environment and way of life. But like all scientific theories, it was limited by the knowledge available at the time. Similarly, Gregor Mendel's 19th century insights into genetics were also limited by available knowledge and were greatly advanced by 20th century studies by Morgan and his colleagues. In the chapters preceding this one, we have already seen important new perspectives that do not supplant classical Darwinian theory, yet do provide complementary insights.

Chapter 9

From the "Modern Synthesis" to Molecular Genetics: What Was Missing?

In 1868, for example, a Swiss investigator isolated deoxyribonucleic acid, DNA, from a cell's nucleus, but he had no idea of its function. Not until three-quarters of a century later, at the conclusion of some research directly related to the 1918 influenza pandemic, did anyone even speculate, much less demonstrate, that DNA carried genetic information.

—John M. Barry, in *The Great Influenza* [72]

The Big Picture

The "Modern Synthesis" of the 1940s combined the theory of Natural Selection with the expanded understanding of Mendelian Genetics. The statistical behavior of alleles in natural populations linked Darwin's concept of speciation to changes in the frequencies of specific alleles.

Through the 1950s and 1960s, the abstract concepts of alleles and phenotypes were combined with subsequent discoveries concerning the molecular interactions that translate DNA sequences into specific proteins, ushering in the age of Molecular Genetics.

The discovery that DNA is the genetic material of all living things led to the discovery of its structure. Protein synthesis was shown to take place on ribosomes. Messenger RNA links DNA sequences to amino acid sequences during protein synthesis, and transfer RNA carries individual amino acids to the ribosomes. This led to the discovery of the "genetic code", and by 1966 all 64 triplets (codons) of the "genetic code" were mapped to 20 specific amino acids, as well as three stop signals that terminate protein chains.

Since proteins carry out most of the diverse functions of cells and organisms, an understanding of the diverse roles of particular proteins in generating biological organization is essential for a modern understanding of evolution.

The one-to-one correspondence between sequences of codons and sequences of amino acids explained the various dominance relationships of Mendelian Genetics in terms of gene products and biochemistry. Various types of mutations and the phenotypes that they determine could now be understood in terms of diverse roles played by protein sequences in the inner-workings of cells.

Origin and Persistence of the "Modern Synthesis"

The "Modern Synthesis" combined Natural Selection with Mendelian Genetics and Population Genetics

The two most significant 20th century updates to classical Darwinian evolutionary theory were the "Modern Synthesis"[a] and the subsequent revolution in Molecular Genetics.

Over 80 years had elapsed since publication of the first edition of Darwin's *On the Origin of Species* in 1859. The "Modern Synthesis" combined the classical theory of Natural Selection [20] with the updated **Mendelian Genetics** of Morgan and colleagues [22] (see Chapter 8), plus a new specialty known as **Population Genetics.**[b]

Mendelian Genetics provided classical Darwinian theory with crucial insights concerning the nature and behavior of the hereditary factors, now called **genes**, that are passed from parents to offspring during sexual reproduction.

Population Genetics described the ways that variants of these factors, called **alleles**, change in frequency over many generations in natural

---

[a] https://en.wikipedia.org/wiki/Modern_synthesis_(20th_century)

[b] https://en.wikipedia.org/wiki/Population_genetics

populations. One of the major forces involved in that change is Natural Selection. Over time, reproductive isolation can turn hereditary population into separate gene pools, giving rise to new species.

Population Genetics, gene pools, and speciation mechanisms

Population Genetics, one of the three legs on which the "Modern Synthesis" stood, applied new mathematical and statistical methods to analysis of changes in frequencies of specific **alleles** in natural populations. Populations share a common **gene pool** because they breed freely during **sexual reproduction**. Natural populations are defined as part of a biological species when they breed freely and successfully in nature and produce **viable offspring**. The Population Genetics paradigm, stated briefly, is that biological species may diverge into distinct species when **reproductive isolation** of a natural population disrupts gene flow, and causes gene pools to diverge. Some of this divergence arises by means of Natural Selection, while other forces that change **allele frequencies**, such as genetic drift (random changes in allele frequencies) also play important roles.[c] In other words, reproductive isolation can lead to the origin of species. This led to a greater understanding of speciation mechanisms—a subject that is certainly relevant to evolutionary theory.

However, critics of the "Modern Synthesis" correctly point out that Population Genetics does *not* help us to understand how specific combinations of alleles interact to generate particular phenotypes. When individual genes, alleles and mutations are treated as proxies for phenotypes, this obscures the molecular, cellular and developmental links between genetic information and biological organization [70]. Today, we understand quite a bit more about those links. This topic will be discussed further in Chapters 12 and 13.

The scientific literature on the "Modern Synthesis" culminated in the publication of several definitive books in the 1930s and 1940s, including *Systematics and the Origin of Species from the Viewpoint of a Zoologist*[d]

[c] https://en.wikipedia.org/wiki/Hardy-Weinberg_principle

[d] https://en.wikipedia.org/wiki/Systematics_and_the_Origin_of_Species

by Ernst Mayr [73], *The Genetical Theory of Natural Selection*[e] by Ronald Fisher [74], *Genetics and the Origin of Species*[f] by Theodosius Dobzhansky [75], *Tempo and Mode in Evolution*[g] by George Gaylord Simpson [76], and Huxley's *Evolution: The "Modern Synthesis"* [23].

Genes Are Linked to Proteins, Including Enzymes

The one gene, one enzyme hypothesis

One important reason for the persistence of the "Modern Synthesis" is the success of the **one gene, one enzyme hypothesis** proposed by Beadle and Tatum in 1941.[h] In 1941, Beadle and Tatum published their landmark experimental paper [77] illustrating their success of deducing biochemical pathways in eukaryotic organisms by analysis of Mendelian mutations. This was made possible by clever experiments that relied on the fungus *Neurospora* as a model system for controlled experiments. This represented a considerable advance over analysis of biochemical pathways in prokaryotes. In the cautious style appropriate for reporting the results of pioneering studies, Beadle and Tatum wrote:

> The preliminary results outlined above may offer considerable promise as a method of learning more about how genes regulate development and function.

This was followed with their important insight:

> For example, it should be possible, by finding a number of mutants unable to carry out a particular step in a given synthesis, to determine whether only one gene is ordinarily concerned with the immediate regulation of a given specific chemical reaction.

These preliminary results were borne out by subsequent experiments.

---

[e] https://en.wikipedia.org/wiki/The_Genetical_Theory_of_Natural_Selection
[f] https://en.wikipedia.org/wiki/Genetics_and_the_Origin_of_Species
[g] https://en.wikipedia.org/wiki/Tempo_and_Mode_in_Evolution
[h] https://embryo.asu.edu/pages/george-w-beadles-one-gene-one-enzyme-hypothesis

From one gene, one enzyme to one gene, one protein chain

Following the discovery that DNA was the genetic material, subsequent discoveries in Molecular Genetics the 1950s and 1960s led to a concrete understanding of the structure of DNA, the "genetic code", and protein synthesis. Genes were shown to contain sequences of DNA that determine the sequence of amino acids in a variety of protein chains, not just those found in enzymes. As discussed in Chapter 4, each protein chain is a linear sequence of **covalently** bonded amino acids, where each position in the sequence is one member of a set of 20 possible amino acids. Each **codon** is one of 64 possible nucleotide (base) triplets in the DNA strand ($4 \times 4 \times 4$ possible nucleotides per triplet).

With the new understanding of Molecular Genetics, the one gene, one enzyme hypothesis—a powerful tool for deducing biochemical pathways in eukaryotes with genetic experiments—was expanded to the broader **one gene, one protein-chain** paradigm that is now well established.[i]

The significance of protein synthesis and protein chains as timeless innovations was discussed in Chapter 4. Protein chains represent fundamental units of structure and function. They are involved in virtually every molecular interaction that governs the inner-workings of cells.

New molecular techniques link genes to DNA, RNA, and proteins

The "Modern Synthesis" of the 1940s took place prior to the revolution in Molecular Genetics that began with the discovery of DNA in 1944 but did not get underway until the discovery of the structure of DNA in 1953, the cracking of the "genetic code" in the 1960s, and the recombinant DNA revolution in the 1970s.

---

[i]Actually, we now know that only certain segments of eukaryotic genes, called exons, are incorporated into DNA, and in some cases, alternative splicing of these exons can sometimes generate a variety of protein chains containing different amino acid sequences.

Proving that DNA is the genetic material of living things

Prior to 1944, biologists suspected that proteins, rather than DNA, constituted the genetic material of living things. Both DNA and proteins are found in chromosomes, but proteins are far more complex, since they consist of sequences of 20 different amino acids rather than four different nucleotides. Proteins seemed to be a better candidate for the genetic material, because it seemed likely that genes would require a complex molecule to encode them.

The definitive experiment that proved that DNA—not proteins—is the genetic material of living things was performed by Avery, McLeod, and McCarthy in 1944.[j]

The complex relationship between clinical and basic research

Oswald Avery[k] spent most of his early career as a tireless and meticulous researcher who was determined to find the infectious agent that causes influenza. Influenza, complicated by bacterial pneumonia, was responsible for some 50–100 million deaths worldwide during the flu pandemic of 1918. Although the virus that causes influenza was not identified until 1931 by Richard Shope,[l] Avery's research focused on a **virulent** bacterial strain, a secondary infection of flu victims that was responsible for the majority of deaths in the 1918 pandemic.

The Great Influenza [72], an engaging popular historical account by John Barry, tells the clinical history behind the serendipitous discovery of the phenomenon known as **genetic transformation** by Oswald Avery, McLeod and McCarty.[m] Briefly, they discovered that DNA, taken up by

---

[j] https://en.wikipedia.org/wiki/Avey-MacLeod-McCarty_experiment

[k] https://en.wikipedia.org/wiki/Oswald_Avery

[l] http://www.influenzavirusnet.com/history-of-influenza.html

[m] The Great Influenza is not, as the title might imply, only about the flu pandemic of 1918. The book is also a treasure trove of other information describing the fascinating history of medical research and clinical practice in the United States. The book shows not only how basic research (a.k.a. fundamental research) led to technological advances and clinical applications, but also how clinical research fortuitously led to major advances in basic research, as described above. For those who enjoy audiobooks, Scott Brick provides a masterful narration.

bacterial cells, could turn harmless strains of bacteria into lethal pathogens. This led to critical experiments demonstrating that DNA is the genetic material of all cells, in 1944, resulting in the Nobel Prize.[n] This illustrates the complex, mutually beneficial relationship between basic and clinical research. It also shows that luck and accidental discoveries play major roles in both basic and applied research, and the value of a particular line of research often cannot be predicted in advance. Biomedical science rests on a firm foundation of basic research, and it is necessary to cast a very wide net.

The Nature of DNA and Cracking of the "Genetic Code"

Discovery of the structure of the DNA double-helix

The molecular structure of DNA was determined in 1953 by James Watson and Francis Crick, with data taken from Rosalind Franklin. It is important to realize that Watson and Crick could not have determined the double-helical structure of DNA without relying on the images that Rosalind Franklin was able to produce, using the empirical technique of X-ray diffraction, a method related to X-ray crystallography, which was first applied to biological molecules by Dorothy Crowfoot Hodgkin when she solved the structures of cholesterol and penicillin in 1937 and 1946, respectively. It's worth pointing out that although Hodgkin was awarded the Nobel Prize in Chemistry in 1964 and went on to solve the structure of insulin in 1969, it is only recently that the accomplishments of women such as Franklin and Hodgkin have become more widely known to the general public.

James Watson and Francis Crick did publish one of the most influential papers of all time—in what has become one of the most prestigious scientific journals in biology—*Nature*—and they mention the use of Franklin's unpublished data in that paper.

[n] http://www.jbc.org/content/277/16/13355.full.pdf

The classic paper [25] is remarkably concise—it is only about 1,100 words long—and it contains what is arguably the most famous scientific understatement of all time:

> It has not escaped our notice that the specific pairing we have postulated immediately suggests a possible copying mechanism for the genetic material.

The paper also includes the iconic image of the double-helix. It's worth noting that while biologists often must rely on words and images to demonstrate elegant principles, physicists have the privilege of using simple mathematical expressions to capture the most extraordinary and universal scientific laws—such as the most famous math equation of all time provided by Albert Einstein[o]: $E = mc^2$.

Messenger RNA, proteins, and the "central dogma" of molecular biology

Once DNA was shown to be the genetic material of all cells and organisms, a series of extraordinary discoveries concerning the connection between DNA and proteins could follow. Each required the use of empirical laboratory techniques to perform appropriate experiments, and in many cases, these techniques had not been invented yet.

In 1957, Francis Crick published "On Protein Synthesis" [78], in which he laid out both the rationale and the experiments demonstrating the connections between DNA sequences and the synthesis of protein chains, which are specific sequences of amino acids:

> It is an essential feature of my argument that in biology proteins are uniquely important… Watson said to me, a few years ago, 'The most significant thing about the nucleic acids is that we don't know what they do.' By contrast the most significant thing about proteins is that they can do almost anything… Almost all chemical reactions in living systems are catalyzed by enzymes, and all known enzymes are proteins… I shall also argue that the main function of the genetic material is to control (not necessarily directly) the synthesis of proteins…

---

[o] https://en.wikipedia.org/wiki/Mass-energy_equivalence

Many of the key experiments were performed by using T4, a virus that replicates by infecting cells of the common gut bacterium T4, injecting a copy of its DNA, and using the host cell machinery to replicate its DNA and to synthesize the viral proteins, resulting in the replication of infectious virus particles.

To make a long story short, in 1958 Crick, relying on the collective works of many brilliant laboratory scientists who shared their findings in peer-reviewed scientific publications, published what came to be known as the "central dogma" of molecular biology:[p]

> The Central Dogma. This states that once 'information' has passed into protein it cannot get out again. In more detail, the transfer of information from nucleic acid to nucleic acid, or from nucleic acid to protein may be possible, but transfer from protein to protein, or from protein to nucleic acid is impossible. Information means here the precise determination of sequence, either of bases in the nucleic acid or of amino acid residues in the protein.

Today, we know that information flows not only from DNA to RNA to protein, but also in many other directions, and at a variety of levels of complexity. Transcription factors, for example, are proteins that bind to specific DNA sequences and regulate their expression.

Cracking the "genetic code"

By using proflavin—a chemical **mutagen** that acts by inserting itself between the bases of DNA—to generate single-base insertions and deletions in the DNA of the bacterial virus T4, Crick and colleagues were able to determine that DNA contains a **genetic code** consisting of groups of three DNA nucleotides (bases) known as **codons**.

By the mid-1960s, Marshall Nirenberg, Har Gobind Khorana, Robert Holley and colleagues were able to decipher the entire "genetic code", and also to determine the structure of transfer RNA (tRNA), the RNA molecule that delivers amino acids to the ribosomes, where they are assembled into protein chains.[q] The specific sequence of amino acids in each protein

[p] https://en.wikipedia.org/wiki/Central_dogma_of_molecular_biology

[q] http://www.dnaftb.org/22/bio.html (15 March 2018).

chain corresponds to the sequence of codons in a sequence of messenger RNA (mRNA). The sequence of codons in mRNA corresponds to the sequence of nucleotides in the DNA template.

There are 64 possible codons in each triplet of DNA, since there are four different nucleotides (A, C, G and T): $4 \times 4 \times 4 = 64$. 61 codons specify which of the 20 possible **amino acids** will be placed at each position in a newly synthesized **protein chain**. Some amino acids are specified by more than one codon. Three of the 64 codons are stop signals, which terminate synthesis of a protein chain at the ribosome.

The molecular interpretation of Mendelian gene mutations

The central importance of proteins to virtually every aspect of the life of the cell was discussed in Chapter 4. In the context of **Mendelian Genetics**, the phenotypes of **protein-coding**[r] genes should generally involve the presence or absence of specific protein chains, or the control of their synthesis (see Chapter 10).

Mutations in a protein-coding DNA sequence should usually either eliminate synthesis of a functional protein chain, or modify the protein chain that is synthesized. The cracking of the "genetic code" provided concrete explanations for the phenotypic consequences of a variety of **mutations** observed by Mendelian geneticists. Readers without a background in biology will find an accessible introduction to genetics in Wikipedia.[s]

Here are several concrete examples of the ways in which Molecular Genetics has helped to clarify the phenotypes of Mendelian mutations, in terms of their visible effects:

(1) When mutations result in changes in the sequence of nucleotides in protein-coding DNA sequences, most result in the loss of function by interfering with synthesis of a specific protein chain.

---

[r]Although the concept of protein-coding genes or DNA sequences is a useful one, and although biologists often refer to a "genetic code," the metaphorical use of the word "code" can be misleading, as discussed in Chapter 12.

[s]https://en.wikipedia.org/wiki/Introduction_to_genetics

(2) Prokaryotic cells usually contain one copy of the genome, so mutations in critical biochemical pathways are often lethal, and can be used experimentally to deduce the individual steps in biosynthetic pathways.
(3) Most multicellular eukaryotes spend the majority of their life cycle in the diploid state, which means that they carry two copies of the genome. Diploid genomes of sexually reproducing organisms contain two sex chromosomes and two similar sets of autosomes. This means that the individuals carry two similar sets of autosomal genes.
(4) Homozygous diploid individuals carry two identical alleles of each autosomal gene. Heterozygous diploid individuals carry two different alleles.
(5) For protein-coding genes, loss of expression of a viable protein chain in a homozygous individual may not have a deleterious effect, because the other unmutated copy of the gene can still produce a normal protein.
(6) In organisms with X and Y sex chromosomes, mutations on the X chromosome are more likely to show up in the phenotype of males, because males carry only one copy of those sex-linked (X-linked) genes.
(7) Since protein-coding genes are read in groups of three nucleotides, frameshift mutations (insertions or deletions of nucleotides that are not multiples of three nucleotides, Figure 1) will change the way that the codons are translated at the ribosomes. This may result in a premature stop codon which terminates the protein chain (and usually destroys protein function), or it may change the amino acid sequence of a protein chain for every codon that is downstream from the point of insertion. Often this will destroy protein function, but not always.
(8) A recessive mutation often destroys one of the two copies of a protein chain encoded by a pair of autosomal genes. Their effects are often masked by the presence of a normal protein chain that is encoded by the other nonmutated copy of the gene on the other autosome.
(9) Incomplete dominance often involves pairs of autosomal alleles that code for different protein chains, or for protein chains that require two copies to be fully expressed. For example, if one allele codes for

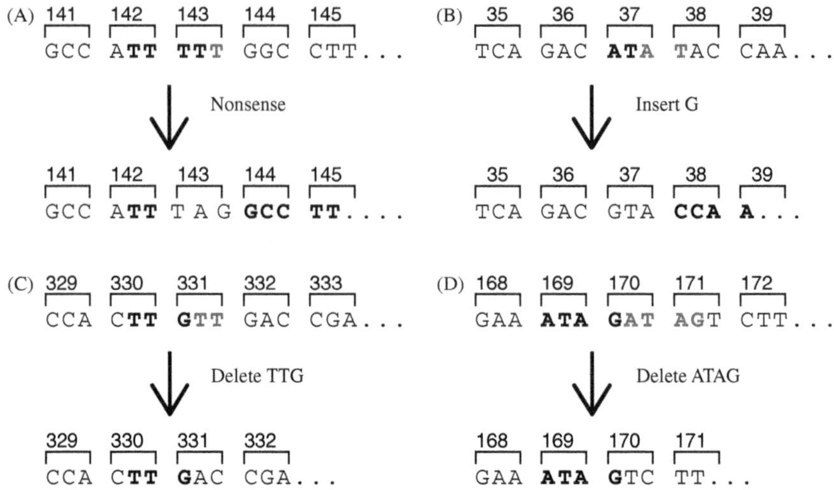

Figure 1. Diagram illustrating how various types of mutations can affect the way that codons are translated at the ribosomes. This can lead to a variety of changes in the amino acid sequences of protein chains. A: Changing a single base (nucleotide) may change a single amino acid; this is known as a nonsense mutation. B: Insertion of a single base represents a frameshift that alters the codon reading frame and changes several amino acids that are downstream. C: Deletion of a multiple of three bases does not change the reading frame, but can remove and/or change one or more amino acids in the chain. D: Deletion of bases that are not multiples of three cannot only remove amino acids but can also cause a frameshift and change several amino acids that are downstream. Not shown: a variety of other mutational consequences are possible, depending on the context of the sequence in which they occur. For example, an insertion or deletion can either produce a premature stop codon that truncates the protein chain, or can remove a stop codon and lengthen the chain.

Source: Page-link: https://commons.wikimedia.org/wiki/File:Frameshift_mutation.jpg. File-link: https://upload.wikimedia.org/wikipedia/commons/0/0a/Frameshift_mutation.jpg. Attribution: By Sumukal [CC BY-SA 3.0 (https://creativecommons.org/licenses/by-sa/3.0)], from Wikimedia Commons. Rendered in B&W.

red flower color and the other codes for white flower color, if heterozygous individuals have pink flowers, then this is an example of incomplete dominance. This phenotype may be determined by the amount of red pigment that is synthesized by one or by two protein chains.

(10) Co-dominance often involves alleles that code for protein chains that have different amino acid sequences but are expressed in the

same cells. For example, the blood type alleles A and B can result in type A, type B, or type AB individuals. These phenotypes are caused by proteins on the surface of red blood cells that are recognized as antigens by the immune system. There is also an "i" allele that is recessive to both A and B, which does not produce a functional protein chain, and is nonantigenic.

(11) Since protein chains are involved in virtually every aspect of the life of the cell (Chapter 4), mutations may affect a variety of cellular functions. Some will affect critical functions and result in lethality. As discussed in Chapter 15, systematic screening for lethal mutations in fruitflies provided important insights into the developmental processes that give rise to the body plan, not only of insects but in a broad range of animals as well.

(12) Many important genes do not code for protein-chains, but instead play critical roles in the expression of particular protein chains in particular cell types under specific circumstances. This will be discussed in more detail in Chapters 10 and 11.

(13) Especially for processes involving development, phenotypes often depend on complex interactions between multiple protein chains that are products of several genes. In such cases, the complex ways that these proteins interact with other molecules, at a variety of levels of organization, must be described. Fortunately, modern laboratory techniques have made it possible to systematically analyze such complex functions in an empirical manner, and many complex functions are now well-understood.

From protein chains and DNA to higher levels of organization

Although Mendelian and Molecular Genetics represented important conceptual advances for evolutionary theory, the remarkable success of this conceptual framework also had a downside. The problem can be briefly stated as follows:

(14) Beyond biochemical pathways, most multicellular adaptations involve higher levels of organization that arise from complex interactions of macromolecules, not just single gene products.

(15) Additional principles of molecular, cell, and developmental biology are required to explain the relationship between the structure of genes and proteins and their useful functions. Unfortunately, the essential features of this information did not become available until later.

Recombinant DNA and the Biotechnology Revolution

One of the themes of this book is the close relationship between evolutionary biology and an empirical understanding of the inner-workings of the cell. Advances in both biology and medicine have played a major role in advances in evolutionary theory. Recombinant DNA technology is especially relevant, since it provides insights into DNA, RNA, proteins, and various functions of the genetic apparatus.

The power of recombinant DNA technology

The molecular structure of DNA, RNA, and proteins consists of **polymers** (chains of subunits (**monomers**) that are linked together in long chains by **covalent** bonds). Since similar sets of monomers (four different **nucleotides** for DNA and RNA, and 20 different amino acids for proteins) are used by virtually all species, DNA (or RNA) molecules transferred from a donor species to a host species will often perform similar functions in the host. This is also true for DNA and RNA sequences from **viruses**, which depend on host cells for their replication. A primer on the steps involved in protein synthesis can be found on Wikipedia.[t]

This also means that **recombinant DNA** sequences—DNA polymers derived from two different species—can be transferred into host cells, where they will be replicated, and where they can often be designed to synthesize mRNA and proteins in the host cell.

The potential power of recombinant DNA technology was recognized by Herbert Boyer and colleagues at UCSF and Stanley Cohen and colleagues at Stanford University. These pioneers developed recombinant

---

[t]https://en.wikipedia.org/wiki/Protein_biosynthesis

DNA technology[u] in the early 1970s, which dramatically accelerated the pace of discovery in both biology and medicine.

Briefly, recombinant DNA is prepared by inserting a desired sequence of DNA into a **vector**—usually either a **plasmid**, a small circular piece of DNA that can replicate in a harmless strain of ***E. coli*** bacteria, or the DNA of a virus (**bacteriophage**) that replicates in *E. coli.* Special enzymes, known as **restriction enzymes** are used to cut the DNA sequences, and **ligase** enzymes are used to seal the DNA sequences together. The vectors are then grown in bacterial cell cultures, resulting in **molecular clones** of specific DNA sequences that can then be analyzed in a variety of ways.

DNA sequencing, databases, and sequence alignments

Recombinant DNA technology has extraordinary power as a tool of discovery in evolutionary biology. As may be expected, the power of these laboratory techniques has been combined with several other technological advances—including **DNA sequencing** technology[v] and sophisticated bioinformatics[w] software for cataloging, searching, comparing and analyzing massive amounts of sequence data. These sequences are stored in public databases, along with analysis tools that are easily accessed on the Internet,[x] and educational resources geared to high school students are also now available.[y]

From cDNA and PCR to CRISPR

Three other powerful biotechnology tools should be briefly mentioned here: **cDNA**, **PCR**, and **CRISPR**.

(16) Complementary DNA technology (cDNA) is now routinely used for analysis of messenger RNA sequences—the processed mRNA

[u] https://en.wikipedia.org/wiki/Recombinant_DNA
[v] https://en.wikipedia.org/wiki/DNA_sequencing
[w] https://en.wikipedia.org/wiki/Bioinformatics
[x] https://www.ncbi.nlm.nih.gov/
[y] https://geneed.nlm.nih.gov/

sequences that are ultimately translated into protein sequences. cDNA uses an RNA template to synthesize a DNA sequence, which can then be cloned with recombinant DNA technology, sequenced, or used in any other way.

(17) In 1987, Kary Mullis invented PCR, which earned him the Nobel Prize. PCR[z] is an extremely powerful method that is much faster, easier, and less expensive than recombinant DNA technology. Today, PCR is a multi-billion-dollar industry. The technique is routinely used in both basic research and for a variety of clinical applications, as well as forensic analysis, paternity, and to determine human ancestry. It can be used to create large numbers of copies of specific DNA or RNA sequences, and all that is required is that about 40–60 nucleotides of the flanking sequences of the target DNA sequence be known in advance. This means that virtually any target sequence of DNA can be amplified (copied in large numbers, even from total genomic DNA). The technique is so sensitive that the genomic DNA of even a single cell can be used as starting material. PCR now plays a major role in evolutionary DNA sequence comparisons used to determine relationships between distantly related organisms.

(18) CRISPR is a very powerful and new technique[aa] for targeting and changing specific DNA sequences in genomic DNA of living organisms. Human beings can now artificially alter the nature and evolution of various species, including humans. CRISPR has been quite useful as an experimental tool, and has great promise for advances in molecular medicine. But there are several reasons to be concerned about ways that this technology could be abused to create "designer' babies and to destroy the natural balance of the human gene pool [79]. Unfortunately, human greed rather than human need is often the driving force in biotechnology. Powerful technologies such as CRISPR could profoundly impact the future of our own species, and the planet that sustains us. We need to be proactive in our decision-making to assure positive outcomes.

---

[z] https://en.wikipedia.org/wiki/Polymerase_chain_reaction

[aa] https://en.wikipedia.org/wiki/CRISPR

The Fruits of the Molecular Genetics Revolution

The nature and function of the genome is a topic of obvious relevance to evolutionary biology, and new technologies have dramatically increased the pace of discovery. This chapter has briefly surveyed some limitations of the conceptual frameworks of the "Modern Synthesis" and Population Genetics, while introducing the extraordinary power of recombinant DNA technology. New techniques led to the conceptual revolution of Molecular Genetics. Interested readers can find literally dozens of additional articles at the evolutionary biology portal on Wikipedia.[ab]

In the chapters that follow, we'll explore the molecular, cellular, and developmental mechanisms that generate complex organization. The new molecular perspectives transcend both classical Darwinian theory and Mendelian Genetics. Several important new paradigms will be introduced in the following chapters that have taken our understanding of evolutionary biology to a whole new level.

Beyond the "Modern Synthesis"

The unfortunate moniker "Modern Synthesis"—intrinsically unlikely to improve with the passage of time—came from Julian Huxley's book titled *Evolution, the "Modern Synthesis"* [23]. The title was catchy, however. The continued use of the term causes considerable confusion, because what was modern in 1942 is now past history. Strictly speaking, the "Modern Synthesis" took place before the discovery of DNA and cracking of the **genetic code**.

Many evolutionary biologists now use a broader term, **neo-Darwinism**, to describe both the "Modern Synthesis" and subsequent discoveries in Molecular Genetics. Our understanding of variation and the inner-workings of cells has grown by leaps and bounds since 1942.

In other words, the "Modern Synthesis" is no longer "modern". It did represent a major milestone in the history of evolutionary theory, however. Some 75 years later, it is *still* the textbook account of evolution taught in

---

[ab] https://en.wikipedia.org/wiki/Portal:Evolutionary_biology

many introductory biology courses in middle school, high school, and even college, despite the fact that it is incomplete and out of date.

Updating the Language and Conceptual Framework of Evolutionary Theory

How history and language have caused confusion that extends beyond the "Modern Synthesis" to neo-Darwinian theory

In his clear and insightful 2015 synthesis[ac] of ideas put forward in his earlier books and lectures, Denis Noble pointed out how the "web of interpretation" that represents the "whole conceptual scheme of neo-Darwinism" has created significant difficulties for evolutionary theory:

> Each concept, simile or metaphor reinforces the overall mind-set until it is almost impossible to stand outside it and appreciate how beguiling it is. As the "Modern Synthesis" has dominated biological science for over half a century, its viewpoint is now so embedded in the scientific literature, including standard school and university textbooks, that many biological scientists may not recognize its conceptual nature, let alone question incoherencies or identify flaws.

Noble correctly points out that many of the problems with the "Modern Synthesis" arise from the way that neo-Darwinism represents the information, rather than the empirical findings *per se*:

> These forms of representation have been responsible for, and express, the way in which 20th century biology has most frequently been interpreted…The concepts therefore form a biased interpretive veneer that can hide those discoveries in a web of interpretation.

No single word has caused greater problems for evolutionary theory than the word "gene". To understand this, we will need to look more

[ac] Noble D. (2015) "Evolution beyond neo-Darwinism: a new conceptual framework". Originally published in *J. Exp. Biol* 218, 7–13. Corrected online version available as pdf: http://jeb.biologists.org/content/jexbio/218/1/7.full.pdf

closely at the gap between revolutionary empirical discoveries about the nature of the gene and the entrenched conceptual framework of the "Modern Synthesis".

Noble provides an insightful analysis of the conceptual and semantic dichotomy between the Mendelian concept of the gene and the term $gene_J$ to describe the conceptual dichotomy it has created:

> The word 'gene' was introduced by Johannsen (Johannsen, 1909). But the concept had already existed since Mendel's experiments on plant hybrids, published in 1866 (see Druery and Bateson, 1901), and was based on 'the silent assumption [that] was made almost universally that there is a 1:1 relation between genetic factor (gene) and character' (Mayr, 1982). Of course, no-one now thinks that there is a simple 1:1 relation, but the language of direct causation has been retained. I will call this definition of a 'gene' $gene_J$ to signify Johannsen's (but essentially also Mendel's) meaning…

The $gene_J$ definition subtly implies that there is a 1:1 relationship between a genetic factor and a specific visible characteristic of an individual—consistent with the visible correlation between the presence of a specific genetic allele and the visible phenotype (character) that is observed when that allele is present:

> $Gene_J$ referred to the cause of a specific inheritable phenotype characteristic (trait), such as eye/hair/skin colour, body shape and mass, number of legs/arms/wings, to which we could perhaps add more complex traits such as intelligence, personality and sexuality.

Having summarized some of the key empirical discoveries of molecular biology (or Molecular Genetics) from the 1960s, we can now return to Denis Noble's discussion of the conceptual baggage of the "Modern Synthesis" and neo-Darwinism. The empirical discoveries of Molecular Genetics revolutionized our understanding of the nature of the gene, and Noble draws a distinction between the older definition of $gene_J$ and a new one:

> Following the discovery that DNA forms templates for proteins, the definition shifted to locatable DNA sequences with identifiable beginnings and endings. Complexity was added through the discovery of

> regulatory elements (essentially switches), but the basic cause of phenotype characteristics was still thought to be the DNA sequence as that forms the template to determine which protein is made, which in turn interacts with the rest of the organism to produce the phenotype. I will call this definition of a 'gene' $gene_M$.

The "M" subscript in $gene_M$ represents a "molecular definition of the gene". The problem, however, is that the conceptual framework of the "Modern Synthesis"—which was closely tied to the Mendelian Genetics of the *Drosophila* experiments—continued to associate genetic variation with the "random mutations" described by Julian Huxley.

How do more recent empirical discoveries change the meaning of the "gene"?

Empirical discoveries related to the "gene" have continued to challenge the way that we define the word—so much so that at the present time, about 18 years into the 21st century, the molecular definition is insufficient, and we require a new one. For now, let's follow Noble's lead and give it a new name: we can call it $gene_{2019.}$ Here are some essential features:

(19) A $gene_{2019}$ is a unit of heredity usually transmitted from generation to generation—or from cell to cell—as a DNA sequence, and sometimes by means of an RNA intermediate. This is consistent with the molecular definition ($gene_M$).

(20) The information contained in the DNA sequence can carry both *explicit* information—such as the sequence of amino acids in a protein chain—as well as *implicit* information—such as a variety of regulatory roles—that depend on a variety of molecular and cellular interactions. This already extends beyond $gene_M$.

(21) The meaning and significance of a $gene_{2019}$ is context-dependent, where "context" refers to a broad range of other entities such as chemical modifications, the presence or absence of other DNA sequences in the cell or individual, the cell type, the presence or absence of a variety of other molecules—including but not limited to transcription factors, the developmental and physiological status of the cell, and more.

(22) Replicated copies of the DNA sequence—or portions thereof—can change in a variety of different ways in different cells, different individuals, and different generations—which include, but are not limited to: point mutations, deletions, insertions (which may or may not result in frameshifts or other effects in proteins), chromosomal rearrangements, gene duplications, cross over events, slipped-strand mispairing events, methylation or acetylation, and more.

(23) The phenotypic effects of the DNA sequence—the ways that it influences or helps to determine the shape and form of cells, organs, individuals, or interactions between individuals—can be defined both in terms of its generative effects—a subject discussed in Chapter 4—as well as its ecological impact on the survival and reproduction of the individual.

Our Updated Evolutionary Synthesis (UES) incorporates all of these new distinctions of $gene_{2019}$. It also has the flexibility to accommodate a variety of empirical discoveries that are yet to be made.

In the chapters that follow, we'll begin to explore the higher levels of organization that represent the true significance of alleles and protein chains. We'll begin to focus on the molecular, cellular, and developmental interactions from which phenotypic characteristic emerge. We'll go beyond genes or cells in isolation, and begin to explore the ways that biological organization is reproduced in each generation. We'll begin by exploring the new evolutionary opportunities made possible by the evolution of multicellularity.

Chapter 10

The Origins of Multicellularity and Embryonic Development

Cell differentiation is based almost certainly on the regulation of gene activity, so that for each state of differentiation a certain set of genes is active in transcription and other genes are inactive.

—Roy Britten and Eric Davidson[a] [80]

The Big Picture

Despite the extraordinary combined explanatory power of the "Modern Synthesis" and Molecular Genetics, neither can explain how cells—which all carry a complete set of identical DNA sequences—give rise to the diverse and complex structures and functions that arise, in each generation, during the development of multicellular individuals.

In multicellular species, useful elements that are modified and reused in novel ways are aptly described as members of "developmental toolkits". These toolkits include flexible and reusable genetic components that regulate complex interactions between cells and macromolecules during development. The toolkit metaphor is also useful for exploring the flexibility and reuse of duplicated genes involved in prokaryotic metabolic networks.

Throughout the entire range of carbon-based life forms that have ever existed on planet Earth, a common theme is that their evolution has been

[a]https://embryo.asu.edu/pages/gene-regulation-higher-cells-theory-1969-roy-j-britten-and-eric-h-davidson

driven by adaptations that help them to adopt particular ways of life (*niches*) that facilitate capture of energy and nutrients in novel ways. This leads to complex ecosystems of producers and consumers, including predators and prey.

In terms of evolvability and innovations that facilitate capture of energy and nutrients, multicellularity has several distinctive and useful features. Among these are the production of embryos, and the robust and flexible genetic determinants that allow emergence of higher levels of interaction between cells and the reproduction of complex structures and functions during their development.

High-speed analysis and comparison of genomic DNA sequences and other cutting-edge techniques show that multicellularity has arisen multiple times in the deep history of eukaryotic evolution, and that multicellular animals share common ancestors with the choanoflagellates.

The Powerful Toolkit Metaphor

Sean B. Carroll [15] has found the metaphor of the **toolkit** (or toolbox) to be quite useful in the relatively new field of **evolutionary developmental biology** (**evo-devo**). Evo-devo is mostly concerned with the interactions and specialization of eukaryotic cells in **multicellular** organisms.

Toolkits for capturing energy and nutrients

Maslov, Krishna, Pang and Sneppen argue [81] that if **enzymes** are viewed as tools, and **metabolic pathways** are viewed as collections of tools that allow use of a new **metabolite** (such as a new carbon source), then genomes need to evolve fewer and fewer new enzymes, because existing enzymes can be reused in new ways. In other words, they become part of the prokaryotic toolkit.

Reuse of existing enzymes often requires the evolution of new regulators, such as **transcription factors** or **operons**. Therefore, Maslov *et al.* argue that the number of new regulators should increase more rapidly than the number of enzymes. They explored this prediction by comparing the results of mathematical models to empirical observations. The authors report that the branching structure of empirically observed prokaryotic metabolic networks is consistent with the predictions of their mathematical models.

Several synonymous terms describe evolutionary *reuse* of metaphorical toolkit elements in new ways. The term **exaptation**[b] is used to describe the way that **adaptations** can be (and frequently are) redeployed in new and useful ways, resulting in new **phenotypes** as well as new ***niches***.

Horizontal gene transfer between prokaryotes dramatically extends the **gene pools** of prokaryotic species, providing numerous opportunities for a given species B to redeploy metabolic innovations that arose (often for different reasons) in species A.

Comparing the biomass captured by various carbon-based life forms on Earth

Considering their limitations in size, shape, and mobility, prokaryotes have done remarkably well in exploiting available sources of energy and nutrients. According to a recent estimate by Bar-On, Phillips, and Milo [83], the total **biomass** of all carbon-based life on Earth is about 550 **gigatons** (550 billion tons) of carbon. Of this total, **bacteria** and **archaea** have captured about 70 and 7 gigatons, respectively, primarily in deep subsurface environments.

Plants, on the other hand, which are **multicellular eukaryotes**, have captured about 450 gigatons of carbon. Most of that biomass is found in terrestrial environments. Clearly, in terms of opportunities for capturing energy and converting available carbon to biomass, plants are living proof of the new opportunities afforded to multicellular, terrestrial eukaryotes. Terrestrial plants have far greater biomass than prokaryotes because they are able to harvest and exploit the vast amounts of energy available from photons of sunlight that pass through the atmosphere. They also can readily take in carbon dioxide (CO_2), required for **carbon fixation** during **photosynthesis**, and release oxygen as a byproduct. Much of the biomass is utilized to build rigid and resilient structures from cellulose, which is a polymer of glucose found in plant cell walls. Especially in dense

[b] https://en.wikipedia.org/wiki/Exaptation

rainforests, trees compete for available sunlight by building tall trunks and branches that thrust their leaves skyward towards the canopy, where sunlight is most abundant.

Other environmental factors, such as water, proximity to the equator, optimal temperatures, nitrates, and phosphates, are limiting resources that are responsible for the distribution of plants among terrestrial biomes.[c]

In the oceans, both carbon dioxide and water are abundant, but sunlight is rapidly filtered out by increasing depths of water, and nitrates and phosphates are usually more limited (except in waters polluted by, for example, agricultural run-off). The largest communities of marine organisms that depend on photosynthesis usually thrive near the water's surface, where unfiltered wavelengths of sunlight are most abundant.

A thought experiment involving a hypothetical prokaryotic terrestrial pioneer

Now consider the following imaginary scenario. Suppose that a hypothetical photosynthetic bacterium evolved that could adhere to virtually any terrestrial surface. Let's call it *Laminatium adherens*. These hypothetical wonders could somehow conserve water while allowing carbon dioxide to get in and oxygen to get out. In this thought experiment, *L. adherens* can utilize sunlight effectively, but is relatively resistant to temperature extremes as well as mutations caused by ultraviolet light. The cells could survive periods of extreme cold, dryness, or absence of sunlight by forming dormant spores that could survive for a year or two until conditions improved.

L. adherens could outcompete multicellular plants, because, in our hypothetical scenario, it could adhere to any surface (including those plants) and be first in line to capture the direct rays of the sun.

The question is, what would happen next? Perhaps, if *L. adherens* could adhere to any surface, then by definition, those cells would evolve ways to adhere to *one another*. But that would mean that the adhering cells that are in the most direct path of the sunlight would have an advantage, because they could absorb the useful wavelengths of sunlight, and the cells covered by them would have less available energy, and probably

[c]https://en.wikipedia.org/wiki/Biome

decreased ability to exchange carbon dioxide and oxygen with the atmosphere. This would presumably lead to an arms race between cells that can adhere to others and cells that could somehow prevent other cells from adhering to them.

Or perhaps *L. adherens* would form aggregates in which the upper layers captured sunlight and fixed CO_2, but Natural Selection would favor cells in the lower layers that could either fuse with the upper cells and extract their nutrients, or somehow attack those cells as consumers or parasites to make use of a ready-made source of energy and macromolecules.

How does this imaginary species fare in relation to known biological principles? In fact, real-life ecologies that involve multicellular terrestrial eukaryotes are far more inventive than this hypothetical prokaryote. In the real-world, life has repeatedly found better ways to exploit virtually every imaginable source of energy and nutrients.

Among real-life prokaryotes, **cyanobacteria** are photosynthetic organisms with species that *do* form filaments, sheets, or hollow spheres. Also, in real-life, terrestrial plants evolved from green algae. Green algae evolved from ancestral eukaryotes that engulfed ancient species of cyanobacteria.

The Transition to Multicellularity

There are two general ways for single-celled species to make the transition from unicellular to multicellular existence. One way would be for individual, formerly separate cells to form aggregates. Slime molds, which are **fungi**, routinely do as part of their life cycle.

However, in terms of **evolvability**, a better way is for daughter cells that share a common **genome** to divide, but remain attached to one another afterwards. This makes it possible for cells with the same set of genetic information to regulate the expression of their genes, so that cells can specialize and cooperate to form higher-level, emergent multicellular structures and functions. These structures can form diverse shapes and forms (**morphologies**[d]) that equip the organisms with the tools required to capture energy and nutrients, survive, and reproduce their own kind.

[d]https://en.wikipedia.org/wiki/Morphology_(biology)

Among the many advantages of multicellular existence is the ability to capture energy and nutrients from other organisms, including **producers** as well as other **consumers**. The dazzling array of ecological dependencies among various life forms has long delighted both nonscientists and scientists alike.

In this chapter, we'll focus on the origins and evolutionary opportunities of multicellularity, with an emphasis on animals.

Multicellular species are closely related to distant common ancestors of choanoflagellates

The origins of animal multicellularity can be traced to ancient single-celled eukaryotes that are closely related to common ancestors of predatory **choanoflagellates** that are still abundant in creeks, lakes, and oceans today. As told by Sean B. Carroll [15]:

> Choanoflagellates are voracious single-celled predators. The beating of their long flagellum both propels them through the water and creates a current that helps them to collect bacteria and food particles in the collar of 30 to 40 tentaclelike filaments at one end of the cell...
>
> ...recent studies suggest that these obscure organisms are among the closest living single-celled relatives of animals. In other words, choanoflagellates are cousins to all animals in the same way that chimpanzees are cousins to humans. Just as the study of great apes has been vital to understanding human evolution, biologists are now scrutinizing choanoflagellates for clues about one of the greatest transitions in history—the origin of the animal kingdom.

In terms of capturing biomass, animals are over two orders of magnitude less successful than plants, since they account for about two gigatons of carbon, most of which is found in marine environments. This does not surprise ecologists, however, because every time an animal consumes either a plant or another animal, only about 10% of the energy from the consumed materials can be used for the production of new macromolecules.

Perhaps we humans are biased, but by most criteria, the ecological opportunities and evolutionary innovations of animals generate higher levels of complex organization. In both plants and animals, complex organization depends on multitudes of cooperating, specialized cells. In terms of cellular specialization and communication, animal embryos are more complex and more diverse than plant embryos. Evolutionary innovations make it possible for animals to adapt to a broader range of distinct *niches* or ways of life than plants. They (and we) are, however, almost entirely dependent on plants for their continued existence. Without plants, virtually every form of animal life on Earth would quickly go **extinct**.

Before considering the evolution of multicellular animals, let's quickly survey some of the *niches* that are available to eukaryotic species that are usually found as unicellular or colonial species.

The Solitary Lifestyle of Single-Celled Eukaryotes

The extraordinary diversity of the protists

Eukaryotes that are not animals, plants or **fungi** fall into a catch-all category known as **protists**. Protists consist of diverse, mostly single-celled organisms that do not represent a natural group or lineage. They obtain energy and nutrients in a variety of ways. Some are autotrophs and rely on photosynthesis. Some are heterotrophs and survive by engulfing or otherwise capturing microscopic prey. Some survive by parasitizing larger hosts, while others gather energy and nutrients from decaying organic matter. Some have flagella or cilia that enable them to propel themselves through the water or to sweep prey into a mouth-like structure. Some have both animal-like and plant-like characteristics. **Euglenids,**[e] for example, are not only able to perform photosynthesis, but are also able to detect light with their photosensitive **eyespot apparatus,**[f] and use their flagella to move towards the light, to increase their rates of photosynthesis.

Protists evolved by endosymbiosis that occurred independently on multiple occasions, leading to a diversity of separate lineages in the deep history of eukaryotic evolution.

---

[e]https://en.wikipedia.org/wiki/Euglenid

[f]https://en.wikipedia.org/wiki/Eyespot_apparatus

Yeasts evolved from multicellular fungi

Yeasts,[g] which are usually found as single cells, represent only about 1% of fungi. Yeasts evolved from multicellular fungal ancestors. Although multicellularity has several potential advantages—such as the opportunity to form diverse shapes and forms, develop specialized cell types, and diverse **life-history strategies**, single-celled yeasts have one important advantage that is similar to prokaryotes: they can divide very quickly. Yeast can divide about every 90 minutes, whereas prokaryotes can divide even more quickly—about every 20 minutes. Presumably, the ability to divide independently and relatively rapidly provided the survival advantages that led to the evolution of single-celled yeast lineages, in the deep history of life.

The Origins and Advantages of Multicellularity

We have already seen that terrestrial plants evolved from ancestors closely related to **green algae**, eukaryotic species that live both as **unicellular** and **colonial** organisms. Green algae originally evolved from prokaryotic cells by **endosymbiosis**.

Ecologists often refer to plants as **producers**, while animals usually occupy ***niches*** available to **consumers**. Plants are usually **autotrophs,** which means that they can use available energy (from sunlight) to fix carbon from the atmosphere to build complex organic chemicals, whereas most animals are **heterotrophs**, which means that they consume other organisms (often plants or other animals) to obtain their energy and nutrients.

Multicellularity makes it possible for each individual to develop complex structures and functions that are not available to unicellular or colonial species. Multicellular individuals obtain energy and nutrients by exploiting a variety of additional ***niches*** that usually involve consumption of other organisms. Many of these *niches* involve complex relationships with other species that are routinely studied and well-understood by ecologists. There is a close affinity between ecologists and evolutionary biologists.

---

[g] https://en.wikipedia.org/wiki/Yeast

Where does multicellular evolution fit into the timeline for the deep history of life?

In *The Vital Question* [11], Nick Lane discusses the special challenges required for eukaryotic cells to evolve from prokaryotic ancestors. This is why, he argues, it took a long time (about two billion years) for eukaryotic cells to arise on planet Earth, whereas prokaryotic cells appeared much earlier. Although estimates are continuously changing in response to new empirical findings, a timeline[h] of the evolutionary history of life on Earth shows that it took roughly 500 million years for the earliest life forms to appear after the Earth was formed, about 500 million years for photosynthesis (in bacteria) to appear, and 1,500 million years longer for the first eukaryotic cells to arise, some two billion years ago.

About 500 million years after that, the first multicellular life forms appeared: this was about 1.5 billion years ago.

The choanoflagellate connection

Regarding the deep ancestry and evolution of multicellular organisms[i] in general and animals in particular, a series of reviews and original scientific reports authored by Nicole King and colleagues [84, 85] provide both historical background and recent insights that are based on state-of the-art techniques.

Evolutionary theories regarding the deep ancestry of multicellular organisms in general, and **animals**[j] in particular, date back to the 19th century. They relied on comparisons of visible structures and **morphology** (shape and form), as well as similarities of **embryonic development**.

About 130 years ago, Ernst Haeckel proposed that all animals develop by means of "repeated self-division of [a] primary cell", based on observations that a single fertilized egg cell (**zygote**) divides repeatedly to give rise to animal embryos [86].

---

[h] https://en.wikipedia.org/wiki/Timeline_of_the_evolutionary_history_of_life

[i] https://en.wikipedia.org/wiki/Multicellular_organism

[j] https://en.wikipedia.org/wiki/Animal

As early as 1867, James-Clark described the morphological similarity between collar-shaped cells that are found in both **choanoflagellates**[k] (mentioned above), and **sponges**.[l] This led to the early hypothesis that all animals are descended from choanoflagellates [84]. Both cell types have **flagella**,[m] long appendages that can beat rapidly and capture potential food particles (including bacteria or other microorganisms) inside. Choanoflagellates are eukaryotic **protozoans**[n] that are found in both unicellular (single-celled) and colonial (multicellular) forms. Sponges are relatively simple multicellular organisms consisting of a few specialized cell types.

This early hypothesis has been confirmed by striking and recent scientific observations. This was made possible by innovations in rapid whole-genome DNA sequencing, combined with methods for systematically comparing genomes that share a common ancestry—a general approach known as **phylogenomics**.[o] By comparing two existing genomes of choanoflagellates to 19 new genomes that they sequenced, and to genomes of various groups of animals, Richter, Fozouni, Eisen and King found a treasure-trove of new information regarding the deep-ancestry of these gene families [85]. Some of their conclusions can be summarized as follows:

(1) All animals ranging from sponges and reef-building corals to elephants and humans are descended from a common ancestor that lived over 500 million years ago.
(2) Although the first animals consisted of soft-bodied creatures that rarely left any traces in the fossil record, they shared many gene families with modern organisms that allow detailed lineages to be traced.
(3) Comparisons of about 1,000 sequenced animal genomes to 21 choanoflagellate genomes reveals clear and striking patterns regarding

[k] https://en.wikipedia.org/wiki/Choanoflagellate
[l] https://en.wikipedia.org/wiki/Sponge
[m] https://en.wikipedia.org/wiki/Flagellum
[n] https://en.wikipedia.org/wiki/Protozoa
[o] https://en.wikipedia.org/wiki/Phylogenomics

the origins and conservation (or loss) of various gene families among diverse animal lineages.

(4) Genes that were essential for the development of the **innate immune system**, which allows animals to detect (and eliminate) **pathogens**,[p] arose before the first animals evolved, in the last common ancestor of animals with choanoflagellates.

(5) Among the 1,944 animal-specific gene families that were identified, genes that are now known to play well-characterized roles in development, including **transcription factors** and **signaling proteins** such as **TGF-beta**, **Hedgehog**, **Pax** and **Sox**, were found.

(6) 153 of these animal-specific gene families were found in all but 10% of all animals compared in the study.

(7) The origins of animals, choanoflagellates, and the single lineage that includes both, were accompanied by the evolution of distinctive gene families.

(8) About 372 of the gene families previously thought to be restricted to animals have **homologs** (ancestrally related genes) in the 19 newly sequenced choanoflagellate genomes.

(9) Gene families in pathways critical for biosynthesis of seven essential amino acids that are present in choanoflagellates were lost in animals that share common ancestors. These amino acids are commonly provided by prokaryotes (such as *E. coli* in humans) that have symbiotic relationships with animals.

These findings, based on phylogenomic analysis, provide abundant molecular evidence for the origin of species and biological complexity based on common ancestry and mechanisms such as Natural Selection. They not only confirm 19th century hypotheses proposed on the basis of morphological comparisons, but they also represent more rigorous and compelling quantitative evidence. In these data, we find numerous examples for innovations that repurpose existing gene families as well as the appearance of new families. The loss of gene families such as those for biosynthesis of amino acids provided by symbiotic organisms provide additional evidence that the genes that are conserved, as well as those that

---

[p] https://en.wikipedia.org/wiki/Innate_immune_system

are lost, reflect the diverse and changing ecological requirements for capture of energy and nutrients among animals that occupy diverse *niches* and that have evolved diverse structures that enable them to survive in various ways of life.

Multicellularity evolved multiple times independently

In an insightful summary titled "The momentous transition to multicellular life may not have been so hard after all," Elizabeth Pennisi offers an updated perspective based on the work of King and colleagues as well as numerous other scientists.[q] As Pennisi's title implies, recent evidence suggests that the evolution of **multicellularity** was not particularly challenging. Although the multicellular lineage leading to both plants and animals arose only once, multicellularity has arisen independently about 46 times,[r] including about a dozen separate events in the evolution of **fungi**. Separate leaps to multicellularity also occurred in other groups, such as red, brown, and green algae.

The advantages of using cell division as the main path to multicellularity

As mentioned above, there are two general ways to generate multicellular organisms—by aggregation, in which separate free-living cells come together, and by cell division (**cleavage**) of a single cell. The latter has the advantage that the lineages descended from a single cell can share the same genome, which is more consistent with a shared strategy for survival based on cooperation and division of labor. This also tends to discourage "cheating" by individual cells that might benefit individually by "going rogue", at the expense of the remaining cells that rely on cooperation [84, 87].

---

[q] http://www.sciencemag.org/news/2018/06/momentous-transition-multicellular-life-may-not-have-been-so-hard-after-all

[r] https://en.wikipedia.org/wiki/Multicellular_organism

Divide and Conquer: Evolutionary Opportunities Provided by Coordinating Multitudes of Specialized Cells

For **multicellular organisms** such as **animals**, it is difficult to appreciate the beauty—or to accept the validity—of 21st century evolutionary theory—without a deep understanding of the ways that complex organization is routinely reproduced in each generation. In the sections below, as well as the following chapters of Part 2 of this book, we'll take a closer look at the molecular, cellular, and developmental mechanisms that give rise to complex biological organization.

We have seen that while prokaryotes are masters of biochemical innovation, multicellular eukaryotes have the potential to evolve a broader range of creative adaptations that can capture energy and nutrients in novel ways.

In each generation of every multicellular species, the reproduction of complex structure and functions depends on cooperation between multitudes of specialized cells. In this section, we'll focus on the general capabilities of animal cells that make this possible.

Eukaryotic cells are remarkably versatile units of structure and function. Most animal cells are found at higher levels of organization known as **tissues** and **organs**. According to Wikipedia[s]:

> In biology, tissue is a cellular organizational level between cells and a complete organ. A tissue is an ensemble of similar cells and their extracellular matrix from the same origin that together carry out a specific function. Organs are then formed by the functional grouping together of multiple tissues.

The **extracellular matrix** consists of **secreted** materials in tissues that provide structural and biochemical support to the individual cells, while allowing them to communicate in various ways with other cells, including both nearby and distant cells.

---

[s]https://en.wikipedia.org/wiki/Tissue_(biology)

Biologists specializing in evo-devo study three broad categories of innovations that permit cells to cooperate and generate higher levels of organization in each individual **embryo**:

(1) **Regional specification:** Specifically, particular regions of the embryo respond to signals and change their patterns of gene expression in coordinated ways.
(2) **Cellular differentiation:** Specifically, the chromosomes undergo changes that regulate the expression of the genes in complex ways; these changes are passed on during cell division, and are known as **epigenetic** changes.
(3) **Morphogenesis:** Specifically, the ability to generate new multicellular structures and functions by communicating with other cells and expressing specific sets of genes, changing cell shapes, moving around, and growing and dividing in a coordinated way.

All of these capabilities have been studied in great detail, and many of these mechanisms are now well-understood. In general, they depend on **multigene families** consisting of both **protein-coding** DNA sequences as well as a variety of kinds of **regulatory sequences**. Some of these families are closely related to those found in choanoflagellates [85], while others evolved later in the deep history of life.

The MADS-box

Multicellular organisms include animals, plants, and fungi. One particular type of genomic sequence motif—called the *MADS-box*—can be found in all three multicellular kingdoms. Genes that have *MADS-box* domains often function as transcription factors. The *MADS-box* plays important roles in growth and development, especially in plants, animals, and even in fungi such as **cellular slime molds**.[t]

MADS-box in fungi

Cellular slime molds form multicellular fruiting bodies as part of their life cycle—a process that depends on signaling and aggregation of individual

[t]https://en.wikipedia.org/wiki/MADS-box

cells that are able to crawl around and ultimately stick together and form collective 3D multicellular structures.[u]

A *MADS-box* transcription factor (*SrfA*) plays multiple roles during **sporulation** of the cellular slime mold *Dictyostelium*, including the organization of **actin** filaments in the **cytoskeleton** as well as stabilization of spore coats [88].

MADS-box in plant development

In plants, the *MADS-box* gene family is organized into several different groups whose evolutionary history appears to reflect their distinct functional roles in flower development [89]. *MADS-box* gene family members play a variety of major roles in the development of plants, including embryos, seeds, roots, flowers, fruits, and male gametophytes [90, 91].

MADS-box in animal development

In animals, *MADS-box* transcription factors play roles in muscle development, including cell proliferation and differentiation [92].

Comparing plant and animal development

In **sexual reproduction** in plants, development begins with a fertilized egg that gives rise to a tiny plant embryo. Plant cells have rigid cell walls and limited motility, which places significant constraints on **morphogenesis**. However, both growth and development of plant cells is free to continue throughout the life of the plant in the **meristematic** cells found at leading edges of both roots and shoots.[v]

In sexual reproduction in animals, a fertilized egg divides into a multitude of specialized cells during embryogenesis. Much of the development of the organism takes place during embryogenesis, but some animals undergo metamorphosis in which larval tissues are replaced by newly developed adult tissues. Developing animal cells can often change their shape and form and sometimes migrate from one embryonic location to

---

[u] https://en.wikipedia.org/wiki/Slime_mold

[v] https://en.wikipedia.org/wiki/Meristem

another. Growth and development of **stem cells** continues through the life of the individual, and wound healing and repair and replacement of damaged or dead cells continues to varying degrees among various animal species.

The roles of specialized genes and cell differentiation in multicellular organisms

Prokaryotic and eukaryotic cells share a number of conserved metabolic pathways, such as those required for energy production, transport, and the breakdown and synthesis of macromolecules. In multicellular organisms, genes that are expressed in virtually all cell types that are required for basic cellular functions are called "housekeeping genes".

But during development, cells of multicellular organisms can also **differentiate**, which means that they can specialize and play distinct roles in different tissues. One general strategy that multicellular organisms use to accomplish this specialization is to control the expression of tissue-specific genes as the cells mature. For example, in mammals, liver cells will make enzymes that detoxify molecules in the blood, blood cells will make oxygen-carrying hemoglobin, and muscle cells will make contractile proteins. But development adds new layers and new levels of complexity. Cellular differentiation and specialization provides several evolutionary opportunities to multicellular species that are not available to single-celled organisms.

Multicellularity makes it possible for the organism to go after new sources of energy and nutrients, both in the seas and on the land (and even beneath the surface of both).

Multicellular plants evolved several innovations that enabled them to invade the land [93–95]. These included conservation of moisture, growth of rigid supports for leaves to reach the sunlight, transport of minerals and water from the roots, and transport of sugars and other complex molecules from the leaves.

Plants have also excelled in production of secondary plant products that play defensive roles against consumers and other infringing species. Flowering plants have also excelled in creating mutualistic opportunities for the **coevolution**[w] of pollinating insects and other animals.

---

[w] https://en.wikipedia.org/wiki/Coevolution

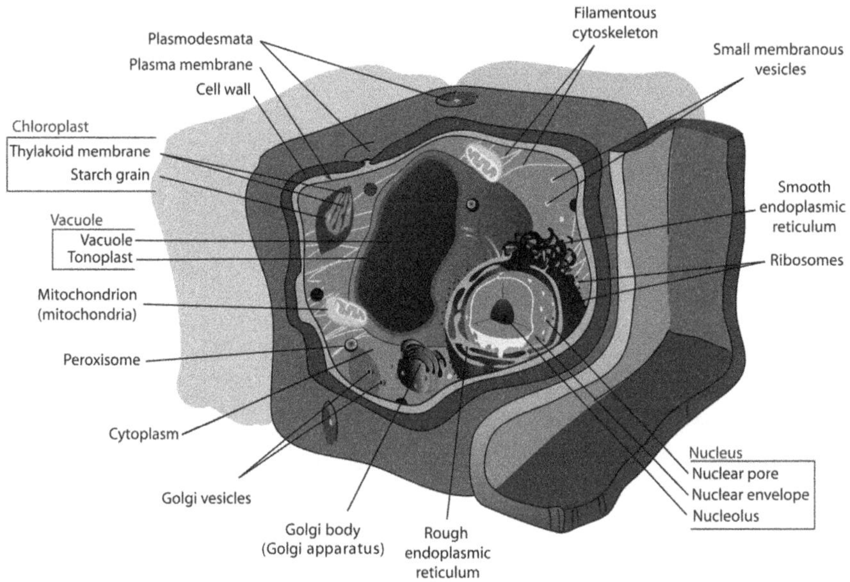

Figure 1. Diagram of a plant cell. Like all eukaryotic cells, plants have mitochondria that use proton gradients to transform glucose into **ATP**, a major cellular energy currency. Plants are producers, and plant cells have chloroplasts that capture the energy of sunlight and use it to transform carbon dioxide and water into glucose. Plant cells are constrained by their rigid cell walls that contain cellulose, a resilient polymer of glucose.

Source: Page-link: https://commons.wikimedia.org/wiki/File:Plant_cell_structure-en.svg. File-link: https://upload.wikimedia.org/wikipedia/commons/d/d8/Plant_cell_structure-en.svg. Attribution: By LadyofHats [Public domain], via Wikimedia Commons. Rendered in B&W.

Despite these opportunities, however, even the carnivorous plants[x] that so fascinated Darwin[y] are constrained by their dependence on sunlight and photosynthesis as primary sources of energy and carbon skeletons. Plant cells (Figure 1) are constrained by the rigid cell walls that limit motility both of individual cells and also of the entire organisms.

Animals are equipped to occupy a variety of *niches* unavailable to plants, including the various mutualistic relations, the roles of consumers at various levels in the food web, and as predators and prey. Animal cells

---

[x] https://en.wikipedia.org/wiki/Carnivorous_plant

[y] http://pages.britishlibrary.net/charles.darwin3/insectivorous/insect01.htm

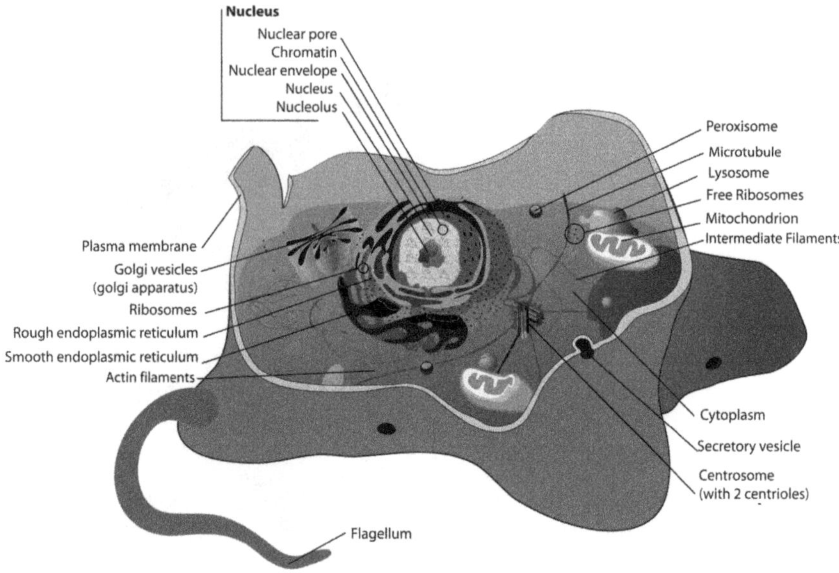

Figure 2. Diagram of an animal cell. Like all eukaryotic cells, plants have mitochondria that use proton gradients to transform glucose into **ATP**, a major cellular energy currency. Animals are consumers, and animal cells must obtain energy and nutrients that are captured from other organisms. Animal cells have flexible shapes that are determined by cytoskeletal elements such as actin filaments and microtubules. Some animal cells have cilia or flagella for motility.

Source: Page-link: https://commons.wikimedia.org/wiki/File:Animal_cell_structure_en.svg. File-link: https://upload.wikimedia.org/wikipedia/commons/4/48/Animal_cell_structure_en.svg. Attribution: By LadyofHats (Mariana Ruiz) [Public domain], via Wikimedia Commons. Rendered in B&W.

(Figure 2), with flexible cell membranes and various cytoskeletal elements, have mastered motility (mobility and ability to change shape and form), at the level of individual cell shape and collective shape and motion of tissues and organs, and even in cellular migration during embryogenesis.

For animals, the struggle for existence—and Natural Selection—gave rise to the evolutionary innovations of mobility and propulsion. The need for awareness of real-time events, such as those involving potential predators or prey, gave rise to sense organs, central nervous systems, and instinctive behaviors, as well as the evolution of intelligence.

Once brains and behaviors had evolved, this provided opportunities for our own primate ancestors to evolve. Awareness of past events, social evolution and language, abstract thinking and imagination, tool-making, planning and goal-seeking: these all became possible once animals with larger brains had the opportunity to evolve.

Capabilities of embryonic cells

Embryonic cells have evolved a variety of useful capabilities that go far beyond the capabilities of single-celled individuals. A partial list of these capabilities are as follows:

(1) The ability to become regionally specified.
(2) The ability to regulate growth and **cell division** in a coordinated fashion.
(3) The ability to undergo **programmed cell death** (**apoptosis**) in a coordinated way, which is useful in morphogenesis as well as controlling infections and cancer.
(4) The ability to routinely express a specific set of genes, so that they synthesize a specific set of proteins.
(5) The ability to exchange energy and nutrients with neighboring cells, often by means of cellular structures such as the circulatory and lymphatic systems, and by means of specialized cell junctions that connect the cytoplasm of adjacent cells.
(6) The ability to differentiate, so that they take on permanent specialized roles, both during development and in the adult individual.
(7) The ability to form stem cells that serve as reservoirs for specialized cell lineages.
(8) The ability to routinely send and respond to signals, to and from other cells. These capabilities depend on protein chains that serve as **cell surface receptors,** as well as a variety of **signal transduction pathways** in both **cytoplasm** and **nucleus**.
(9) The ability to regulate their gene expression, shape, and state of differentiation in response to signals from other cells.
(10) The ability to change shape and move around in a coordinated fashion (cell motility).

(11) The ability to adhere to neighboring cells by a variety of mechanisms known as **cell adhesion**.

Specialization of differentiated cells

In addition, many cell types are capable of exiting from the **cell cycle** and becoming highly specialized adult cells that are integrated into tissues and organs. In part because of their medical importance, such specialization in human tissues has been studied in great detail by **cell biologists** and **cell physiologists**. Such cells often undergo major structural changes and express proteins that carry out highly specialized functions. Here are just a few examples, found in a variety of **vertebrates** (animals with backbones):

(1) Nerve cells become highly specialized for generating and transmitting electrical signals to adjacent cells, often by means of cellular extensions (axons and dendrites) that rapidly transform electrical signals into chemical ones, and vice versa.
(2) **Schwann cells** function as electrical insulators for nerve cells by wrapping around the cells to form **myelin sheaths**.
(3) Red blood cells (erythrocytes) become specialized for transport of oxygen and carbon dioxide by losing their nuclei and becoming packed with hemoglobin molecules.
(4) **B cells** and **T cells** rearrange their DNA to generate an extremely diverse set of antibodies and T cell receptors, respectively, that can recognize, and help eliminate a broad range of toxins and pathogens with exquisite specificity.
(5) **Skeletal muscle cells** synthesize **actin**, **myosin**, and other proteins that use the energy of ATP to contract. They fuse with adjacent cells to form long myotubes that connect to other skeletal elements.
(6) Liver cells synthesize a variety of specialized enzymes that are routinely used to detoxify materials in the blood.

Evolutionary robustness of embryonic cells

In the late 19th century **experimental embryologists** attempted to solve the mysteries of development with cellular manipulations in model organisms such as sea urchins or amphibians. They discovered that embryos are remarkably resilient, but in the absence of any real understanding of the inner-workings of cells, these experiments could only lead to qualitative, phenomenological explanations, and provided many more questions than satisfying answers.

The behavior of developing sea urchin embryos was especially puzzling to Hans Driesch. When he separated the two embryonic cells (called **blastomeres**) resulting from the first division of the fertilized sea urchin egg, he found that they both developed into complete individuals. This was also true after the second division: all four cells could separately develop into four complete individuals. By 1885, Driesch had found that the embryo even developed normally after shuffling or recombining blastomeres of separate embryos.

As an undergraduate student at UC Berkeley, my fellow classmates and I reproduced these sorts of remarkable results in academic exercises in the laboratory. For example, the blastomeres of sea urchin embryos fall apart in artificial seawater lacking calcium and magnesium ions. When normal concentrations of calcium and magnesium are restored, cells from separate individuals reaggregate, and the chimeric individuals subsequently undergo normal development.

In 1885, however, much to the chagrin of his colleagues, Driesch began to part ways with the objectivity of empirical science, and instead began to dabble in **vitalism**, a mystical approach to philosophy. He invoked the concept of **entelechy**, a term borrowed from Aristotle, to try to explain his puzzling experimental findings. He proposed that some form of nonlocalized, mind-like life force was responsible for guiding the development of the sea urchin embryos.

It's important to distinguish vitalism and mystical forms of **emergentism**, which generally invoke some sort of consciousness or supernatural guidance, from the scientific definitions of emergence, potential, EEP and ***res potentia*** that were discussed in Chapter 2 of *Rethinking Evolution*. To be clear: The Updated Evolutionary Synthesis (UES) does *not* include any form of mystical or supernatural consciousness. **Consciousness** arises from material and natural causes, such as the activity of the brain. Apart from concrete mechanisms such as the **Baldwin Effect**, there is currently no evidence that consciousness plays a significant role in the evolution of developmental mechanisms.

Beyond engineering, design, intelligence, and instructions

When seen from the **anthropomorphic** perspective of engineering, design, intelligence, or instructions, the multicellular reproduction of complex structures and functions during animal development seems unimaginably challenging. In the absence of conceptual frameworks that are now widely accepted, the robustness of development remained a black box to experimental embryologists for about a century after Darwin published his *On the Origin of Species*. Peter Lawrence and Michael Levine provide a lucid explanation [96]:

> From the mid-1800s and for about a hundred years, mainstream embryologists ignored genetics and tried to understand the mechanisms of animal development without it. The attempt was a brave one, but it became increasingly foolish. In the early 20th century, the gifted embryologist Thomas Hunt Morgan realised the importance of genes and took what he thought would be a temporary diversion into genetics. But it was a long detour; only at the end of his life was he able to return to his beloved embryos. But most other embryologists continued to work as if genes were irrelevant…
>
> …many of their results were so counterintuitive and conflicting that their hypotheses became abstract and ornate. The philosophical and the whimsical found this attractive. By contrast, it was the mathematical and the rigorous who joined the new science of genetics. Naturally enough, the two types of scientist failed to understand each other and embryology drifted off into metaphysical swamps while genetics explored the dry savannahs of statistics…

Escape from both the "metaphysical swamps" such as vitalism and entelechy and the "dry savannahs" such as **Population Genetics** became possible when several brilliant scientists, including Ed Lewis, Christiana NüssleinVolhard, and Eric Wieschaus applied Mendelian and Molecular Genetics to a systematic analysis of the development of fruitfly embryos. By relying on state-of-the-art methods such as ***in situ* hybridization** and other methods for visualizing the patterns of expression of RNA and proteins, these pioneers provided deep insights into the molecular, cellular, and developmental mechanisms of embryonic development.

These mechanisms are determined by complex interactions between protein-coding and regulatory DNA sequences that are now well-understood. This has revolutionized our understanding of both evolution and development. More about this will be discussed in Chapter 15.

In this context, the vitalism and entelechy of embryologists such as Driesch can be seen as well-meaning but misguided attempts to explain the astounding robustness of embryonic cells to form a complex and coherent individual. The tendency to seek supernatural explanations for phenomena that we do not understand is deeply embedded in the human DNA. But we also have the capacity to find scientific explanations for these phenomena, and, once the facts are available, such supernatural explanations are often supplanted by scientific ones.

Development from Egg to Organ Systems

"Mosaic" or "regulative" embryos: Two sides of the same coin

Sydney Brenner once quipped that animal development can proceed according to either the European or American plan [97]. Under the European plan, you are defined by your ancestry. Under the American plan, what you do depends on your neighbors.

These are actually two ends of a conceptual spectrum known as known as **autonomous specification** and **conditional specification**, respectively. This is quite similar to the conceptual spectrum that distinguishes **mosaic embryos** from **regulative embryos**. Right up to the present day, textbooks often emphasize this distinction, but as pointed out by Lawrence and Levine [96], these are just "two sides of the same coin":

> To oversimplify: mosaic development depends on agents, such as transcription factors, being placed locally in the egg by the mother. Regulative development depends in part on long-range gradients of positional information, such as that provided by the *Hedgehog* protein, that can pattern many cells at once. Regulative development can also be driven by short-range signals that trigger changes in cell identity in nearby neighbours.

Among multicellular organisms, complex organization arises during development. The evolution of complex organization therefore involves the evolution of genes that control the process of development.

In many animals and plants with sexual reproduction, we often define "the beginning of development" or "the next generation" as the process that begins when an egg is fertilized and then divides into a multitude of specialized cells.[z] In reality, both parents and offspring represent a continuum of growing and dividing cells that span a multitude of generations. Only living cells give rise to other living cells. This is one of the core principles of cell theory.[aa]

A morphological description of the major stages of embryogenesis

Long before the major advances brought about by fruitfly genetics, embryologists described the stages of embryo development in terms of visible attributes of interacting cells. This process is a fascinating one that is routinely studied from an anatomical point of view by students of various medical professions. A very brief description of some of these stages is as follows:

(1) **Fertilization:** The fusion of a sperm cell from the father with an egg cell from the mother to form a fertilized egg, known as a **zygote.**
(2) **Cleavage:** Division of the zygote. This takes various forms in various animals. In insects such as fruitfly, the egg subdivides internally into a layer of cells known as a syncytial blastoderm. In sea urchins and amphibians, the zygote forms a hollow fluid-filled ball of cells known as the blastula. In animals such as birds and mammals, the dividing cells form a disc that floats on top of a mass of yolk, known as a blastocyst.
(3) A layered structure, essentially a tube within a tube, is formed by the inward movement of cells. This process is known as **gastrulation**. The inner tube generally gives rise to the gut or digestive system.

---

[z] https://en.wikipedia.org/wiki/Cell_division
[aa] https://en.wikipedia.org/wiki/Cell_theory

(4) Through a variety of cellular interactions, the formation of integrated tissues and organs takes place as the various cells differentiate to take on specialized functions. For example, in vertebrates, a process known as **neurulation** gives rise to the brain and spinal cord. In general, the formation of tissues and organs is referred to as **organogenesis**.

These specialized cells communicate and interact in complex ways to reproduce the structures and functions that depend on accumulated genetic information that has been passed from generation to generation.

The multitudes of specialized cells in a developing embryo all carry the same genome—the same set of genes—but they express them in different ways to form complex multicellular structures and functions involving tight coordination between neighboring cells. This requires elaborate

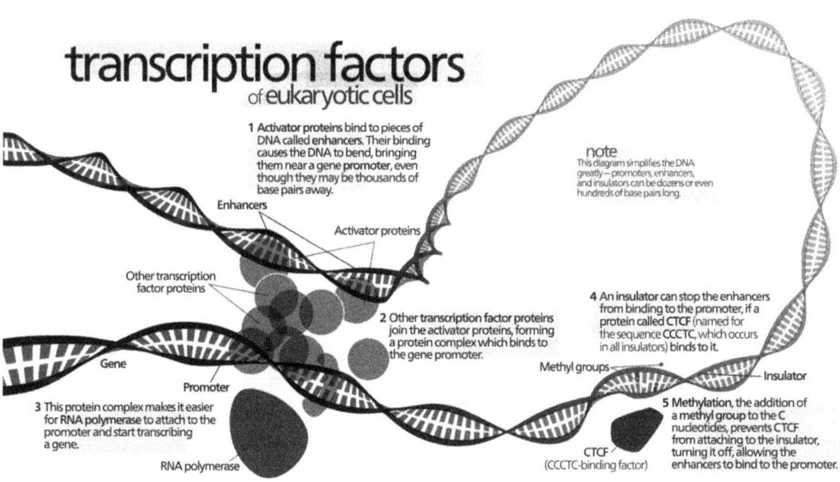

Figure 3. Transcription factors regulate the expression of eukaryotic genes by controlling production of RNA. Transcription factors play major roles throughout development, and often interact in complex and nuanced ways by binding to DNA, RNA, and other proteins, including other transcription factors.

Source: Page-link: https://commons.wikimedia.org/wiki/File:Transcription_Factors.svg. File-link: https://upload.wikimedia.org/wikipedia/commons/8/80/Transcription_Factors.svg. Attribution: By Kelvin13 [CC BY 3.0 (https://creativecommons.org/licenses/by/3.0)], from Wikimedia Commons. Rendered in B&W.

patterns of intercellular signaling. Some of the major ways that this is accomplished include the following innovations in multicellular eukaryotes:

(1) Distribution of **cytoplasmic determinants** during cleavage of the embryo.
(2) **Regional specification**, which embryologists have referred to as **positional information, pattern formation**, and **morphogenic fields,** by molecular mechanisms such as ***Wnt*** **signaling pathways** or **gradients** of proteins or other molecules known as **morphogens**. When concentrations of morphogens reach critical thresholds during development, they act as genetic switches to change the developmental fates of particular cells in specific regions.
(3) **Cellular differentiation** through a variety of mechanisms, including cytoplasmic determinants, various short and long-range cellular interactions, **transcription factors** (Figure 3), and other mechanisms involving **control of gene expression**.
(4) Modifications of eukaryotic chromosomes, known as **epigenetic** changes, that change the ways that genes are expressed in particular lineages of specialized cells.

Our understanding of these mechanisms has advanced by leaps and bounds in recent years, and will be explored in more detail in the following chapters.

Chapter 11

Embryos Are Not Computers, and DNA Is Not Software

It is not birth, marriage, or death, but gastrulation which is truly the most important time in your life.

—Lewis Wolpert[a]

The Big Picture

Embryonic development is a process that is shaped by genetic determinants. Chromosomes contain those determinants, and each chromosome includes a long DNA molecule, where each position of the sequence is represented by one of four possibilities. These qualities lend themselves to a variety of metaphorical descriptions involving codes, information, computers, and digital data files.

But as pointed out by Gunther Stent,[b] although the genome defines the unique characteristics of each species and each individual, the genome cannot be usefully considered to be a 1D representation of the organism's multidimensional phenotype (or phenome).

A critical analysis of the pros and cons of various metaphors for development and for genomic DNA, and their evolution, leads to a better approximation of reality more suitable for the Updated Evolutionary Synthesis (UES).

---

[a] https://en.wikipedia.org/wiki/Lewis_Wolpert

[b] https://onlinelibrary.wiley.com/doi/full/10.1038/npg.els.0005468

The Pros and Cons of Various Metaphors of Development, DNA, and Evolution

Introduction

In each generation, reproduction of complexity, manifested within individual offspring, arises in a sequential, but massively parallel way, during embryonic development. The complexity of the **fertilized egg (zygote)** increases dramatically during this process. The zygote divides into a multitude of **cells** which give rise to increasingly complex **multicellular** structures and functions, which interact in increasingly complex ways. These increases in complexity can be observed at several levels of organization, and we can focus on various molecular, cellular, and developmental events that involve individual cells as well as groups of cells. At any given time, each cell is poised to respond to appropriate signals from both nearby and distant neighbors, and to send out appropriate signals to its neighbors. Cells specialize, control cell division, control gene expression, control shape, form, and adhesion to other cells, and sometimes move to new locations in the embryo.

All of these events are driven by molecular interactions, including specific binding-interactions between two or more macromolecules as well as catalyzed chemical reactions involving enzymatic activities. Each embryo inherits genetic determinants that help to drive and shape the process of embryonic development. Therefore, the nature and evolution of those genetic determinants of development are of obvious importance and relevance to our understanding of evolution.

The evolution of intelligent, extraterrestrial life forms

Gazing into the past with 20–20 hindsight, it's difficult to imagine the depth of our ignorance even a few decades ago. In the 1970s, our speculations concerning the evolutionary origins of biological intelligence relied heavily on metaphors drawn from our own cultural history. Empirical data regarding the inner-workings of cells and the origins of life was severely limited.

Even more profound, of course, was the depth of our ignorance when it came to speculations regarding the search for extraterrestrial intelligence. Today, we understand far more about the molecular, cellular, and developmental origins of complexity than we could possibly understand over four decades ago. We also have a better grasp on the conditions that may be necessary and sufficient for the origin of life, and the inherent barriers to life forms more complex than microbes (Chapter 5). We should expect the quality of our informed scientific speculations to improve considerably in the light of recent discoveries, but there is always a significant lag between the cutting-edge of empirical discovery among specialists and the widespread dissemination of that knowledge among the broader community. Consequently, both scientific-minded professionals and nonscientists often wander through the bewildering maze of outdated conceptual frameworks and metaphorical perspectives.

The genetic blueprint metaphor

It is all too easy to extrapolate from familiar experiences to biological metaphors. Most are misleading. One of the more prevalent is the concept of the **genetic blueprint**. According to Wikipedia[c]:

> It is widely believed that genes provide a "blueprint" for the body in much the same way that architectural or mechanical engineering blueprints describe buildings or machines. At a superficial level, genes and conventional blueprints share the common property of being low dimensional (genes are organised as a one-dimensional string of nucleotides; blueprints are typically two-dimensional drawings on paper) but containing information about fully three-dimensional structures. However, this view ignores the fundamental differences between genes and blueprints in the nature of the mapping from low order information to the high order object.
>
> In the case of biological systems, a long and complicated chain of interactions separates genetic information from macroscopic structures and functions. The following simplified diagram of causality illustrates this:
>
> Genes → Gene expression → Proteins → Metabolic pathways → Subcellular structures → Cells → Tissues → Organs → Organisms.

[c]https://en.wikipedia.org/wiki/Common_misunderstandings_of_genetics

The 3D printer rendering metaphor

A modern-day variation of the genetic blueprint metaphor is to imagine an organism being reproduced by transmission of a 1D stream of digital data to a 3D printer which renders that data.

Long before 3D printers had been invented, however, scientists and communicators were already thinking of the genome as a linear sequence of homogeneous, coded digital information.

Distinguishing explicit 1D DNA sequences from implicit multidimensional determinants of development

I fondly recall a brilliant lecture [98, 124] by Gunther Stent, one of my professors during my undergraduate days at UC Berkeley (UCB). Professor Stent, whose intellectual journey had taken him from the cutting-edge of **Molecular Genetics** to the exciting new field of **developmental neurobiology**, took issue with the widely-used metaphors of genomic DNA as a blueprint, a coded set of instructions, a digital computer code, or a 1D data stream.

Even the most brilliant among us reflect the zeitgeist in our conceptual frameworks

Carl Sagan is best known as the prime mover behind *Cosmos*[d] [99] and his Pulitzer Prize-winning *The Dragons of Eden* [4]. But in the 1970s, Sagan was also heavily influenced by the conceptual framework shared by those who attempted to bring cosmology together with the biological sciences. This fusion of the physical and biological sciences created the highly speculative field known as **Astrobiology**.

Perhaps in the tradition of Erwin Schrodinger,[e] Sagan had suggested, at a conference on the search for extraterrestrial intelligence, a thought experiment involving a housecat. According to Stent, Sagan had argued, partly in jest, that:

---

[d] https://en.wikipedia.org/wiki/Cosmos_(Carl_Sagan_book)

[e] https://en.wikipedia.org/wiki/Schr%C3%B6dinger%27s_cat

> to transmit via radio signals the DNA nucleotide sequence of a cat to a Distant Alien Civilization is equivalent to sending the Aliens the cat itself…

Like many of his colleagues, Stent, who was searching for a window into the black box of developmental genetics, took issue with the idea that the "essence" of a housecat, and the means of reproducing one, could be understood by somehow translating the explicit 1D DNA sequence of the cat genome (which was yet to be determined).

> …On the contrary, what *is* obvious is that the Alien intelligence, even if it possessed a table of the terrestrial "genetic code", would not be able to reconstruct the cat from its DNA nucleotide sequence. To make this reconstruction, the Aliens would have to know a good deal more about terrestrial life than the formal relations between DNA nucleotide base sequences and protein amino acid sequences.

The digital computer code metaphor

In the 1970s, it seemed reasonable—especially to physicists and astronomers, who had begun to rely heavily on computers—to compare DNA sequences to digital computer code. Although each position of the DNA sequence has four possible values instead of two, those values are discrete, and when DNA is replicated without copying errors (as it usually is), the copies remain identical to the original, just as they are in binary computer code.

But Stent argued, correctly, that this should not imply that the *meaning* or *significance* of DNA can be understood simply by translating its *explicit* nucleotide sequence.

Stent was pointing out what embryologists had already known for decades: namely, that there is more to the making of an embryo than a simple catalog of protein-coding sequences of DNA.

In fact, today we know that only a tiny fraction of the genomes of mammals such as cats [98], dogs, or humans consists of sequences that are transcribed into mRNA and translated into amino acid sequences of proteins.

Not only that, today we know that **regulatory DNA sequences**, which account for at least twice as much DNA sequence data as protein-coding sequences, contain important pieces of the puzzle of how the complex organization of a cat is reproduced during development.

Stent went on to argue that understanding the many-sided *implicit* significance of DNA sequences is critical to understanding the ways that genetic determinants interact during embryogenesis, in the reproduction of complexity.

Although Carl Sagan was one of our most gifted science communicators, he was not immune to the influence of the zeitgeist of the 1970s, any more than any of us could be. But even though Stent could not anticipate the revolutionary discoveries that would eventually lead to **evolutionary developmental biology**, he *did* know enough about the nature of animal development to pinpoint the limitations of metaphors based on any explicit nucleotide sequence.

Carl Sagan's deeper legacy

But if Carl Sagan could not magically conjure knowledge that had to await later discoveries, one of the many things he *did* offer, as a science communicator, was a deep understanding of both history and the tenuous nature of democracy, truth, and the scientific method.[f] The storyline of *Contact* (especially the 1997 film outlined by Sagan and Ann Druyan[g]) dramatized several recurrent themes that have influenced the social history of science right up to the present day, including greed, political power, religious zealotry, and tribalism.

---

[f]With Cosmos: A Personal Voyage, Sagan alerted millions of readers and television viewers to the dangers of historical anti-intellectualism and its violent attacks on science and truth itself. In Part 4, we'll review his account of the staggering and irreparable losses that came with the sacking of the ancient library of Alexandria. Only now, in the United States, have his warnings concerning the fragility of science and inconvenient truths become painfully evident in our daily lives through President Donald Trump and his G.O.P enablers in Congress (see Part 4. Online description of the television series at https://en.wikipedia.org/wiki/Cosmos:_A_Personal_Voyage.

[g]https://en.wikipedia.org/wiki/Contact_(1997_American_film)

The phenotype of the Siamese cat

I was rapt as I listened to Stent's lecture at UCB, when he explained the Molecular Genetics behind the **phenotype** of the Siamese cat, a breed that carries a **mutation** in an **enzyme** that produces dark **pigmentation**. The cats have an unusual phenotype: the tips of their nose, paws and ears tend to have black fur, while the rest of the body is white.

Not only that: the cats' eyes are blue, and, apart from the black pigmentation of the extremities, the cats look very much like albinos. The reason is that the mutation that eliminates dark pigmentation is **temperature-sensitive**. When the developing kittens are in the womb, at body temperature, the protein for the tyrosinase enzyme required for pigment production is inactivated by the body heat of the mother (and her blood supply), so the fur of the developing kittens remains white (rather than pigmented) and the eyes remain blue.

In contrast, the peripheral extremities, such as the paws and parts of the face remain slightly cooler in the womb, so the enzyme can function normally and the pigment can be produced. The result is the striking phenotype of the Siamese breed.

The phenotype of the Siamese breed cannot be understood without understanding the higher-level emergent properties that arise during development. But we have not even come to the most interesting part of the story.

Why are Siamese cats cross-eyed?

Once the kittens are born, they become cross-eyed, and they stay that way as adults (Figure 1). As Stent pointed out, the biological reason behind this strange phenomenon is fascinating. According to Wikipedia:

> The same albino allele that produces coloured points means that Siamese cats' blue eyes lack a tapetum lucidum, a structure which amplifies dim light in the eyes of other cats. The mutation in the tyrosinase also results in abnormal neurological connections between the eye and the brain. The optic chiasm has abnormal uncrossed wiring; many early Siamese were cross-eyed to compensate, but like the kinked tails, the crossed eyes have been seen as a fault and due to selective breeding, the trait is far less common today.

Figure 1. Photograph of a Siamese cat, illustrating the phenotypic traits of dark pigmentation on the extremities but not the main part of the body, and the light-colored crossed eyes. A temperature-sensitive mutation in pigmentation affects both the fur color and the eyes. The eyes, which lack normal pigmentation, cannot properly absorb stray light, and during the early development of the kitten, this results in a skewing of the connections between the retina and the brain. The musculature of the eyes compensates as it develops in the young kitten, assuring overlapping visual fields for binocular vision, but also resulting in crossed eyes. This process illustrates the implicit nature of the phenotypic determinants carried by the genome.

Source: Page-link: https://commons.wikimedia.org/wiki/File:Siamese_Cat_Cross-Eyed.jpg. File-link: https://upload.wikimedia.org/wikipedia/commons/a/a7/Siamese_Cat_Cross-Eyed.jpg. Attribution: By Emery [CC BY-SA 2.5 (https://creativecommons.org/licenses/by-sa/2.5)], from Wikimedia Commons.

Normal binocular vision in the kittens involves neural connections that map specific regions of the retina onto the brain. This is completed after the kittens are born. Correct, overlapping visual fields required for binocular vision involve an active, light-dependent process. This assures that correct connections are hooked up properly, and incorrect ones are eliminated.

There is no "blueprint" for these correct connections: rather, evolution has led to a flexible, adaptive process that is robust and that coordinates the vision in both eyes, despite individual variations in each newborn kitten.

Normal binocular connections depend on elimination of stray light by a pigmented layer attached to the retina. But in the Siamese kittens, this pigment has been inactivated by the mutation. Consequently, stray light causes the connections to become skewed. The muscles and alignment of the eyes compensate for this abnormality and assures that the kittens develop binocular vision. There is no intelligence or design responsible for this. Rather, this is just one example of the numerous robust and flexible developmental mechanisms that coordinate the interactions of cells, tissues, and organs, not only in the embryos but also during the early life of each juvenile animal.

Explicit and implicit informational content of the genome

Why did Stent choose to focus on this particular story? Because it clearly illustrates that there is, as he would put it, "implicit informational content in the genetic code". That implicit information includes all of the normal mechanisms of development. In each generation, complexity is reproduced during development (including embryogenesis as well as the early life of juveniles such as kittens). Complexity consists of hierarchical layers of complex molecular and cellular organization that emerge in a step-wise fashion.

An understanding of the phenotype of any animal in general, and the Siamese cat in particular, requires detailed empirical knowledge about the inner-workings of cells, and the molecular and cellular interactions that reproduce complex levels of organization during development of the individual. *Any* intelligent being would need to know about these things in order to understand their developmental (and evolutionary) origins. The idea that a super-intelligent being could deduce this information from the "genetic code" alone is, by 21st century standards, preposterous. As Stent put it:

> What they would have to know, above all, is that the actual cat, or feline *phenome*, arises from a fertilized egg containing the feline DNA nucleotide sequences by an *epigenetic* process of embryonic [and juvenile] development. Moreover, the Aliens would have the understand the

> epigenetic relation between phenome and genome… the Aliens could not, regardless of their level of intelligence and technical sophistication, work out this relation by a semantic analysis of the feline DNA, for the reason that epigenetic relations cannot be found in the DNA.

The complex organization of the individual must be reproduced, in each generation, by a sequential series of interactions. At each stage of development, higher levels of organization emerge from previous interactions. These, in turn, lay the groundwork for still higher levels of emergent organization that depend on the scaffold provided by previous interactions.

Any reasonable metaphor for the reproduction of complexity during development—and the evolution of the determinants of development—must take implicit informational content into account. But the appeal of explanations based on mathematical expressions, probability, and determinism is very strong, especially among physical scientists and mathematical theorists.

The deterministic or probabilistic digital information metaphor

According to Wikipedia, in physics and cosmology, **digital physics**[h] is:

> a collection of theoretical perspectives based on the premise that the universe is describable by information. It is a form of digital ontology about the physical reality. According to this theory, the universe can be conceived of as either the output of a deterministic or probabilistic computer program, a vast, digital computation device, or mathematically isomorphic to such a device.

The deterministic and probabilistic features of this family of perspectives are problematic, to say the least. This is inconsistent with *res potentia* and EEP (Chapter 2). Such perspectives also assume, incorrectly, that mathematical expressions can take the place of knowledge concerning higher-level interactions that emerge from previous interactions. Such processes—such as the development of a cat—are *not* properly described

[h] https://en.wikipedia.org/wiki/Digital_physics

as an unfolding of preexisting information. Unlike the theory of preformation, described in the following section, the embryo does not exist as a tiny individual inside the head of a sperm cell!

Preformation: The "miniature human inside a sperm cell" metaphor

As preposterous as it may seem today, **preformationism**—which argued that human beings develop from tiny miniatures, carried by sperm cells, was accepted by philosophers and physicians for centuries. According to Wikipedia:

> Pythagoras is one of the earliest thinkers credited with ideas about the origin of form in the biological production of offspring. It is said that he originated "spermism", the doctrine that fathers contribute the essential characteristics of their offspring while mothers contribute only a material substrate. Aristotle accepted and elaborated this idea, and his writings are the vector that transmitted it to later Europeans…Later, European physicians such as Galen, Realdo Colombo and Girolamo Fabrici would build upon Aristotle's theories, which were prevalent well into the 17th century.

As pointed out by Robert Wright in *The Moral Animal* [32], **evolutionary psychology (evo-psych)** explores the persistent biases concerning human sexuality that have been with us for thousands of years, and are still very much with us today. It should not be surprising that those biases color the theories put forward by ancient philosophers and early scientists, even though many of those theories seem ludicrous today. One of the more bizarre ones is preformation,[i] in which tiny fully-formed human beings were thought to be transmitted to the unfertilized egg in the head of a sperm cell.

From sexist fantasies to cutting-edge empirical science

Powerful male "authorities" have long assumed that males play a larger role in shaping the characteristics of human beings, both before and after

[i]Illustration on the web at https://commons.wikimedia.org/wiki/File:Preformation.GIF

birth. In fact, preformationism is a biological travesty. In mammals, there can be little doubt that mothers contribute far more to the development of the fetus and the nurturing of infants and juveniles than fathers.

The fact that mothers invest *far more* resources in their young is a fundamental principle taught in elementary biology courses. In fact, the mother also contributes far more molecular, cellular, and developmental information—in the form of **cytoplasmic determinants** (from **maternal genes**), to say nothing of the inherent complexity of the egg cell. Apart from both genetic and **epigenetic** information, the mother also contributes the **mitochondria** (which carry their own genomic DNA). Human mothers contribute more genes than human fathers, because the X chromosome inherited from the mother is much larger than the Y chromosome contributed by the father. Males are also more prone to recessive genetic diseases, because they only have the one maternal X chromosome, whereas females have two. Males do, however, determine the gender of the offspring. On balance, mothers play a much bigger role in perpetuating the species than males. Clearly, the "miniature human inside a sperm cell" metaphor is misguided, to say the least.

The software motif metaphor

The "implicit informational content" of the genome is in some ways analogous to tasks that can be accomplished by running a computer program. In both cases, a sequence of steps must be completed before the significance of the explicit information can be manifested in the real world.

While it is true that a knowledgeable software developer can look at a computer program and recognize patterns that are known to bring about certain kinds of consequences, it is also true that a skilled biologist can find numerous motifs in the genome that are known to bring about certain kinds of molecular, cellular, and developmental consequences.

But in both computer science and in biological science, that knowledge represents extensive previous collective efforts of discovery, experimentation (including many trials and many errors), and empirical observations and experiments by thousands of researchers. Explicit, recognizable sequence motifs in the DNA can be linked to known cellular processes that generate complexity, but only because those higher levels of complexity have already been observed, analyzed, documented, and communicated to the community at large.

The genetic determinants of development are quite different than software. The fact that experts can find meaningful motifs in both—before the software program runs, or before development takes place—reflects the fact that in both of these domains of human activity, collective knowledge has been mapped onto the explicit sequences of bits or nucleotides. But the real-world significance of those sequence motifs can only be glimpsed in the context of those maps.

The coding metaphor and the mythical "genetic code"

Only a small fraction (around 1–2%) of typical animal genomes "code for" amino acid sequences of protein chains. In other words, only a small fraction consists of **codons** that are **transcribed** into RNA, spliced together during RNA processing to make mature mRNA, and translated into amino acid sequences at the **ribosomes**.

Many DNA sequences, called **introns**, are removed (spliced out) during processing of RNA. The retained portions found in mRNA are called **exons**.

Large stretches of noncoding flanking sequences play a variety of regulatory roles that involve interactions with other molecules such as transcription factors.[j] Most of the typical animal (or plant) genome contains vast stretches of noncoding DNA that can play a variety of useful roles,[k] can serve as a genomic scrapyard, and also includes metaphorical **selfish DNA** elements such as repetitive sequences that can expand, diversify, and move around the genome by a variety of mechanisms (see Chapter 16).

Other problems with the "coding" metaphor

The coding metaphor carries conceptual baggage that should be recognized as such and excluded from the Updated Evolutionary Synthesis

[j] https://en.wikipedia.org/wiki/Transcription_factor

[k] https://www.encodeproject.org/

(UES). A code implies a code designer as well as a code breaker, and neither of these human mathematical or computer-assisted attributes apply to Natural Selection. There was no intelligent design involved in protein-coding sequences in particular nor in genomic evolution in general.

This should in no way diminish the profound historical importance and brilliance of the creative conceptual leaps by Francis Crick and the pioneers who documented the detailed Molecular Genetics of the "genetic code" and protein synthesis.

The quest to "crack the code" led to the current understanding that there is a many-to-one correspondence between the sequences of 61 of the 64 possible codons (DNA/RNA triplets) in protein-coding sequences, and the 20 possible amino acids at each position in the sequence of a protein chain.

Fortunately for the history of science, and for our present-day understanding of biology, these pioneers made their discoveries largely because they *were* searching for a *code*. The metaphor was quite useful in that context. But it is time to move on to 21st century descriptions that represent a better approximation of the truth.

Molecular binding interactions, not metaphorical information

Interactions between organic molecules always depend on their physical and chemical properties—including their 3D contours and the ways that they can fit and bind together. The various complementary, weak chemical binding interactions that take place between a variety of DNA sequences, RNA sequences, protein sequences, and other molecules have been well-characterized and cataloged in textbooks [126, 127] as well as the scientific literature. They all share in common the fact that no intelligence or design is responsible for their interactions. They are all governed by principles of organic chemistry, biochemistry, and biophysics that can be both qualitatively and quantitatively described in great detail, using modern laboratory methods and instrumentation.

Organic molecules in general, and **macromolecules** such as DNA, RNA, and protein **polymers** in particular, represent enormously diverse sets of highly-specific 3D shapes and chemical binding potential. When these binding events occur in useful ways, they generate an extraordinary range of biological functions. Many of these functions involve **emergent**, complex properties and interactions involving numerous subunits, units, and levels of organization (see Chapter 2).

Although it is sometimes useful to describe the highly specific organic structures as containing implicit *information*, this can also be misleading, because most of the meanings of the word *information* imply more than the chemical or physical properties and potential of organic molecules which interact in specific (and often useful) ways.

The metaphors of running programs, reading computer files, or decoding streaming digital data

Most types of computer programs, software and digital data files have something in common: they are stored, read, written, and transmitted as long, linear sequences of **binary bits** that have a starting bit and an ending bit. There is a one-to-one correspondence between the unique sequence of bits and the unique discrete information that it contains.

We frequently think of "running" this sequence—i.e. sequentially reading a computer file or stream of bits—from one end to the other—in order, so that its potential applications or meanings can be manifested. An MP4 computer file, for example, can be decoded such that a sequence of video frames can be viewed and an audio track can be heard.

The metaphor of the fertilized egg that "runs" a developmental "program" to reproduce a multicellular individual

As a fertilized egg develops, molecular binding events trigger other molecular binding events which trigger still other molecular binding events. As a result of these events, molecules are synthesized, modified, and/or broken down into smaller products in a variety of ways. New levels of intracellular and intercellular complexity emerge as a consequence of previous events, and these new levels of organization set the stage for still higher (more complex) levels of organization to arise.

The determinants of these events are robust and flexible, so that even genetically identical individuals will follow slightly different developmental sequences but generate functional individuals.

During development, a sequences of events trigger other sequences of events. However, these sequences take place in parallel—not in series—in a multitude of molecules and in a multitude of cells. Many simultaneous binding events occur, both dependently and independently, in different locations, and researchers can describe some of these binding events in terms of cause and effect, while others have no mutual interactions.

When fertilization triggers the "running" of a developmental program, this process is fundamentally different than reading a linear sequence of bits from a computer file. It is also different than simultaneously reading several computer files in parallel in different threads. The shapes and structures of the molecules, their ability to bind and undergo mutual shape changes, and their **catalytic** potential (i.e. their ability to facilitate particular chemical reactions) does not involve linear sequential digital information.

Biological development should not be viewed as running a metaphorical developmental "program". Although this language does express certain aspects of biological development, it is misleading in many ways.

The metaphors of the watchmaker and the blind watchmaker

In Chapter 10, we began to see how discoveries of concrete mechanisms of molecular and cellular interaction help dispel metaphors such as engineering, design, or intelligence in biological evolution.

Prior to such detailed knowledge, Richard Dawkins' 1986 metaphor of *The Blind Watchmaker* [5] provided an accessible popular explanation of why Natural Selection is inconsistent with such **anthropomorphic** or **creationist** explanations.

Evolution of complexity and the origin of species are driven by the struggle for existence. The deciding factors are competition for survival and reproduction. Each variant that occurs is tested against both constraints and opportunities presented by the environment and available *niches*. Adaptations are naturally "selected" in this way.

Even the word "selection" carries conceptual baggage, because it implies a conscious choice. But that is an illusion. That is why the metaphor of the blind watchmaker is a powerful one: it acknowledges our all-too-human tendency to attribute complex organization to intelligent design. In Darwin's time, this was exemplified by William Paley's watchmaker analogy that was based on Natural Theological precepts. Today, Paley has been replaced by more cynical approaches such as attempts to force the teaching of so-called **Intelligent Design pseudoscience** in the public schools. Such attempts are readily exposed by genuine logic and science, yet are still promoted by powerful political and economic interests and appeals to religious fundamentalism.[1]

When do metaphors outlive their usefulness?

In this chapter, we've explored several metaphors that have, in terms of scientific progress, been both a blessing and a curse, including metaphors of:

(1) genetic blueprints,
(2) 3D printers,
(3) digital computer code,
(4) deterministic or probabilistic digital information,
(5) miniature humans inside the head of a sperm cell,
(6) software motifs,
(7) coding and the "genetic code",
(8) reading or writing or running linear sequences of digital bits in computer files,
(9) watchmakers and blind watchmakers.

We've explored the critical importance of implicit informational content that manifests itself during both normal and abnormal development, and gives rise, in a sequential fashion, to increasingly complex, emergent layers or levels of complexity, as observed in the Siamese cat.

[1] https://en.wikipedia.org/wiki/Intelligent_design

We've also seen that the metaphor of the blind watchmaker is a useful remedy for our human tendency to attribute design or intelligence to the complexity and **efficacy** of **adaptations** that arise by Natural Selection.

Metaphors can be quite helpful in making hidden events more tangible and unfamiliar concepts comprehensible. Metaphorically, they bring challenging blue-sky concepts down to Earth. They often play important roles that motivate groundbreaking research (as in the quest for the "genetic code"). You will also find them peppered throughout this section in particular and *Rethinking Evolution* in general. Metaphorical language is frequently, and deliberately, used in the popular science and peer-reviewed scientific literatures and media, because it is a useful rhetorical communication tool, so long as it is understood to represent poetic license rather than literal truth.

But as new discoveries are made and new conceptual frameworks become available, certain metaphors outlive their usefulness. Despite this, they continue to be carried by the community as intellectual baggage.

In the next chapter, we'll consider two interdisciplinary insights that are firmly grounded in empirical science (rather than metaphor) and provide a deeper understanding of the evolution of genetic determinants as well as the ways that those determinants help reproduce complex organization during the process of development.

Chapter 12

Shape-Specific Molecular Interaction and Binding Events (SSM-IBE)

Biochemistry is the science of life. All our life processes—walking, talking, moving, feeding—are essentially chemical reactions. So biochemistry is actually the chemistry of life, and it's supremely interesting.

—Aaron Ciechanover[a]

The Big Picture

Several metaphors, such as those that treat DNA as software code or embryos as computers that interpret digital information, have outlived their usefulness for evolutionary theory.

A more useful 21st century perspective is that biological organization depends on shape-specific molecular interactions and binding events (SSM-IBE), which obey predictable physical and chemical principles.

When applied to the origins of life and the emergence of the genetic apparatus, SSM-IBE illustrates how simple molecular interactions lay the groundwork for more complex levels of structure and function to evolve.

[a] https://www.ynetnews.com/articles/0,7340,L-4730699,00.html

Introduction

In Chapter 11, we reviewed several metaphors that have been useful in the past, such as describing DNA as software, or describing cells or embryos as computers that "run" or interpret digital information, instructions, or blueprints. One metaphor in particular—the genetic "code"—led to systematic empirical studies designed to "crack the code".

For a 21st century, Updated Evolutionary Synthesis (UES), however, several metaphors have outlived their usefulness and instead stand in the way of a deeper understanding of how biological organization evolves.

Today, we have a deep and detailed understanding of the genetic apparatus, including the precise molecular interactions involved in each stage of protein synthesis, and the mapping between 64 **codons** and 20 amino acids plus start and stop signals in a particular protein chain.[b]

This mapping is commonly referred to as the "genetic code", but few people realize that "code" is actually a metaphorical description with subtle conceptual and semantic limitations. Human brains create codes, but evolution does not.

Today, conceptual frameworks of codes, information, software, blueprints, or computers present obstacles that get in the way of a deeper understanding of real-time molecular interactions, and the way that higher levels of organization can evolve from them.

In this chapter, we'll first replace these metaphorical descriptions with down-to-earth, concrete descriptions that are firmly grounded in well-established principles of chemistry, physics, and biology.

Next, we'll bring several key concepts together, and consider, in general and hypothetical terms, the ways that complex molecular organization can emerge from the evolutionary process.

Replacing Outdated Metaphors with Molecular Interactions

Transcending the coding metaphor

What biologists routinely call the **genetic code** is not actually a code at all. According to Wikipedia, a code[c] is:

---

[b] https://en.wikipedia.org/wiki/Protein#Synthesis

[c] https://en.wikipedia.org/wiki/Code

> …a system of rules to convert information—such as a letter, word, sound, image, or gesture—into another form or representation, sometimes shortened or secret, for communication through a communication channel or storage in a storage medium.

The Wikipedia entry on the "genetic code", despite its clarity and pedagogical value, fails to recognize the semantic limitations of the coding metaphor[d]:

> The "genetic code" is the set of rules used by living cells to translate information encoded within genetic material (DNA or mRNA sequences) into proteins.

This sentence semantically implies that cells understand and actively consult a set of rules to translate information. Obviously, the sentence was not intended to be taken literally. But the language is encumbered with the conceptual baggage of goal-directed engineering or intelligent design. This does not help us to understand how a blind, natural evolutionary process—such as Natural Selection—can have sufficient organizing power to give rise to the genetic apparatus in the first place.

Now contrast that sentence with a sentence taken from the Wikipedia entry on biological translation[e]:

> Translation is accomplished by the ribosome, which links amino acids in an order specified by messenger RNA (mRNA), using transfer RNA (tRNA) molecules to carry amino acids and to read the mRNA three nucleotides at a time.

This sentence is a concrete and useful brief description of the molecular interactions that routinely take place during protein synthesis. The link between codons and amino acids emerges from the action of enzymes known as aminoacyl tRNA synthases (aaRS). A more specific Wikipedia page describes them as follows[f]:

> An aminoacyl-tRNA synthetase (aaRS or ARS), also called tRNA-ligase, is an enzyme that attaches the appropriate amino acid onto its tRNA…

---

[d] https://en.wikipedia.org/wiki/Genetic_code

[e] https://en.wikipedia.org/wiki/Translation_(biology)

[f] https://en.wikipedia.org/wiki/Aminoacyl_tRNA_synthetase

> ...In humans, the 20 different types of aa-tRNA are made by the 20 different aminoacyl-tRNA synthetases, one for each amino acid of the "genetic code".

Each aaRS chemically links a specific amino acid to one of the **tRNA** molecules that will carry it to the ribosome. As is typical of enzymes, the **active site** of the aaRS has a complementary shape and binding potential that specifically allows the appropriate amino acid and tRNA molecules to bind. The enzyme then catalyzes the synthesis of a strong chemical bond between the amino acid and the appropriate tRNA, forming an amino acid-tRNA complex.

When this complex moves to a ribosome that is actively synthesizing an amino acid sequence, a specific RNA base triplet known as an **anticodon** will **base-pairs** to its complementary **codon** on the mRNA that is undergoing translation. The sequence of codons on the mRNA, transcribed from a particular gene, determines the sequence of amino acids in the growing protein chain. The ribosome has active sites that break the bond between the amino acid and the tRNA that is base-paired to the mRNA, and that synthesize a strong chemical bond, known as a peptide bond, between each sequential amino acid in the growing protein chain.[g]

Replacing coding metaphors with molecular binding interactions

Today, the detailed biochemistry of each of the steps in translation has been thoroughly characterized in great detail. For *Rethinking Evolution*, the important point is that higher-level structures and functions that link specific DNA sequences to specific amino acid sequences arise from well-characterized molecular interactions that obey the fundamental laws of chemistry and physics. The 3D shapes of the aaRS, the amino acid, and the tRNA bind together with exquisite specificity, as do the active sites of the ribosome and the base-pairing between codons and anticodons. This

[g] https://en.wikipedia.org/wiki/Ribosome

specificity depends on the 3D shapes of the molecules as well as the physical and chemical properties of the parts of the macromolecules that fit together in a complementary fashion. Such specific interaction and binding potential are intrinsic properties of macromolecules. Without this potential, carbon-based life forms, with all of their complexity and diversity, could never have evolved.

Transcending the lock-and-key and molecular recognition metaphors

Among macromolecules, the potential diversity and specificity of complementary shape-specific interactions is breathtaking. An enzyme, for example, depends on the precise fit—the complementary 3D shape—of the enzyme's active site and the reactants that bind to it. Both shape and chemical stability make the binding exquisitely specific. Precise fit and gradual refinement of that precision are certainly consistent with Darwinian concepts of incremental change in DNA sequences. Subtle changes in key amino acid residues of protein chains should be reflected in the 3D surfaces that fit together.

When two or more 3D molecular surfaces fit together, the classical textbook metaphorical description is that of a **lock and key**. Another metaphor is **molecular recognition**, but this suffers from the same problem as coding metaphor: it implies that the complementarity is derived from some sort of conscious choice, awareness, or intelligence, which is clearly not the case.

Modern organic chemistry and biochemistry provide detailed, concrete predictions and descriptions of complementary binding potential between macromolecules. These phenomena are described quantitatively in terms of free energy, attraction or repulsion to the surrounding water molecules, and similar parameters, and the 3D shapes can be observed routinely with proper instrumentation and can be predicted by computer programs.

The shapes of macromolecules such as proteins are not actually static (like a lock and key). Instead, the shapes of the molecules change as a consequence of the binding process, which is referred to as **induced fit** (Figure 1).

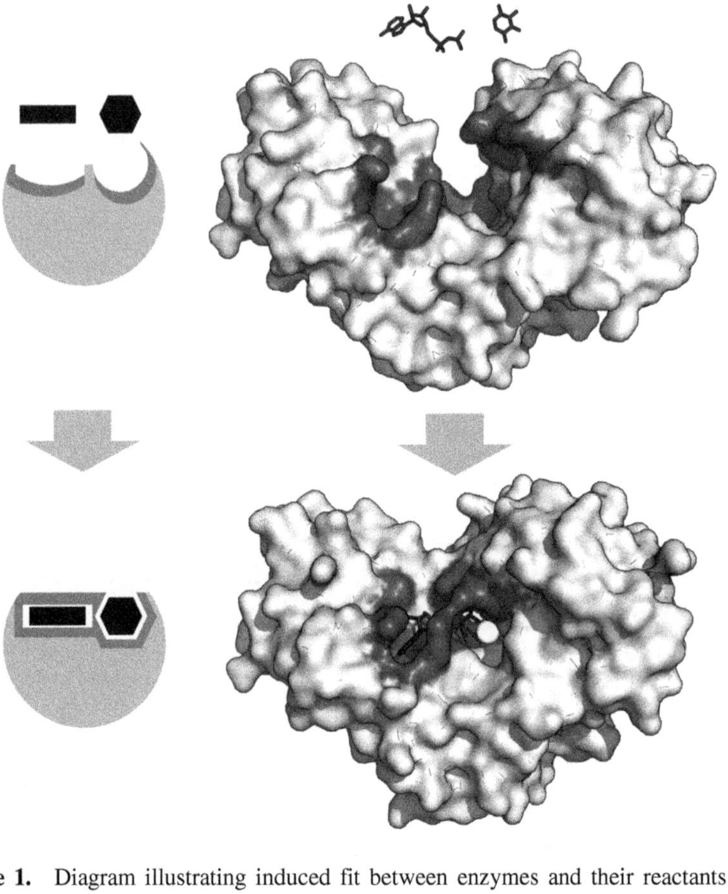

Figure 1. Diagram illustrating induced fit between enzymes and their reactants (also called substrates). Enzymes facilitate specific chemical reactions by providing complementary shapes and other physical properties that only permit binding of specific reactants at their active sites. When the correct reactants are in place, the enzymes often change shape so that reactants are held tightly in place, a process known as induced fit. Once the reaction has taken place, the enzymes change shape again as products are released. Enzymes lower the activation energy and speed up chemical reactions that otherwise would be unlikely to occur. Amino acid side chains at the active site are often involved. Often, enzymes are arranged in specific sequences within biological membranes, so that they can facilitate a series of chemical reactions in a biochemical pathway.[h]

---

[h] Page-link: https://commons.wikimedia.org/wiki/File:Hexokinase_induced_fit.svg. File-link: https://upload.wikimedia.org/wikipedia/commons/f/f5/Hexokinase_induced_fit.svg. Attribution: By Thomas Shafee [CC BY 4.0 (https://creativecommons.org/licenses/by/4.0)], from Wikimedia Commons. Rendered in B&W.

Shape-specific molecular interaction and binding events (SSM-IBE)

The intrinsic properties of organic molecules, and their potential to fit together in a complementary fashion can be seen repeatedly throughout the biosphere. These interactions govern the inner-workings of cells, the cooperation between cells during development, and the reproduction of complexity in each generation. Let's give these interactions a more descriptive name: let's call them **shape-specific molecular interaction and binding events**, or **SSM-IBE** for short.

When coding metaphors are replaced with SSM-IBE, it's easier to understand how complex organization can evolve.

The implications of SSM-IBE for the evolution of complexity

At the instant that two or more macromolecules bind together, they form a higher-level structural unit. This is an intrinsic emergent property of molecules in general. Their ability to bind together specifically—as well as any changes that may take place as a consequence of that binding, including shape-changes as well as making or breaking strong chemical bonds—represents the structural foundation of the functional roles that they can play.

In aqueous (water-based) solutions, the specific shapes, binding potential, and enzymatic potential of macromolecules are set by the laws of chemistry and physics. Biochemists and molecular biologists routinely rely on these laws, in a quantitative fashion, to predict the events that may or may not occur. There is nothing subjective about any of this. The laws of nature are the same throughout the universe, and are predictable by well-established principles of chemistry, physics, and biology.

What this means for evolution is that SSM-IBE defines the specific organized structures and functions that can arise. Fortunately for us and other complex life forms, the potential diversity of complementary shapes and interactions among protein chains and other macromolecules is astronomical.

To see this, just consider the theoretical diversity of a random sequence of 20 possible amino acids. A short random chain of just ten amino acids has a potential diversity of 20 raised to the tenth power, or over ten trillion possible sequences, and each has a unique shape and physical and chemical properties. These shapes become highly specific when longer protein chains fold into characteristic 3D structures, and when multiple folded structures self-assemble into more complex 3D structures.

Sequences of 200 or more amino acids are quite common in protein chains, and their theoretical diversity far exceeds the number of atoms in the known universe.

Metaphors implying active choices have outlived their usefulness

Human beings *translate* a code to something that has a clear or plain meaning by using some sort of map or table. But the biological process that we refer to as *translation* does not use any map or table whatsoever. Therefore, it is not actually a code at all, but rather, yet another example of the emergent properties of organic molecules and their potential to interact.

Is this just nitpicking? Not at all. The difference is a crucial one, because it demonstrates the clear distinction between information and SSM-IBE. Information is something that intelligent beings exchange. For the purists among you, the term information also has other scientific and technical definitions related to entropy, signals and noise that can be expressed in mathematical terms—as described by Claude Shannon and others.[i] But those definitions of information do *not* provide the useful insights into the evolution of development that SSM-IBE does.

Thought Experiments Concerning the Evolution of the "Genetic Code"

The origin and evolution of complex molecular interactions

In terms of **evolvability** and the generation of complex biological organization,[j] the potential value of linking DNA triplets to amino acid

[i] https://en.wikipedia.org/wiki/A_Mathematical_Theory_of_Communication
[j] https://en.wikipedia.org/wiki/Evolvability

sequences cannot be overstated. But how did this linkage evolve in the first place?

Exploring the evolutionary potential of hydrothermal mounds for the evolution of the "genetic code"

There is no shortage of peer-reviewed papers that rely on indirect empirical evidence to propose hypothetical ways that the "genetic code" may have evolved [100]. Here, I'd like to consider some logical but imaginary evolutionary scenarios that combine the principles of EEP, SSM-IBE, and Natural Selection. These represent an extension of principles discussed in Chapters 5 and 6.

Setting the stage

The setting is any one of the ancient hydrothermal mounds (**UAHM**) that were discussed in Chapter 5 and further explored in Chapter 6. The process of **serpentinization** continues to vent hydrogen and carbon dioxide gases. Mounds of hollow, porous tubes are **precipitated**. Proton gradients are generated in those mounds where **alkaline** fluids from serpentinization meet the acidic seawater. Abundant, inorganic catalysts trapped by the hollow, porous tubes catalyze a variety of organic (carbon-containing) **redox** reactions that are powered by the potential energy of the proton flux. These inorganic products include, but are not limited to, a limited set of amino acids, and RNA nucleotides, as well as compounds containing high-energy phosphate bonds.

Selective pressures for synthesis of diverse organic compounds in the hydrothermal mounds

Efficient production of diverse organic materials

At the origin of life, efficient production of a diverse range of organic materials has emergent evolutionary potential (EEP). Obviously, if no organic materials were produced, life could not evolve. As the number of different organic compounds increases, the potential for these compounds to bind, interact, and facilitate new chemical reactions also increases. By a sort of molecular trial and error, greater diversity of organic compounds increases the chances that something potentially useful may emerge.

The ability of small organic molecules to interact and react with each other depends on their physical and chemical properties, such as the presence or absence of particular **functional groups**. As compounds increase in size, shape and weak binding potential play larger roles in the specificity of the interactions that take place.

Template-free, cell-free molecular proliferation

Prior to the origin of cells, and prior to the origin of RNA or DNA templates, organic molecules may have been exchanged between the porous, tube-like structures that arise at hydrothermal mounds. Such exchanges would not require more than the ability of the molecules to flow from one location to another. We can easily list examples of chemical and physical factors that might facilitate such exchanges:

(1) ability of molecules to dissolve,
(2) buoyant density of molecules,
(3) ability of molecules to be carried by flows or currents,
(4) ability of molecules to attach to other molecules,
(5) ability of attached molecules to form **aggregates**,
(6) ability of molecules or aggregates to attach to porous tubes, or to other molecules that are already attached to the tubes,
(7) ability of molecules or aggregates to be cleaved off or otherwise released from their attachments to particular tubes,
(8) efficiency of synthesis and quantity of production of organic molecules, which depends on factors such as porosity of the tubes, presence of mineral or organic catalysts, ability to harness the energy of proton fluxes, presence or absence of particular inorganic or organic compounds or functional groups, etc.

Note that both passive mobility (ability to detach, attach, dissolve, flow) and efficiency of production are among the factors that could influence molecular proliferation among multiple locations in the hydrothermal mounds. Note also that in this cell-free, template-free environment at the origin of life, organic processes and interactions can freely proliferate if the organic compounds are able to move around and attach to other objects.

EEP and evolvability

In the context of cell-free, template-free environments such as the ancient hydrothermal mounds, EEP and evolvability can depend on the rate of production of particular organic compounds with potentially useful chemical or physical properties, their ability to move from one location to another, and their ability to combine with and interact with other materials at other locations. Higher-level structures and interactions are emergent phenomena. As new structures and interactions arise, these in turn lay the groundwork for new potential interactions in the future.

The origin of primitive linkages between amino acid sequences of proteins and nucleotide sequences of RNA

As mentioned above, we can replace the conceptual framework of a metaphorical "genetic code" with a better representation of reality—one that is based on shape-specific molecular interactions and binding events (SSM-IBE).

Based on its widespread similarity among both prokaryotic and eukaryotic organisms, the present-day genetic apparatus for protein synthesis has **deep homology** with the earliest known cellular ancestors.

At the molecular level, the correspondence of codons with specific amino acids is enforced by SSM-IBE between the active sites of specific amino-acyl tRNA synthetase (aaRS) enzymes and the complementary 3D contours and binding capabilities of specific amino acids and tRNA molecules. The tRNA molecules carry the correct amino acids to the ribosomes, and the triplet anticodons of those amino acids base-pair with complementary codons in the mRNA sequence. In this way, as sequential codons of the mRNA molecule bind to their anticodons, the amino acid sequence of a specific type of protein chain is assured.

Although the metaphor of a "genetic code" was quite useful during the discovery phase of the genetic apparatus, that metaphor does not help us to understand how that genetic apparatus evolved.

Genetic determinants for simple interactions lay the groundwork for more complex interactions to evolve

Evolutionary theory must account for the origins of the genetic apparatus from simpler forebears. Contrary to the pseudoscientific belief in "irreducible complexity", concrete empirical studies show that complex structures and functions do not arise suddenly.[k] Complex functions do not arise because of a single opportunity or selective pressure in the struggle for existence. Rather, they are blindly cobbled together by what might be described metaphorically as Rube Goldberg machines,[l] tinkering, opportunism, or trial and error. This is metaphorical because there is no intelligent design involved. Natural Selection acts, as Richard Dawkins aptly puts it, as a blind watchmaker [5]. Often, adaptations that arise in one ecological context are reused for different purposes in different contexts, a process known as **exaptation**.[m]

With this in mind, can we imagine some hypothetical early steps in the evolution of the enforced correspondence of amino acids with particular codons in the genetic apparatus?

If the genetic apparatus gradually evolved from the coalescence of simpler units of structure and function, what are some of those simpler units, and how could their emergence be useful in the struggle for existence at the origin of life?

Useful molecular structures and functions in the hydrothermal mounds

The following is a list of the sorts of structures and functions that could play useful roles in capture of energy and nutrients and dispersal throughout primitive hydrothermal mounds. The hypothetical structures and functions are listed in an arbitrary order, and it is assumed that these structures and functions evolve in parallel as initially independent units. As these independent units arise, they have EEP that leads to their coalescence into

---

[k] https://en.wikipedia.org/wiki/Irreducible_complexity

[l] https://en.wikipedia.org/wiki/Rube_Goldberg_machine

[m] https://en.wikipedia.org/wiki/Exaptation

higher levels of structure and function, gradually leading to a primitive and then a more refined genetic apparatus by Natural Selection acting at the molecular level:

(1) synthesis of amino acids,
(2) linkages between amino acids and other organic compounds,
(3) synthesis of compounds with high-energy phosphate bonds,
(4) synthesis of ribonucleotides and ribonucleotide triphosphates,
(5) synthesis of peptide bonds, oligopeptides and polypeptides,
(6) synthesis of polyribonucleotide sequences (RNA),
(7) coupling of inorganic catalysts to organic compounds,
(8) formation of organic aggregates with complementary functions,
(9) tethering of organic molecules or aggregates to porous tubes,
(10) release of organic molecules or aggregates from porous tubes,
(11) dispersal of organic molecules or aggregates in hydrothermal mounds,
(12) formation of RNA chains with catalytic capabilities,
(13) formation of secondary structure and base-pairing in RNA sequences,
(14) synthesis of polymers of sugars such as glucose (primitive carbohydrates),
(15) synthesis of glycerol and fatty acid chains (primitive lipids),
(16) synthesis of short and long polypeptides,
(17) synthesis of polypeptides with secondary structures that bind to other inorganic or organic structures,
(18) diversification of side chains of amino acids,
(19) synthesis of long polypeptides with inorganic catalytic cofactors,
(20) synthesis of long polypeptides with organic catalytic capabilities,
(21) folding of long polypeptides into 3D shapes (tertiary structure),
(22) binding of complementary 3D polypeptides into aggregates with **quaternary structure**,
(23) RNA and/or protein aggregates that bind specifically to smaller RNA molecules and specific amino acids and link them together with strong chemical bonds, leading to primitive tRNA and primitive forms of aaRS,
(24) refinement of primitive aaRS specificity as amino acid side chains and primitive tRNA diversify,

(25) synthesis of simple RNA templates that function as primitive codons,
(26) primitive ribosomal functions,
(27) refinement and coalescence of primitive ribosomes, codons, and other molecules and evolution of the genetic apparatus,
(28) evolution of DNA as the complementary storage genetic material,
(29) synthesis of primitive mRNA.

The success criterion and the struggle for existence in the hydrothermal mounds

Strictly speaking, classical Darwinian Natural Selection does not apply to molecular evolution prior to the existence of cells and genomes. Darwin was concerned with the origin of species that transmitted hereditary factors from generation to generation. He was concerned with competition among individuals that vary among themselves and must compete with one another, and with other species, in the struggle for existence in a particular environment and way of life. Evolutionary success could be measured by relative numbers of offspring that carry a specific set of hereditary variants, and the ability of those offspring to reproduce those variants. In order to succeed, the structures and functions of the offspring must enable them to excel at capturing energy and nutrients, maintaining their integrity, and reproducing their own kind, relative to other individuals that compete with them.

Prior to the evolution of cells and genomes, evolutionary success has a somewhat different meaning, but there are also some similarities.

The items listed in the section titled *Template-free, cell-free molecular proliferation* illustrate how to see that evolutionary success can be achieved by efficient production and dispersal of a broad range of organic molecules that have the capacity to interact in various ways.

The items listed in the section titled *Useful molecular structures and functions in the hydrothermal mounds* illustrate how more complex levels of organization can arise from simpler ones, because the simpler functions lay the groundwork for the evolution of more complex organization.

In the second list, we see some hypothetical ways that a linkage between amino acid sequences and RNA and DNA sequences can arise from SSM-IBE. This represents the early origins of genetic determination of complex structure and function. Even a primitive genetic apparatus would be expected to dramatically accelerate subsequent evolution, because it lays the groundwork for classical Natural Selection as an evolutionary force.

Later, when innovations such as long, folded protein chains acquire catalytic capabilities, and when their interactions are determined by their ability to fit together and find in a complementary and specific fashion, the extraordinary evolutionary potential of proteins, and interactions between proteins and genomic DNA, become possible. The general expectation is that highly complex multi-chain proteins, such as the ATP synthase[n] and the flagellar motor complex[o] would evolve from the coalescence and modification of simpler functional assemblies, and there is some empirical evidence for this. We'll explore some of the widespread and diverse ways that **deep homology** and reuse of both genomic and developmental **toolkits** have led to complex levels of organization by evolutionary "tinkering" in subsequent chapters.

---

[n] https://en.wikipedia.org/wiki/ATP_synthase

[o] https://en.wikipedia.org/wiki/Flagellum

Chapter 13

The Genetic Algorithm (GA): Blind Production of the Combinatorial Phenotype

Can data generated by an idealized computer model provide useful insights about natural evolutionary systems?

—Stephanie Forrest, 1993 [101]

The Big Picture

Metaphors can be useful tools that help to illustrate the enduring core of Natural Selection in the context of the Updated Evolutionary Synthesis (UES), even when they have teleological overtones. Such metaphors include the blind watchmaker, the tinkerer, the Rube Goldberg device, the opportunist, the scrapyard, the toolkit, the black box, and trial-and-error.

These metaphors help to illustrate the power of a software technique known as the Genetic Algorithm (GA), which was inspired by Natural Selection.

The GA has also been used in simulations that demonstrate the creative power of genetic recombination, which does more than simply reshuffle genetic determinants.

Metaphors for Natural Selection That Are Still Useful

The absence of awareness, consciousness, goal-seeking, sentience, intelligence, choices, and design are fundamental aspects of Natural Selection. They are also among the most challenging for human beings to understand, because of the way that our brains are wired [102]. In the fast mode of human thinking,[a] we are wired to react to real-time, actual events—usually those we can see or hear—because that was essential during the evolution of our ancestors. In the slow mode of human thinking, we are goal-seeking storytellers and intelligent, goal-seeking designers. We understand the world by analyzing and attaching labels, and we plan strategies in an intentional fashion.

Natural Selection does not work in the way that our brains are wired. Yet Natural Selection has extraordinary organizing power, and there is no shortage of examples where the complexity and efficacy of adaptations far exceed human capabilities.

Metaphors can be useful in providing a deep understanding and appreciation of the ways that Natural Selection differs from human designs. Seven kinds of useful metaphors are briefly summarized as follows:

The metaphor of the blind watchmaker

Richard Dawkin's clever reply to William Paley's metaphor of the watchmaker [71] is captured in the title of his popular book, *The Blind Watchmaker* [5]. Clearly, Natural Selection routinely generates remarkably complex and intricate adaptations, but it does so in a blind fashion. As we'll see shortly, however, metaphorical "blindness", in this context, is a feature, and not a bug.

---

[a] Holt, J. Two brains running. *New York Times* Sunday Book Review, November 25, 2011. Online at https://www.nytimes.com/2011/11/27/books/review/thinking-fast-and-slow-by-daniel-kahneman-book-review.html

The metaphor of the tinkerer

The Wikipedia page for "tinkering" redirects to one titled "bricolage"[b]:

> In the arts, bricolage (French for "DIY" or "do-it-yourself projects") is the construction or creation of a work from a diverse range of things that happen to be available, or a work created by mixed media.

In the context of Natural Selection, creation of useful adaptations most definitely arises from a diverse range of things that happen to be available. They consist of genetic determinants, prior innovations, and structures and functions that depend on EEP (Chapter 2) and SSM-IBE (Chapter 12).

The metaphor of the Rube Goldberg device[c]

The Wikipedia description of a **Rube Goldberg** device is:

> …a machine intentionally designed to perform a simple task in an indirect and overcomplicated fashion. Often, these machines consist of a series of simple devices that are linked together to produce a domino effect, in which each device triggers the next one, and the original goal is achieved only after many steps.

It is important to recognize that there is no "intention" involved in Natural Selection. Nevertheless, the results are often reminiscent of a domino effect, in which each device triggers the next one, and the useful result (in the life of the cell or organism) is achieved only after many steps. **Signal transduction pathways**[d] often involve these sorts of cascades of events.

The reasons that the products of Natural Selection resemble Rube Goldberg machines can be found in the emergent ways that SSM-IBE arises from the emergent properties of macromolecules, and the way that previous innovations pave the way for new levels of complexity to arise, often by coalescence.

---

[b] https://en.wikipedia.org/wiki/Bricolage

[c] https://en.wikipedia.org/wiki/Rube_Goldberg_machine

[d] https://en.wikipedia.org/wiki/Signal_transduction

The metaphor of the opportunist

Since selection acts on the phenotypic gestalt—which is often created by a diverse range of things that happen to be available—it appears opportunistic. In this case, opportunism is an entirely blind and unconscious natural result.

The metaphor of the scrapyard or toolkit

The metaphors of the genomic scrapyard [38] or toolkit, and the developmental toolkit have already been mentioned in earlier chapters. Natural Selection often involves **exaptation** (redeployment, with or without modification, of an element that evolved for a different reason), and reuse of genetic elements is widespread and commonplace.

The metaphor of the black box

The metaphor of a black box is useful in the sense that selection acts on the phenotypic gestalt, and useful phenotypic gestalts can arise from a diverse range of things that happen to be available. The combinations and sequences of events that lead to a useful result are like a black box to Natural Selection, because they arise in a blind fashion, without any "awareness", by natural processes.

The metaphor of trial-and-error

Since in each generation, individuals must compete in the struggle for existence, and since variation is extensive, only a few combinations of variants will be selected because they generate the most useful phenotypes. Selection acts at the level of real-time, actual events that involve both the individual and its environment. Therefore, repeated rounds of variation and selection resembles a trial-and-error operation. As long as we realize that this process does not involve deliberate choices or awareness, the metaphor is apt.

All of these metaphorical attributes of Natural Selection demonstrate why Natural Selection became the inspiration for a powerful problem-solving software technique known as the Genetic Algorithm, which will be discussed in the following sections.

How Natural Selection Inspired a Powerful New Software Technique: The Genetic Algorithm (GA)

Origins and characteristics of the GA

In 1975, John Holland described the GA[e] in his aptly titled book, *Adaptation in Natural and Artificial Systems* [103].

An algorithm can be defined as an unambiguous set of steps that lead to the solution of a problem.[f] By combining certain well-characterized aspects of heredity and Natural Selection with the concept of an algorithm, we come up with a powerful software engineering tool that borrows heavily from biological evolution. This is known as the **Genetic Algorithm (GA)**.

Unlike computer programs that solve problems by an orderly sequence of steps leading to the solution, the Genetic Algorithm simply varies a large number of parameters in a random fashion, and then selects the combinations that fortuitously provide the best solution.

The GA helps to illustrate the blind and unconscious way that evolution generates complex phenotypes. Holland's stroke of insight was to abandon the traditional logic of computer programming, in which engineers consciously solve a problem by an orderly and logical sequence of specific steps. Instead, he borrowed Darwin's brilliant concept, in which repeated cycles of random variation, followed by selection, lead to useful results. He did this by representing a multitude of random choices for elementary programming steps, and then allowing the computer to blindly select for the combinations of steps that happened to result in the best solutions. As described by Holland in a popular online article[g]:

---

[e] Holland, J.H. Genetic Algorithms: Computer programs that "evolve" in ways that resemble Natural Selection can solve complex problems even their creators do not fully understand. Online at http://www2.econ.iastate.edu/tesfatsi/holland.gaintro.htm

[f] https://en.wikipedia.org/wiki/Algorithm

[g] http://www2.econ.iastate.edu/tesfatsi/holland.gaintro.htm

> Computer programs that "evolve" in ways that resemble natural selection can solve complex problems even their creators do not fully understand...
>
> ... Living organisms are consummate problem solvers. They exhibit a versatility that puts the best computer programs to shame. This observation is especially galling for computer scientists, who may spend months of years of intellectual effort on an algorithm, whereas organisms come by their abilities through the apparently undirected mechanism of evolution and natural selection...
>
> ... [In the 1960s] I had been investigating mathematical analyses of adaptation and had become convinced that recombination of groups of genes by means of mating was a critical part of evolution...The result was the classifier system, consisting of a set of rules, each of which performs particular actions every time its conditions are satisfied by some piece of information...The conditions and actions are represented by strings of bits corresponding to the presence or absence of specific characteristics in the rules' input and output...

According to Wikipedia,[h] examples of problems solved by GAs include mirrors designed to funnel sunlight to a solar collector, antennae designed to pick up radio signals in space, walking methods for computer figures, and optimal design of aerodynamic bodies in complex flow fields. A common characteristic of these problems is that the optimal solutions represent a combination of parameters that interact in a complex way that defy the logical or linear approaches that are consciously applied by designers or engineers.

A list of key attributes of Natural Selection that loosely apply to the GA

(1) Natural Selection acts upon the advantages of a particular variable phenotype over other phenotypes, in the struggle for existence.
(2) Selection has its own built-in criteria for success, in particular ecological contexts.

---

[h] https://en.wikipedia.org/wiki/Genetic_algorithm#Commercial_products

(3) Selection works by a process that is aptly described as trial-and-error.
(4) Selection is path-independent.
(5) Selection is opportunistic.
(6) Nature has no awareness, intelligence or design—it is like a blind watchmaker that generates organization by any or every possible means.
(7) Selection acts on the combined, integrated effects of a variety of molecular, cellular, and developmental mechanisms.
(8) These combined effects are aptly described as various **phenotypic gestalts**.
(9) **Convergent evolution** selects for useful outcomes, not for the path to get there—there are numerous distinct pathways that accomplish similar complex outcomes in terms of their adaptive significance.
(10) Selection is opportunistic, a metaphorical tinkerer, a builder of metaphorical **Rube Goldberg devices**.
(11) In each generation, offspring are varied, primarily by means of sexual reproduction, in which recombination shuffles the alternative determinants in the **gene pool** (i.e. **alleles**) that are randomly inherited from the parents.
(12) Useful attributes tend to increase the numbers of the individuals that carry them, and thereby increase the proportion of those alleles in the gene pool.

The GA provides deep insights into the blind organizing power of Natural Selection

Among the most doggedly persistent, widely believed and yet patently false ideas concerning speciation and the origin of novelty are the ideas that:

(1) **point mutations** are the ultimate source of diversity in the gene pool
(2) **recombination** can only shuffle existing sequence motifs
(3) new genotypes lead directly to new phenotypes
(4) most mutations are harmful

(5) the disadvantages of sexual reproduction (almost, nearly) outweigh its advantages as a source of diversity

Each one of these misguided beliefs can be traced to false precepts and misunderstandings dating back to **classical Darwinism**, the **Modern Synthesis**, and **Population Genetics**. Each one has been boosted, at one time or another, by the endorsement of distinguished evolutionary theorists. But fortunately, due to the robustness of professional ethics, logical debate, the scientific method, and new methods of empirical discovery, they have been countered by cogent arguments. For the most part, they have been corrected for the record. But that does not mean that they do not persist to some degree in the scientific community, and to a larger degree in textbook accounts and among nonscientists. Here is a brief list of updated concepts that will hopefully replace these five misguided concepts:

1. *Point mutations are not the ultimate source of diversity in the gene pool*

In Chapter 16 of *Rethinking Evolution*, genome evolution will be explored in some detail. Briefly, point mutations—that is, single-base changes resulting in single-codon substitutions—may be the least interesting of several types of well-documented mechanisms and kinds of changes in DNA sequences, during genome evolution. Also, these various types of genomic changes often synergize in consequential ways.

2. *Recombination can do more than reshuffle existing sequence motifs*

Recombination represents an important source of sequence novelty, because it generates new combinations of shorter sequence motifs and can generate profound changes in both protein-coding and regulatory DNA. This will be discussed further in the following section.

3. *New genotypes often do not lead directly to new phenotypes*

The concept of the **developmental toolkit**, a central concept in evo-devo, needs to be expanded to the concepts of the **genomic toolkit** (i.e. reusable

DNA sequence motifs of all kinds, including duplicated genes) as well as the concept of the **genomic scrapyard** [38], which is a suitable 21st century replacement for the earlier misguided but powerful metaphor of **junk DNA**. DNA sequences that are not explicitly used, at any given moment, in any particular genomic background or ecological context, represent a potent source of materials that can be used at later times or in other contexts, and contribute in many ways to the evolvability of any lineage. Lineages that are evolvable tend to persist for many millions of years, whereas those that are not are more likely to go extinct.

4. *Most mutations are not harmful*

The idea that most mutations represent random changes to a complex, metaphorical fine-tuned machine—and are therefore more likely to be destructive than useful—is a point well-taken. However, this perspective is a limited one because it (a) only considers the phenotypic consequences of mutations in isolation from other genomic changes; (b) assumes that the mutations alter specific protein-coding **codons**, whereas mutations actually represent a much broader range of changes to genomic DNA that are not usually referred to as mutations, including, for example, **gene duplication** (Chapter 16); and (c) it fails to consider the multiple weak, subtle, and combined effects of changes to regulatory sequences of DNA.

5. *The advantages of sexual reproduction do outweigh its disadvantages*

Sexual reproduction is the main source of variation among offspring. This variation arises during the production of sperm and egg cells, primarily by means of segregation of maternal and paternal **chromosomes** during **meiosis**; by random fertilization of one egg cell by one sperm cell; and by means of random **recombination** events, including exchanges between **homologous chromosomes** (paired contributions from the mother and the father) and other chromosomal variations.

In addition to its importance as a potent source of new combinations of useful genes in each individual offspring, the presence of two sets of chromosomes also provides a powerful way of maintaining diversity in the gene pool while masking most harmful recessive mutations in individuals.

In addition to recombining protein-coding genes as well as combinations of protein-coding sequences with various regulatory elements, recombination also represents a potent source of genetic novelty within genes and even at the level of individual codons. This will be discussed in the following section.

Crossovers generate nonrandom recombinants under Darwinian selection

The title of this section is the title of one of my earlier published contributions to evolutionary theory [104, 105]. In addition to helping to clarify the nuances of the roles played by recombination and point mutations in DNA sequence evolution, the paper exemplifies the mutually beneficial ways in which basic research, applied research, and theoretical research interact, leading to new experimental designs and surprising results. I'll begin with an excerpt from the paper's **Abstract** (summary):

> Genetic diversification is not the only consequence of recombination. When recombination occurs in a gene pool under selective pressure, the result is non-random linkage of selected sequence elements with high potential fitness. Computer simulations based on the Genetic Algorithm show that populations of DNA sequences evolve more rapidly *via* homologous recombination than by point mutations alone, especially under mild selective pressure.…[i]

The paper illustrates the way that a technical innovation (the GA) inspired by neo-Darwinian variation and selection led to a computer simulation designed to learn more about the natural system that inspired the technical innovation in the first place.

Briefly, I systematically compared the effects of simulated point mutations (single nucleotide changes) to recombination (reciprocal or unidirectional exchanges between sequences) as well as simulations in

---

[i] http://www.researchgate.net/publication/320554333_Crossovers_generate_non-random_

which both point mutations and recombination were allowed to take place. Various degrees of selection were simulated by selecting sequences for the presence of codons for an arbitrary amino acid (lysine), and then randomly replicating the selected sequence fractions to restore the original populations in each round.

Interested readers are urged to download and review the original paper.[j] Briefly, these simple simulations (carried out on an antiquated desktop computer with a tiny fraction of the computing power of a smartphone) demonstrated why and how, contrary to the precepts of the "Modern Synthesis", recombination can represent a potent creative force in Natural Selection. In fact, the combined effects of recombination and point mutation (along with other sources of DNA sequence variation) play a variety of synergistic roles in actual DNA sequence evolution that has been thoroughly documented in numerous empirical studies (Chapter 16).

---

[j] https://www.researchgate.net/publication/320554333_Crossovers_generate_non-random_recombinants_under_Darwinian_selection

Chapter 14

The Dual Character of Complex Adaptations

While the current neo-Darwinian theory is valid in its proper domain, it represents an abstraction from a much larger implicate and generative order, and its main significance is to be found in its relationship to this latter order...chance is assumed to play an absolute role in evolution as well as in the origin of life...[However,] major evolutionary changes require the coordinated development of many different pieces of the "genetic code".

—David Bohm and F. David Peat [106]

The Big Picture

The relationship between genotype and phenotype is a multifaceted question that has dominated evolutionary theory throughout the 20th century and up to the present day. Genotype refers to the genetic determinants, whereas phenotypes refers to the actual organization, the structures and functions that are determined.

In most cases, the mapping between genes and biological structure and function cannot be understood solely by studying genes in isolation. Multiple genes and gene products must interact in complex ways to generate diverse biological structures and functions. The result is an organized whole which can be called the phenotypic gestalt.

But it is not enough to recognize the phenotypic whole. For multicellular organisms, it is also essential to realize that many genetic determinants reproduce biological organization during the process of development. This involves modular genetic elements that are modified and reused in various ways. The metaphor of a developmental toolkit is useful in this context.

To help clarify the distinction between actual phenotypes that result in particular individuals in particular ecological contexts, it is useful to introduce some new terms that distinguish toolkit elements from actual phenotypes. The former are referred to as the Generative Phenotype, and the latter are referred to as the Ecological Phenotype.

Introducing the Phenotypic Gestalt

Evolutionary forces such as Natural Selection preferentially select for individual characteristics that promote survival and reproduction. Development of the animal embryo reproduces the shared characteristics of the species and the unique set of variations inherited by that individual. The fertilized egg divides into a multitude of cells that sequentially become more specialized. The cells arrange themselves so that they are juxtaposed into layers (the **germ layers**) and coordinate their activities in a robust fashion to generate functional multicellular structures and functions.

We refer to the variable determinants inherited by each individual as its individual **genotype**. The resulting structures, functions, and capabilities of the developed individual are called the individual **phenotype**.

The result is a **phenotypic gestalt**: an organized, multicellular whole with emergent properties and levels of complexity that are greater than the sum of its parts.

Darwin's classical concept of Natural Selection described evolutionary change as a gradual process involving selection of slight, incremental changes that are useful. Over long periods of time, small changes add up to the larger, visible changes that serve to adapt various species to particular ways of life.

But how, exactly, do we reconcile our detailed 21st century knowledge of the molecular and cellular determinants of development with the concept of accumulated, incremental change? Do these developmental determinants gradually change in a concerted, incremental fashion? That might account for phenotypic changes such as the gradual elongation of the limbs, perhaps, but it does not address the ways that the cells cooperate to generate the overall body plans of diverse species, nor can it tell us much about the relationship between genetic determinants and the developmental process. When it comes to linking developmental processes to classical Darwinian selection, we find that there is a disconnect between the description of the final shape, form, and capabilities of the developed individual, and the concerted actions of the cellular mechanisms that reproduce that complex organization in each generation.

Drawing a distinction between Ecological and Generative Phenotypes

For this reason, I propose that we recognize the distinction between the phenotypic gestalt of the fully-developed individual and the cellular mechanisms and toolkits that generate that phenotypic gestalt. To help clarify this distinction, I suggest that we refer to the phenotypic gestalt—the integrated product of development—as the **Ecological Phenotype**, and the molecular, cellular, and developmental determinants of that phenotype as the **Generative Phenotype**.

Why Darwin focused on the Ecological Phenotype

Darwin was naturally focused on the Ecological Phenotype rather than the Generative Phenotype, because molecular, cellular, and developmental biology had not yet come of age. This begs the following question: at what level does Natural Selection act? Does it act on the Ecological Phenotype, the Generative Phenotype, or both?

Clearly, Natural Selection must act on the products of development—that is, the capabilities and adaptations of developed individuals that adapt them to a particular way or life or ***niche***—but *also* must act on the generative processes that organize the multicellular individual, and that can lead to distinctions in their complex shapes, forms, and capabilities.

Unlike the Ecological Phenotype, the Generative Phenotype refers to both **EEP** (Chapter 2) and **SSM-IBE** (Chapter 12), and the ways that the metaphorical blind watchmaking, tinkering, Rube Goldberg machines, genomic scrapyards and developmental toolkits, black boxes, opportunism, and trial-and-error (Chapter 13) link genetic determinants to the reproduction of complex organization.

The short-term selection of the Ecological Phenotype is a real-time generational phenomenon

The Ecological Phenotype determines the **fitness** of each individual during its lifetime—its ability to survive and reproduce its own kind—in a particular environmental ***niche***. The Ecological Phenotype of each individual is under selective pressure that is determined by interactions with its particular environmental circumstances. This selective pressure involves actual real-time events that take place throughout its lifetime. This is the relatively short-term time frame of selection that takes place in each generation. But selection is not limited to the real-time ecological events that affect the life of the individual, and its ability to reproduce its own kind. Selection also acts upon the generative genetic determinants responsible for multicellular biological organization.

The long-term selection of the Generative Phenotype transcends generations (and therefore transcends time and space)

Selection also acts in a long-term way on each **lineage**. Each individual that has ever lived has a chain of ancestry consisting of both ancestors and descendants—unless it has gone extinct. Sometimes genetic information may also be shared *via* **horizontal gene transfer**. In addition to active protein-coding and/or regulatory DNA sequences that are transferred from generation to generation, a larger set of DNA sequences—the genomic scrapyard—is also transferred. Developmental determinants—toolkit elements described by evolutionary developmental biologists—are also transferred. These DNA sequences and toolkit elements have enormous **EEP** that affect the long-term capabilities, adaptations, and survival of every lineage. Often, developmental toolkit elements exhibit **deep homology**—highly conserved elements that are shared by even distantly related species, and that are often deployed in novel ways. This topic is further discussed in Chapter 15.

The long-term survival of any lineage depends on the availability, flexibility, robustness, and efficacy of the Generative Phenotype. Long-term capabilities and survival, in other words, depend on the **evolvability**, and a big part of the evolvability contributed by the **genomic scrapyard** and redeployment of developmental toolkit components can be attributed to the Generative Phenotype. The evolutionary significance of the Generative Phenotype involves EEP and transcends the real-time selective pressures on individuals, and extends over a multitude of generations.

Beyond group selection

The concept of the Generative Phenotype calls for a broader appreciation of the object of selection. In the past, passionate debates among proponents of the "Modern Synthesis", concerning the validity of the group selection, have persisted for decades. Today, there is widespread acceptance of the concept.[a,b] In the UES, we need to go *beyond* the concept of group selection, because EEP calls for a broader understanding of the nuanced ways in which selection acts over many generations.

Also, among prokaryotes, the concept of the individual genome, the individual gene pool, and even the concept of a species, have been supplanted by widespread acceptance of horizontal gene transfer and group phenomena such as quorum sensing and biofilms.

Since genetic information is regularly shared among multiple species, and since individual fitness often depends on group interactions and group behaviors, the question of whether Natural Selection acts at the level of individuals, populations, or species is rendered moot.

What about among multicellular eukaryotes? In Chapter 2, the transcendent quality of emergent evolutionary potential (EEP) was discussed. EEP is not limited to actual events that take place in time and space, but also includes constantly changing logical relationships that transcend space and time.

---

[a] https://en.wikipedia.org/wiki/Group_selection

[b] https://blog.oup.com/2015/01/kin-group-selection-controversy/

The Generative Phenotype is not only subject to the consequences of EEP, but provides illustrative and perhaps quintessential examples of it. Developmental mechanisms and the genomic scrapyard represent toolkits that are reused in various ways in various lineages. The flexibility, robustness, and efficacy of developmental mechanisms and their genetic determinants—and their ability to be reused in different ways—can make the difference between long-term survival of highly branched lineages over many millions of years and rapid extinction. Extinction is a normal aspect of evolution, and extinct species greatly outnumber extant ones. Part of the reason for this is that the Generative Phenotype affects evolvability. Evolvability represents an aspect of Natural Selection that transcends individuals, populations, and species, but cannot always respond quickly enough to rapid environmental changes—such as those caused by human activities. Global warming is, for example, a significant factor in the recent bleaching of coral reefs on a global scale.[c,d]

Co-option and exaptation are just two of the widely recognized aspects of deep homology that are now central concepts in evolutionary developmental biology. The Generative Phenotype represents EEP that is certainly subject to Natural Selection, but is more transcendent than the Ecological Phenotype.

Why distinguish between the Ecological Phenotype and the Generative Phenotype?

The Ecological Phenotype refers primarily to the fitness of the phenotypic gestalt of the developed individual. The efficacy of the structures and functions of the individual, within a particular *niche*, affects its survival and reproduction. This usually involves actual environmental events that take place following the embryonic development of the individual.

The Generative Phenotype, in contrast, refers primarily to the ways that toolkit elements are deployed in particular individuals, to produce that gestalt. This distinction helps to clarify the relationship between evolution, development, and the phenotype of the individual.

---

[c] https://en.wikipedia.org/wiki/Environmental_issues_with_coral_reefs
[d] https://www.chasingcoral.com/

Both concepts help us to understand both the short-term, real-time events as well as longer time frames responsible for evolution. These various time frames have long been recognized by evolutionary theorists in the context of the "Modern Synthesis". But prior to our present-day understanding of the molecular, cellular and developmental basis of the Generative Phenotype, the longer time frame of Natural Selection has been enigmatic. The **Updated Evolutionary Synthesis (UES)** provides a deeper understanding, primarily thanks to the new and active field of **evolutionary developmental biology (evo-devo)**. This is elaborated further in Chapter 15.

The GA demonstrates the evolutionary significance and origins of the phenotypic gestalt

As discussed in Chapter 13, the **Genetic Algorithm** (**GA**) is a software algorithm that emulates Natural Selection, because selection is blind and path-independent: it does not depend on the way that the combinatorial parameters are recombined. Similarly, biological selection acts on the final combinatorial product, the phenotypic gestalt, the product of numerous interacting elements of the Generative Phenotype. Thanks to **SSM-IBE**, the generative elements of the biological phenotype have extraordinary organizing power. Multiple variables contribute to the gestalt that is the target of selection. The multiplicity of variables is *especially* pronounced when Natural Selection acts on determinants that influence development, simply because the molecular, cellular, and developmental toolkits—that is, the numerous contributors to the Generative Phenotype—are complex, redundant, and robust. In virtually every case involving the Ecological Phenotype, there are numerous *overlapping*, *synergistic*, *cooperative*, and *combined* molecular, cellular, and developmental interactions that ultimately contribute to the Ecological Phenotype.

Given the sheer complexity and diversity of elements that contribute to the Ecological Phenotype—and that are part of the Generative Phenotype—it is truly remarkable that we are able to understand as much as we do about the detailed, concrete mechanisms that are involved. That is because biologists have, over many decades, developed extremely powerful empirical techniques for isolating, tracking, and detecting specific interacting components. This is routinely accomplished in literally thousands of biology laboratories throughout the world, on a daily basis.

One of Albert Einstein's most famous quotations—and he had many—is often rephrased[e] as follows:

> The most incomprehensible thing about the universe is that it is comprehensible.

Our ability to understand the universe is astounding. It's important to acknowledge, however, that this comprehensibility has a historical component. In the biology lab, the reason that it is comprehensible is a triumph of the scientific method, of naturalism, and the professionalism of the thousands of unsung scientists who have developed and tirelessly applied powerful empirical techniques, careful judgment, logic, and reason to their experiments and observations, both in the laboratory and in the field.

The processes of fertilization, cleavage, gastrulation, and organogenesis are all high-level cellular descriptions of general stages in the reproduction of complexity by the Generative Phenotype. Cytoplasmic determinants, morphogens, and transcription factors are all involved in molecular interactions required for complex organization to arise in the multicellular embryo (see Chapter 10).

For *Rethinking Evolution* and for the Updated Evolutionary Synthesis (UES), the problem is not that there is not enough detailed empirical information about developmental mechanisms, and the evolution of those mechanisms. The problem is that there is *far too much* detailed information to do it justice. In Chapter 15, we will explore the powerful synergy between developmental and evolutionary biology.

---

[e]Clever rephrasing has probably improved the original quotation with a stickier meme. The original quotation reads, "The eternal mystery of the world is its comprehensibility… The fact that it is comprehensible is a miracle". See https://primemind.com/we-just-cant-stop-misquoting-einstein-19ad4efab26e

Chapter 15

The Creative Forces Behind Evolution, from Darwin to Evo-Devo

Neither Natural Selection nor DNA directly explains how the individual forms are made, or how they evolved. The key to understanding form is development, the process through which a single celled egg gives rise to a complex multi-billion celled animal. This amazing spectacle stood as one the great unsolved mysteries of biology for nearly two centuries. And development is intimately connected to evolution, because it is through changes in embryos that changes in form arise.

The first and still perhaps the most stunning discovery of Evo Devo is the ancient origin of the genes for building all sorts of animals.... Existing genes and structures provide the means for innovation.

—Sean B. Carroll [15]

The Big Picture

Darwin emphasized the origin of species, but the Updated Evolutionary Synthesis (UES) also emphasizes the reproduction of complex biological organization during development.

In multicellular organisms, the storage, transmission, and reproduction of biological organization depends on the inner-workings and coordinated interactions of multitudes of cells. A deep understanding of evolution requires a basic working knowledge of the molecular and cellular interactions that take place during development.

Much of the meaning and significance of the genetic determinants that accumulate by means of Natural Selection is manifested in the useful structures and functions that they generate during embryonic development.

The Creative Forces of Natural Selection

An overarching question of evolutionary biology is as follows: What are the creative forces of Natural Selection?

One reliable and consistent answer is that *natural* forces are responsible: for the origin of species, for accumulation of potentially useful variation, for the efficacy of diverse adaptations, and for the reproduction of complex biological organization in each generation.

This answer extends throughout all domains of life and applies to the biological evolution of human beings as well.

Although the inner-workings of cells and cellular interactions were largely a black box to Darwin and his contemporaries—which he freely acknowledged—he was still able to articulate the theory of Natural Selection which helps to explain many aspects of the inner workings of cells that were subsequently discovered. Natural Selection is a principle and universal factor in biological organization and the origins of species and diverse adaptations. Therein lay his genius. Metaphorically, these natural forces can be viewed as trial-and-error on a grand scale, throughout the living world.

Darwin had a clear and scientific description for the key elements of the natural selective process, thoroughly grounded in empirical facts. Briefly, individuals inherit factors that determine the characteristics of each species, but those hereditary factors vary among the individual offspring in each generation.

Each individual inherits a slightly different set of determinants during sexual reproduction. Darwin freely acknowledged that variation was a black box to 19th century biologists. Today, we would say that such variation consists of sets of alleles—slightly different DNA sequences inherited by the individual. Today, we also recognize that differences are not limited to sequences of DNA that determine the sequences of amino acids in proteins, but also include a variety of sequences that control the synthesis of those protein chains—which we collectively refer to as regulatory sequences.

Variation and randomness

The genesis of the confusing idea that variation is *random*—and the roots of the unfounded conclusion that variation does not play a creative role in Natural Selection—can be traced to "Modern Synthesis", and to well-meaning but misguided efforts to defend Natural Selection against creationist attacks which held that supernatural design, goal-seeking, and intelligence play creative roles.

In the light of well-established principles of sexual reproduction, we can easily clear up some of this confusion by distinguishing between two types of variation.

The first type is the variation that shuffles segments of the parental DNA sequences during sexual reproduction. This shuffling arises from essentially random segregation of parental chromosomes during the meiotic production of sex cells, as well as the largely random recombination events between those parental chromosomes, which also take place during the production of sex cells. Another source of randomness occurs at the moment of fertilization, because only one of numerous possible sperm cells will fertilize only one of numerous possible egg cells.

This first type of variation among individual offspring—**random shuffling of genomic sequences**, and **random fertilization**—is properly viewed as random variation, because the probability of inheritance of any given DNA sequence segment is a function of mathematically calculable probability and chance. This is a blind, undirected natural process that is unrelated to the potential utility or selective value of the possible combinations of genetic determinants that may result.

But there is also a second concept of variation that arose at a time when heredity was attributed to units known as genes and alleles, but before their molecular nature were understood. Changes in these genes, resulting in new alleles, were generally described as **mutations**, and mutations were thought to be random events that are unrelated to the needs of the individual [37, 105].

This second type of variation must be distinguished from the random shuffling of preexisting DNA sequences during sexual reproduction, because this type of variation includes modifications, additions and/or deletions in the pool of available genomic DNA sequences. Unlike the first type of variation, changes in the gene pool will potentially apply to several subsequent generations, rather than one. They are a source of novelty that is essentially *different* than random shuffling during sexual reproduction.

Today, instead of using the term **mutation** to describe this second type of variation, we can more accurately describe the range of possible sequence variations collectively as modification, addition or deletion of genomic DNA sequence elements.

Two key questions distinguish the first type of variation from the genomic sequence elements: are they random, and are they related to their potential utility as adaptations in the struggle for existence? Random shuffling of genomic sequences and random fertilization—the first type of genetic variation—are correctly viewed as random and unrelated to their potential utility.

But the second type of variation—modification, addition or deletion of genomic sequence elements—although it has some random characteristics—is often nonrandom, and is often related to its possible utility. One reason is that Natural Selection acts on both **Generative Phenotypes** and **Ecological Phenotypes** (Chapter 14). Generative Phenotypes include toolkit elements that are modular in nature and that are frequently modified and/or reused in different ways during development. Developmental evolution, which uses toolkit elements, has nonrandom aspects.

Another reason is that the new layers of complexity that can emerge depend on previous layers of complexity. New innovations create new opportunities for new kinds of emergent interactions. This has been previously discussed in the context of EEP (Chapter 2).

A third reason is that when evolution metaphorically "invents", by trial-and-error, a useful entity, it is likely to be useful in a variety of different ways—in a variety of different Ecological Phenotypes. For example, a general amino acid sequence that creates a specific pocket where reactants (substrates) can bind is likely to be useful—often with slight modifications—in a variety of enzymes. A sequence motif that targets newly synthesized protein chains so that they readily insert into, and pass through, biological membranes is likely to be useful for numerous **integral membrane proteins**, such as embedded **cell surface receptors** involved in **signal transduction pathways**. Another broad example, **gene duplication** and divergence and **multigene families**, will be discussed in Chapter 16.

A fourth reason is that evolution will favor robustness, flexibility, and evolvability, and that means that genomic sequence elements which contribute in that way *do* have relationships to their possible utility.

A fifth reason is that in general, exaptation (using innovations for diverse Ecological Phenotypes that are all useful in their own contexts) and co-option are frequently observed. A sixth reason is that certain mechanisms that generate genomic variation—such as repetitive sequences which can expand, diversify, and increase the frequency of recombination events, are often hotspots for genomic changes, and introduce a nonrandom aspect into the variation events that may subsequently occur.

To summarize, the second type of variation—modification, addition or deletion of genomic sequence elements—whether they determine protein sequences or help regulate their expression—are a type of variation that has nonrandom aspects that *can* relate to its potential utility. This should be seen as a major departure from the earlier concept of random variation in the "Modern Synthesis". This is an important change in the UES. Genomic sequence elements *do* play various roles in the creative aspect of Natural Selection. They also help explain how a blind, undirected and natural process can have the extraordinary creative organizing power that it does. In other words, the UES is more plausible than ever before.

Beyond "housekeeping" genes

Apart from the "housekeeping" genes that keep all cells alive, most evolutionary adaptations of multicellular plants and animals depend on the expression of more specialized genes that control cellular and molecular interactions during development. This adds a major dimension to evolutionary biology that goes beyond classical Darwinian Natural Selection, beyond Mendelian Genetics, and beyond Molecular Genetics. The revolutionary fusion of developmental biology with evolutionary biology is called **evolutionary developmental biology** (**evo-devo**), a concept that was discussed to some degree in previous chapters.

The Molecular Genetics of Development

Classical Darwinian selection focused on what is now called the phenotype—specifically, the fully-developed structures and functions that organize the tissues and organs and shape and form of the individual in ways that facilitate survival and reproduction in particular *niches* or ways of life. In Chapter 14, I argued that this should be called the Ecological Phenotype, to distinguish it from developmental toolkits, which also evolve, which I called the Generative Phenotype.

A general trend in evolutionary theory that has taken place since Darwin has been the expansion of evolutionary theory with an increasingly detailed, empirically-based understanding of the inner-workings of interacting cells. None of this invalidates the core framework of metaphorical trial-and-error that Darwin reported in the first edition of *On the Origin of Species* in 1859. But we have seen profound changes in our understanding of hereditary variation.

For plants and animals, however, the most important way that variation in Ecological Phenotypes is manifested is through genetic determinants that control the process of development. Development generates biological organization. In his introduction to *The Making of a Fly* [12], Peter Lawrence puts it this way:

> In my opinion, the heart of development is the step by step allocation of cells to more and more precisely determined fates. These allocations have to occur in the right part of the embryo. This key process can be broken down into two questions: How are the cells chosen by position? What happens inside a cell when it becomes allocated to a specific fate; that is what are the molecular and genetic changes? A good deal of progress has been made in answering these questions.

From cytoplasmic determinants in a single cell to specialized lineages of interacting cells

Focusing on animals, it is convenient to begin our analysis of development, in each generation, with the fertilized egg. However, a critical aspect of development takes place at an earlier time, at a time when the eggs are being produced by the mother. **Cytoplasmic determinants** in the egg are primarily encoded by the maternal genome. These **Generative Phenotypes** are known as **maternal effects.** They include a variety of RNA and protein molecules, including **transcription factors** and **morphogens**.

Bicoid was the first protein identified as a morphogen, and is expressed early in the development of the *Drosophila* embryo.[a] *Bicoid* is a morphogen[b] because it is expressed as a **gradient** along the long axis of the egg, and plays a major role in determining the various regions along the anterior–posterior axis of the developing embryo. *Bicoid* acts as a transcription factor that binds both RNA and DNA targets using its **homeodomain**[c] to regulate their transcription and translation, respectively. *Bicoid* is one of the most important discoveries made by Nüsslein-Volhard and colleagues during their Nobel Prize-winning analysis of mutations that are lethal to embryos and had therefore been previously missed. Development of techniques that made it possible to visualize the *Bicoid* protein gradient in developing embryos is one important example of the ways that advances in empirical techniques have advanced our understanding of both development and evolution. The *Bicoid* protein drives the **differentiation** of embryonic cells along the anterior–posterior axis, and its homeodomain and nucleic acid targets exhibit the deep homology that is one of the most remarkable general principles of evo-devo. More recently, other factors that synergize with *Bicoid* have been described [128].

Cytoplasmic DNA and cleavage

More recently, a powerful gene editing technology known as **CRISPR**[d] has made it possible to trace the entire lineage of the mouse embryo [107]. Briefly, the technique creates a sort of "genetic barcode" that allows investigators to trace the specific patterns of cell division and differentiation that give rise to a multitude of highly specialized cells during embryonic development.[e]

---

[a] https://en.wikipedia.org/wiki/Bicoid_(gene)

[b] https://en.wikipedia.org/wiki/Morphogen

[c] https://en.wikipedia.org/wiki/Homeobox

[d] https://en.wikipedia.org/wiki/CRISPR

[e] https://news.harvard.edu/gazette/story/2018/08/genetic-barcodes-can-record-every-cells-history-in-real-time/

Gastrulation

As implied by the tongue-in-cheek quotation by Lewis Wolpert at the beginning of Chapter 11, **gastrulation** marks an important milestone in life, not only for us humans but for the rest of the animal kingdom as well. For both vertebrates and invertebrates with a central nervous system and an anterior–posterior axis of development, gastrulation establishes the general "tubes within a tube" architecture of the individual. In vertebrates, one of those tubes is the brain and spinal cord, while the other consists of the organs of the digestive system.

Formation of the body plan and organs

Long before the advent of Molecular Genetics and evo-devo, construction of **fate maps**, along with detailed, step-by-step descriptions of the developing tissues of the embryo, were a principal activity for embryologists both in the laboratory and in the classroom. Gastrulation brings particular layers of embryonic cells, known as the germ layers, into proximity, where they communicate extensively in ways that coordinate the gene expression of neighborhoods of cells to produce complex shapes, forms and functional capabilities in tissues and organs. One of the rights of passage of premedical students that continues to this day is to learn the highly-detailed anatomy and physiology of developing embryos—not only of human beings [108], but also of other vertebrates that can be studied in the laboratory.

As an undergraduate student at UC Berkeley, I have vivid memories of the Embryology lectures presented by Richard Eakin.[f] In his well-attended presentations in one of Berkeley's largest lecture halls at the time, Professor Eakin would astound hundreds of us as he would draw, entirely from memory, the most beautiful and detailed pictures describing the various stages of embryogenesis, using a different color of chalk for each of the three germ layers. Beginning with the fertilized egg and then proceeding through gastrulation to organogenesis, Professor Eakin used blue for developing tissues and organs derived from **ectoderm**, red for **mesoderm**, and yellow for **endoderm**.

---

[f]https://en.wikipedia.org/wiki/Richard_M._Eakin

The cellular tools that generate shape, form, and pattern

Developmental biology in general, and evo-devo in particular, are highly interdisciplinary fields that rely on a broad range of specialized research in the biological sciences. For example, since intercellular communication and numerous cellular activities contribute to the Generative Phenotype, and ultimately to the Ecological Phenotype of each individual, many developmental studies fall under the broad rubric known as **cell biology**.[g] The development of shape and form—including gastrulation and the formation of the neural tube, as well as critically important migrations of populations of embryonic cells to particular locations (such as the migration of the **germ line** cells to the developing gonads) involve studies in **cell motility** (changes in shape and form and movements dependent on cellular adhesion and locomotion of cells). All of these aspects of development represent tools that contribute to the Generative Phenotype, and subtle changes in the regulation of gene expression can lead to dramatic differences in the specific Ecological Phenotype that results in a particular species. Cells provide a remarkably robust and flexible toolkit, and the functional efficacy of the coordinated multicellular organs produced during development is breathtaking. All of these remarkable results, however, are consistent with, and can be explained by, the principles of Natural Selection, EEP, and SSM-IBE that have been discussed in previous chapters.

Induction, receptors, and signal cascades

When embryonic cell layers communicate and coordinate their activities to generate unified patterns, they do so by means of a variety of mechanisms, and many aspects of such intercellular communication and signaling are now well understood at a molecular and cellular level.[h] They involve such mechanisms as diffusible proteins, cell-surface receptor proteins, and elaborate signaling pathways that ultimately result in coordination of patterns of cell division, motility, and gene expression.

[g] https://en.wikipedia.org/wiki/Cell_biology

[h] https://en.wikipedia.org/wiki/Cell_signaling

The metaphorical tinkering and Rube Goldberg devices that allow such complex pathways to evolve were discussed in Chapter 12.

Evo-Devo Comes of Age: The Molecular Genetics of Fruitfly Development as a Model System

Ed Lewis takes fruitfly genetics to its premolecular, Mendelian limit

Building on the stunning success of Thomas Hunt Morgan, one of his students and colleagues, Calvin Bridges, discovered the *Bithorax* mutation, which, as the name implies, causes the fruitflies (***Drosophila***) to develop a duplicated thoracic region. Ed Lewis took Mendelian Genetics to the limit when, starting with the Bithorax mutation, he discovered and fully characterized what was arguably the most remarkable and historically important complex locus in the animal kingdom: the **Bithorax Complex** (**BX-C**)[i] (Figure 1).

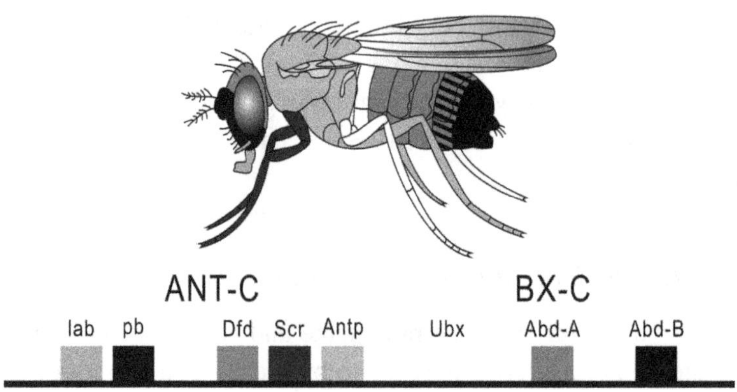

Figure 1. Diagram of the body segmentation and appendages of the adult fruitfly, showing the specific regions of *Hox genes* of the *Bithorax Complex* (BX-C) and *Antennapedia Complex* (ANT-C) that map to the phenotypes of specific segments and appendages that they determine.

Source: Page-link: https://commons.wikimedia.org/wiki/File:Hoxgenesoffruitfly.svg. File-link: https://upload.wikimedia.org/wikipedia/commons/d/da/Hoxgenesoffruitfly.svg. Attribution: By PhiLiP (self-made, based on Hoxgenesoffruitfly.png) [Public domain], via Wikimedia Commons. Rendered in B&W.

---

[i] https://en.wikipedia.org/wiki/Bithorax_complex

Lewis' pioneering work led to the discovery of the ***Hox* genes**[j] that determine the body plans of animals ranging from **invertebrates** such as nematode worms and fruitflies to **vertebrates** ranging from fishes to mammals, including mice and human beings (Figure 2). ***Hox* genes** quickly became the quintessential textbook example of **developmental toolkits** that exhibit **deep homology** throughout the animal kingdom.

Ed Lewis shared the Nobel Prize with Christiana Nüsslein-Volhard and Eric Wieschaus. With their colleagues, these brilliant pioneers took Lewis' work, which relied on experimental techniques that extended the work of Morgan and colleagues, laid the groundwork for the subsequent **Molecular Genetic** studies that ushered in the age of evolutionary developmental biology (evo-devo).

Nüsslein-Volhard and the making of a fly (or a fish, or a mouse, or a human)

Evolutionary theory was taken to a whole new level when they used Molecular Genetics—a powerful combination of techniques from both Mendelian Genetics and Molecular Biology—to perform a systematic screen for **embryonic mutations** in the fruitfly (*Drosophila*). These genes had been previously overlooked by other investigators because they cause mutations that are often lethal in early development, and are therefore easily missed in genetic screens [30].

Techniques for in situ visualization of patterns of gene expression

In the biology laboratory, *in situ* means to examine a phenomenon in the place where it occurs.[k] For example, visualization of a **morphogen**[l]—a substance whose nonuniform distribution governs the pattern of tissue development during embryogenesis—generally requires *in situ* imaging techniques that target specific proteins or RNA sequences and make them visible. The development of *in situ* imaging of morphogens in fruitfly

[j] https://en.wikipedia.org/wiki/Hox_gene

[k] https://en.wikipedia.org/wiki/In_situ#Biology_and_biomedical_engineering

[l] https://en.wikipedia.org/wiki/Morphogen

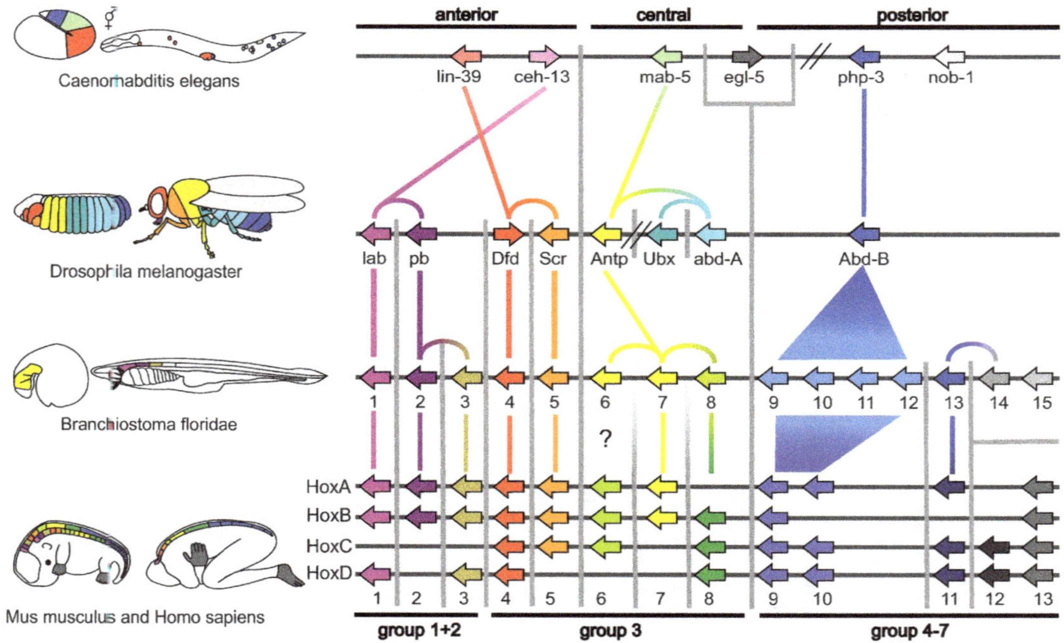

Figure 2. Diagram illustrating how the genomic arrangement of highly conserved *Hox* gene complexes map onto the body plans and segmentation of the animals whose phenotypes they help determine. Highly conserved genes that play major roles in the development of a broad range of distantly related organisms are examples of developmental "toolkit" genes that exhibit "deep homology".

Source: Page-link: https://commons.wikimedia.org/wiki/File:Genes_hox.jpeg. File-link: https://upload.wikimedia.org/wikipedia/commons/0/0c/Genes_hox.jpeg. Attribution: Stefanie D. Hueber, Georg F. Weiller, Michael A. Djordjevic, Tancred Frickey [CC BY 4.0 (https://creativecommons.org/licenses/by/4.0)], via Wikimedia Commons.

embryos, for example, was a turning point that linked Molecular Genetics to development, and paved the way toward major advances in evo-devo [30].

Transcription factors and morphogens

Nüsslein-Volhard and colleagues applied several powerful new techniques. Briefly, they made it possible for the first time to visualize the expression patterns of several key RNA and protein products, including **transcription factors**[m] and **morphogens.**[n] The patterns of expression of these genetic determinants play essential roles in the segmentation and morphogenesis (development of pattern and form) of the embryo. This includes the overall formation of the body plan, segmentation, appendages and local patterning elements of differentiated cells, in fruitfly embryos as well as the bristle patterns and appendages of fruitfly larvae and adults.[o] One of these was already introduced in the section on *Bicoid* above.

Another important example, also discovered by the team headed by Nüsslein-Volhard, was the discovery of **gap genes** [30].[p] Gap genes (Figure 3) encode transcription factors that determine segment formation in the embryo, by directly controlling the expression of **pair-rule genes.**[q]

The "Modern Synthesis" Was Incomplete

Evolution of biological organization depends on development

Sean B. Carroll, in his *Endless Forms Most Beautiful* [15] notes two classically Darwinian concepts in Julian Huxley's *Evolution: The "Modern Synthesis"* [23]:

> First, that gradual evolution can be explained by small genetic changes that produce variation, which is acted upon by natural selection. Second, that evolution at higher taxonomic levels, and of greater magnitude, can

[m] https://en.wikipedia.org/wiki/Transcription_factor

[n] https://en.wikipedia.org/wiki/Morphogen

[o] https://en.wikipedia.org/wiki/Drosophila_embryogenesis

[p] https://en.wikipedia.org/wiki/Gap_gene

[q] https://en.wikipedia.org/wiki/Pair-rule_gene

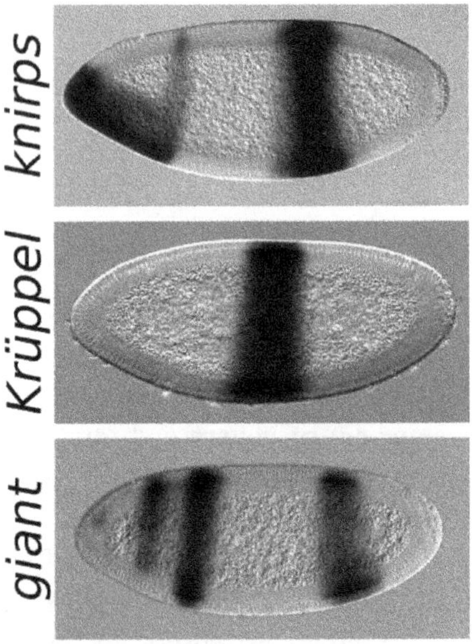

Figure 3. Retouched image of *in situ* hybridization against mRNA of the gap genes *knirps, Krüppel* and *giant* in the early embryo of the fruitfly (*Drosophila melanogaster*) at the stage when new membranes form around nuclei, known as cellularization.

Source: Page-link: https://commons.wikimedia.org/wiki/File:Drosophila_gap_gene_in_situ.png. File-link: https://upload.wikimedia.org/wikipedia/commons/1/13/Drosophila_gap_gene_in_situ.png Attribution: Drosophila_gap_gene_in_situ_brakeless_mutant.png: Haecker A, Qi D, Lilja T, Moussian B, Andrioli LP, Luschnig S, Mannervik, M derivative work: Celefin [CC BY 2.5 (https://creativecommons.org/licenses/by/2.5) or CC BY 2.5 (https://creativecommons.org/licenses/by/2.5)], via Wikimedia Commons Rendered in B&W.

> be explained by these same gradual evolutionary processes, sustained over longer periods. The "Modern Synthesis" established much of the foundation for how evolutionary biology has been discussed and taught for the past 60 years [up to 2006].

Then Carroll succinctly states the problem with this perspective:

> However, despite the monikers of "modern," and "synthesis," it was incomplete. At the time of its formulation and until recently, we could say that forms do change, and that natural selection is a force. But we could say nothing about how forms change…

Mutations have nonrandom features

As will be discussed in more detail in Chapter 16, a variety of mechanisms generate nonrandom changes in genomic DNA sequence elements. If genetic mutations are defined as changes in DNA sequences that are passed on to future generations, then mutations are not simply random changes. More importantly, mutations such as gene duplication events—which are quite common in molecular evolution—create duplicates of genetic elements that have already been subjected to Natural Selection, and have already been proven useful in previous contexts. These duplicates therefore have significant nonrandom aspects. Duplicated genes can and do change by means of classical Natural Selection and generate new structural and functional capabilities. But they begin their evolutionary history in a nonrandom fashion. The "Modern Synthesis" perspective on mutations as purely random events must be tempered with a broader, 21st century perspective that recognizes such nonrandom aspects. Reuse of useful genomic sequence elements implies conservation and/or duplication of those elements, rather than creation from scratch.

The deep homology of genes that determine major features of animal shape, form, and capabilities

In *Endless Forms Most Beautiful* [15], Carroll provides an accessible account of some of the major new principles that evo-devo has added to evolutionary theory:

> Contrary to the expectations of any biologist, most of the genes first identified as governing major aspects of fruit fly body organization were found to have exact counterparts that did the same thing in most animals, including ourselves.

This was not expected, because biologists had assumed that different animals would be genetically constructed in different ways. In the heyday of the "Modern Synthesis", selection was seen as the sole creative force in Natural Selection, and variation was assumed to be random, as discussed earlier in this chapter. If mutations are random and selection is the creative force, then separate gene pools would be expected to evolve according to separate ecological selective pressures, but the ways in which they would change—the particular set of accumulated, random mutations that take place—should have separate histories.

Continuing with some highlights from *Endless Forms Most Beautiful* [15], Sean Carroll points out that:

> The development of various body parts such as eyes, limbs, and hearts, vastly different in structure among animals and long thought to have evolved in entirely different ways, was also governed by the same genes in different animals.

Highly conserved genetic elements that determine development are examples of what is often referred to as deep homology (Figure 4), an important new perspective for evolutionary theory [109–111].

Modularity

Another important principle highlighted in *Endless Forms Most Beautiful* [15] is that animals have modular architecture: they are constructed from metaphorical building blocks:

> A basic theme of animal design becomes obvious when one tries to figure out just what bone or tooth one has found in that shovelful of Florida river gravel. The challenge of the game is both to match the fossil to a species, and also to determine where in the animal it belonged. Why is this so hard? This is one demonstration of a basic fact of animal design. Related animals, such as vertebrates, are made up of very similar parts.

Other questions addressed by evo-devo

Some of the other major questions that can now be addressed with empirical include answers to the following questions, as summarized by Carroll [15]:

(1) What are some of the major "rules" for generating animal form?
(2) How is the species-specific information for building a particular animal encoded?

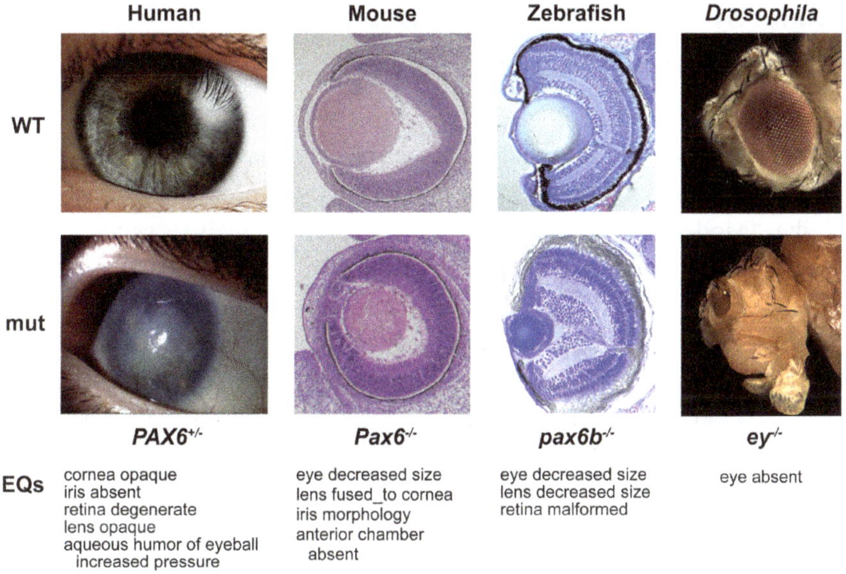

Figure 4. The *Pax* gene family is an example of a developmental "toolkit" gene that demonstrates a high degree of conservation between distantly related animals ranging from fruitflies to humans. Many genes that show such "deep homology" control the phenotypes of homologous segments or organs with related functions, by regulating the expression of multiple genes. This results in distinct phenotypes, but may play similar ecological roles, such as vision in the case of the eye.

Source: Page-link: https://commons.wikimedia.org/wiki/File:PAX6_Phenotypes_Washington_etal_PLoSBiol_e1000247.png. File-link: https://upload.wikimedia.org/wikipedia/commons/1/14/PAX6_Phenotypes_Washington_etal_PLoSBiol_e1000247.png. Attribution: By Washington NL, Haendel MA, Mungall CJ, Ashburner M, Westerfield M, Lewis SE. [CC BY 2.5 (https://creativecommons.org/licenses/by/2.5)], via Wikimedia Commons.

(3) How does diversity evolve?
(4) What explains large-scale trends in evolution, such as the change in number and function of repeated parts?

In addition to deep homology, another important general answer can be found in our newly acquired understanding of regulatory elements such as *Hox* genes and transcription factors, which are briefly discussed in *Rethinking Evolution.*

Model Systems Link Genetic Determinants to the Generative Toolkit

The generative and ecological phenotypic effects of genes and alleles should not be viewed in isolation. Most aspects of development depend on complex interactions between these elements. Unfortunately, one legacy of the "Modern Synthesis" was the tendency to analyze speciation in terms of changes in the frequencies of isolated alleles in the gene pool. But isolated alleles have little to do with phenotypic complexity and change.

A broad interdisciplinary approach is required to understand how Generative Phenotypes generate ecological phenotypic gestalts. Linking Generative Phenotypes to genetic determinants requires new ways to visualize regulation of gene expression during embryonic development.

The ways that the genotypic gestalt *does* change is the subject of active and fruitful recent research. Some of this research is pursued in specific model systems that help to illustrate aspects of both evo-devo that are not easily studied in fruitflies.

Here, we'll focus on three powerful model systems that have been especially illuminating: fruitflies, butterfly wings, and zebrafish.

Although fruitflies have provided some of the earliest and deepest insights into evo-devo, other model systems are sometimes better suited to explore particular aspects of the evolution of development. For comparing the genetic determinants of visible patterns that play important roles in survival and reproduction among different species, butterfly wings have proven quite useful, as described in the following section.

Zebrafish Scales Provide a Model System for Exploring Cellular and Molecular Determinants of the Phenotypic Gestalt

Nüsslein-Volhard, who has served as Director of the Max Planck Institute for Developmental Biology in Tübingen, and leader of its Genetics Department knew, along with her colleagues, that studies of invertebrate (fruitfly) development could only go so far in describing the ways that generative toolkits are deployed in vertebrates. Not long after they published studies covering several major aspects of embryonic fruitfly development, they turned their considerable talents and hard work to laying the groundwork for extensive use of zebrafish as a model system for the Molecular Genetics of vertebrate development.

Like the fruitfly, but to a lesser extent, zebrafish offered ability to maintain strains with various key mutations, and to visualize development *in situ*. Thanks to deep homology throughout the animal kingdom, it was not difficult to find homologous DNA, RNA, and protein sequences shared by fruitflies and fish—even though these toolkits are often deployed in different ways that generate distinct phenotypes.

The molecular determinants of the 3D phenotypic gestalt in zebrafish scales

Most 3D cellular phenotypes are gestalts that are determined by several different generative tools, acting in a concerted fashion. In recent years, it has become possible to bring several lines of evidence, from many different groups, together to describe the regulation of an ancient signaling network that is activated during zebrafish scale development.

A 2018 preprint by Aman, Fulbright and Parichy[r] describes core features of this process by using a modern tool called *in toto* live imaging. Briefly, scale development requires the concerted activity of *Wnt*/beta-catenin,[s] Ectodysplasin (Eda),[t] and signaling by Fibroblast growth factor (Fgf).[u] All of these genes and gene products play widespread roles in animal development. In this case, these genetic elements form a regulatory module that coordinates collective cell migration. In turn, this cell migration depends on another widely used genetic element known as *Hedgehog* (HH).[v]

The authors suggest that a single, ancient system has been adapted in various ways and is responsible for patterning of a variety of other vertebrate skin appendages such as feathers, hairs, and ancient types of scales found in fossilized **tetrapods** (four-legged creatures).

---

[r] https://www.biorxiv.org/content/early/2018/04/02/293233

[s] https://en.wikipedia.org/wiki/Wnt_signaling_pathway; https://en.wikipedia.org/wiki/Beta-catenin

[t] https://en.wikipedia.org/wiki/Ectodysplasin_A

[u] https://en.wikipedia.org/wiki/FGF1

[v] https://en.wikipedia.org/wiki/Hedgehog_signaling_pathway

The Complex Modular Architecture of Butterfly Wings: Exploring the Evo-Devo Toolkit with CRISPR and Comparisons of Genomic DNA

Linking specific genetic determinants to molecular and cellular interactions that generate phenotypic gestalts

Although it is quite challenging to attempt to map the relationships between particular generative elements and Ecological Phenotypes, the powerful new technique of **CRISPR** has been quite revealing. CRISPR makes it relatively easy for investigators to "edit" specific genes and observe the phenotypic consequences[w] [79].

For comparing species-specific, visible patterns, butterflies are a model system of choice. For the first time, it is possible to quickly map the actions of particular genes onto the Ecological Phenotypes of multiple species. This is revealed in the beautiful and highly-variable patterns of butterfly wings.

Briefly, the CRISPR technique makes it possible to cut the genome at a desired location (using a known DNA/RNA sequence) which allows existing genes to be removed or edited, or for new genes to be added.[x] This is a dream-come-true for evo-devo researchers, because it makes it possible to quickly and systematically explore the ways that particular genes are deployed, and how they map onto the Ecological Phenotype in a variety of species.

The results of these experiments clearly demonstrate an important new concept in evolutionary developmental biology (evo-devo): namely, that changes in the same conserved generative gene will have a variety of effects in separate species, because those genes interact in different ways with several other genes. The Ecological Phenotypes in each species arise from complex interactions of multiple genes during the process of development. The meaning or significance of a particular gene depends on the genetic background of numerous factors throughout the genome. In each species, particular genes will play divergent roles.

---

[w] https://en.wikipedia.org/wiki/CRISPR

[x] https://en.wikipedia.org/wiki/CRISPR#Use_for_gene_editing

Science writer Viviane Callier quotes researcher Bob Reed, who says that "single genes can act as extremely discrete switches to completely change morphology". Callier notes that "the genes *WntA* and *optix* are the most important players".

For example, CRISPR studies have revealed that loss of *WntA* function causes a variety of morphological changes in seven species that are members of the Nymphalidae[y] butterfly family. In three species with a conserved wing pattern, *WntA* is required to induce stripe-like patterns, and also activates novel eyespot patterns in one of them. In two other species, *WntA* specifies the boundaries between melanic (dark pigmented) fields and light-colored patterns that they outline. In another species, *WntA* removal has opposite effects of adjacent pattern elements. In yet another species that lacks stripe-like patterns, *WntA* defines distinctive interveinous patterns [112].

Evolution has co-opted conserved genes such as *WntA* and *optix* in a variety of ways [82]. In general, *WntA* defines spatial patterns that generate divergent patterns in various species, whereas *optix* is a **transcription factor** that acts late in development to activate a variety of wing color patterns in the scales of butterfly wings, including both pigmented and textured patterns such as blue iridescence [113, 114].

The Conceptual Framework of Macroevolution and Microevolution

What is the difference between macroevolution and microevolution?

According to Wikipedia, **macroevolution**[z] is:

> evolution on a scale at or above the level of species, in contrast with microevolution, which refers to smaller evolutionary changes of allele frequencies within a species or population.

---

[y] https://en.wikipedia.org/wiki/Nymphalidae

[z] https://en.wikipedia.org/wiki/Macroevolution

But then the article correctly points out that:

> Macroevolution and microevolution describe fundamentally identical processes on different time scales.

This begs the following questions: why is it useful to make this distinction, and how, and why, did this distinction arise in evolutionary theory in the first place?

Macroevolution refers to changes that occur on a geological time scale.[aa] For Darwin, this would have included the significant changes in the shape and form of organisms that served to adapt them to various *niches* or ways of life, such as the wings of birds, bats or insects, or the legs of iguanas, tigers, or human beings.

This implies that microevolution for Darwin would be concerned with incremental changes that take place by means of accumulated **variation** and **selection**.

It follows that for the **Modern Synthesis**, as noted by Ernst Mayr:

> transspecific evolution is nothing but an extrapolation and magnification of the events that take place within populations and species…

This is the reason that the Wikipedia article states that microevolution "refers to smaller evolutionary changes of allele frequencies within a species or population", as noted above. In other words, in this view, microevolution describes the *mechanism* of Natural Selection, as seen through the eyes of proponents of the "Modern Synthesis".

But Mayr also correctly points out that:

> …it is misleading to make a distinction between the causes of micro- and macroevolution.

In other words, microevolutionary changes are the combined and visible results of microevolutionary changes over longer periods of time.

[aa] https://en.wikipedia.org/wiki/Geologic_time_scale

A 21st century view of the limitations of the macro- vs. microevolution distinction

In 19th and 20th century evolutionary biology, the common ancestry of diverse shapes and forms can be seen by comparing their **morphologies**—their visible shape and form, and the way that structure gives rise to function.

But it is one thing to look at the final phenotype of an organism, and quite another to understand the molecular, cellular, and developmental mechanisms that actually generate complex structures from multitudes of specialized, cooperating cells.

The diverse shape and forms of animals—their macroevolutionary adaptations—result from long periods of microevolutionary changes. From Darwin's 19th century point of view, macroevolution was the result of accumulated, incremental changes brought about by means of Natural Selection. From the 20th century point of view of proponents of the "Modern Synthesis", macroevolution was the result of changing frequencies of alleles in natural populations. Later, with the discovery of DNA and the "genetic code", macroevolution was the result of mutations in protein-coding genes. From a 21st century perspective, however, the problem with all of these points of view is that they do not adequately describe the actual molecular, cellular, and developmental mechanisms by which shape and form are now known to arise. In each generation, accumulated genetic information is translated into the complex organization of each multicellular individual. *Many* different mechanisms that involve the inner-workings and complex interactions of cells during development contribute to this complex organization. What evolves is the genetic information that drives the process of development and leads to species-specific outcomes.

Evolutionary change is not always incremental

Presumably, during the **microevolution** of each species, selection can act on the Ecological Phenotype in a way that is reminiscent of classical Darwinian selection—it can generate incremental morphological changes from generation to generation. But that does not mean that the Generative

Phenotype is directly comparable between related species that have been separated for a significant number of generations. The gene pools of those separate species—the collective set of genetic determinants that are unique to that species—will have changed in a variety of ways. This results in divergent generative networks of genes that interact in a variety of ways to produce a variety of Ecological Phenotypes.

When layers of complexity are added to previous ones, we can say that it's the creative force of the environment that's responsible. Certainly, the environment plays an essential role. But without the molecular, cellular and developmental capabilities of cells, none of this could happen.

Not only that, and here's an important point for evo-devo: without development of multicellular organisms, evolution would be limited by significant constraints. It would not be possible to build new complexity on top of the old, or to vary it to any considerable degree, because that complexity depends on the interactive and specialized capabilities of cells. Incremental changes in Ecological Phenotypes do not map directly or in a simple way to incremental changes in the generative phenotypic toolkit.

Toolkit elements are used and reused, and the *ways* that they are used are varied. Some of this variation results from regulation of gene expression.

Changes in morphology are aspects of the Ecological Phenotype. But the cellular and molecular wherewithal to create that morphological diversity and complexity we have defined as the Generative Phenotype.

Evo-devo is a logical outcome of the new insights that are obtained when developmental biology—now greatly improved by empirical advances in molecular and cellular techniques and discoveries—is brought together with evolutionary science, when sophisticated comparisons of genomes obtained via state-of-the-art high-throughput sequencing and comparisons are brought together with visualization of morphogens and deep understanding of the complexities of interacting transcription factors and the like.

Broadly speaking, the genotypic gestalt creates the toolkit of the **Generative Phenotype**, and the ways that toolkit elements are deployed creates the **Ecological Phenotype** (Chapter 14).

Contrasting the 19th century concept of incremental change with 21st century evolutionary developmental biology

The terms "macroevolution" and "microevolution" were not coined until 1927, by Yuri Filipchenko. Nevertheless, we can safely assume that for Charles Darwin, what was later referred to as microevolution referred to the *mechanism* of Natural Selection, as incremental changes brought about by repeated cycles of variation and selection in successive generations. Macroevolution would be seen as the long-term result of those incremental changes.

From a 21st century point of view, however, we need a conceptual framework that describes the relationship between Natural Selection and the generation of complex organization from a molecular, cellular, and developmental point of view. To say that incremental change is responsible for the generation of complex organization is not enough—this is an incomplete description. Today, we know far more about the inner-workings of cells—and the ways that cells interact during the process of development—than Darwin could possibly know. Therefore, we are in a much better position to describe the evolution of complex shape and form in a detailed and meaningful way, in terms of concrete biological mechanisms. In other words, 21st century evolutionary theory provides a more satisfying explanation than classical Darwinian theory or the "Modern Synthesis". However, to this day, nonscientists are more familiar with classical Darwinian theory and the "Modern Synthesis" than with the fascinating principles of molecular, cellular, and developmental biology that are quite familiar to biological researchers.

Chapter 16

Repetitive Sequences, Gene Duplication, and the Varieties of Genomic Variation

Man's creative genius flourished only when his mind, freed from the worry of daily toils, was permitted to entertain apparently useless thoughts.

In the same manner, one might say with regard to evolution that natural selection merely modified, while redundancy created. ... big leaps in evolution required the creation of new gene loci with previously nonexistent functions.

—Susumu Ohno [28]

The Big Picture

Several types of changes in DNA sequences, including production of repetitive sequences, unequal crossing-over, gene duplication, point mutations and more, act on the genome in a synergistic manner. Together, they have extraordinary organizing power.

The UES recognizes that genomes have incorporated several general tools that facilitate, leverage, and accelerate genome evolution.

Darwin's original concept of accumulation of infinitesimal changes must be supplemented with these widespread and now well-known mechanisms of evolutionary change. The "Modern Synthesis" focus on point mutations as the major source of new alleles fails to recognize numerous additional mechanisms of genomic sequence evolution. In many cases, several mechanisms have combined effects on the genome that are far more significant than any one mechanism alone.

Both genotypes and phenotypes can change in a variety of ways that leverage previous innovations. This can result in rapid addition of structures and functions that did not need to arise incrementally, but are often subsequently refined in an incremental fashion.

Sometimes, during DNA replication, the growing strands can mispair with the template strands. This mechanism, known as slipped-strand mispairing or replication slippage, represents a fundamental way that DNA sequences can expand into simple repetitive sequences that are subsequently diversified by other mechanisms such as point mutations.

Gene duplication represents one of the most widespread and significant types of genomic evolution. Duplicated genes played a crucial role in the evolution of diverse forms of life ranging from prokaryotes[a] to plants and animals. Most of the genes in animals are members of multigene families that arose by gene duplication events. Duplicated genes lay the groundwork for new innovations that have far more organizing power.

The keen sense of smell exhibited by dogs is largely a consequence of extensive gene duplication events in olfactory receptors.

The mammalian immune system provides a quintessential example of the sheer organizing power of the combined effects of gene duplication, V(D)J recombination, and somatic hypermutation to generate extraordinary diversity.

Rethinking Mutations

Richard Fortey has poetically captured the significance of genome research for evolutionary theory in a delightful popular book [115]:

> Every new discovery about the genome is consistent with evolution having happened. Whether we find it appealing or not is another question, but personally I like being fourth cousin to a mushroom and having a bonobo as my closest living relative. It makes me feel a real part of the world.

[a] https://bmcgenomics.biomedcentral.com/track/pdf/10.1186/1471-2164-11-588

The **Modern Synthesis** and the **neo-Darwinian** views of genetic variation—that variation (a) arises primarily from **point mutations** in DNA, (b) that DNA sequence variation primarily results in changes in amino acid sequences, and that (c) genetic variation is essentially **random** in nature—are misleading and incomplete, for several reasons:

(1) Genetic information—and complex cellular structures it determines—consists of a metaphorical scrapyard of reusable modules—parts of the developmental toolkit.
(2) Reusable modules are subject to modification, emergent interactions and fine-tuning in various ways brought about by selection—including both sudden and gradual forms of change.
(3) Numerous mechanisms—including metaphorical **selfish DNA** such as **transposable elements** (TE)—modify genomic DNA in various ways and acquire a variety of functions in particular contexts. They are metaphorically "selfish" because, like viruses, the DNA sequences tend to replicate themselves and increase their numbers, independently of the needs of the host organism in which they reside. This can lead to fortuitous innovations that are selected when they happen to prove advantageous in the context of surrounding genomic DNA sequences.
(4) DNA sequences derived from TE comprise substantial portions of many eukaryotic genomes. About 44% of the human genome is derived from TE. In maize (a type of corn plant), TE comprise about 90% of the genome, and are responsible for mutations in the corn plants that behave in a non-Mendelian fashion.[b]
(5) Innovations in the genome—as well as the molecular interactions they determine—can contribute to both generative and Ecological Phenotypes[c] Selection can act on these elements in a variety of ways over multiple generations.
(6) The modular and repetitive nature of determinants of biological organization contribute to much of the variation in phenotypes

[b]https://en.wikipedia.org/wiki/Transposable_element

[c]Generative Phenotypes refer to toolkit elements that play important roles during development, whereas Ecological Phenotypes refer to the adaptations that arise in particular environmental contexts; see Chapter 14.

observed in natural history. Gene duplications are a prime example of reuse and redeployment of elements that arise by duplication and subsequent modification and fine-tuning under Natural Selection.

(7) Self-accelerating mechanisms of genome evolution—such as replication slippage, gene duplication, and unequal crossing-over—can evolve at rates that exceed random point mutations.

(8) The significance of various types of DNA sequence change—such as frameshifts, simple tandem repeats, insertions, and deletions—contribute to a variety of levels of complex gene regulation that are only partially understood at the present time.

(9) Various types of weak molecular interactions are redundant, often are parts of networks of signaling cascades and represent flexible raw material that may be useful in different ways in different cellular and ecological contexts.

(10) Other types of heritable changes in DNA sequences involve rearrangements, insertions or deletions of visible portions of chromosomes, as well as occasional duplications and subsequent evolution of entire chromosomes and even entire genomes.[d]

The UES incorporates a broader spectrum of mechanisms into genomic evolution. It recognizes (a) a variety of types of changes in genomic DNA that are synergistic, (b) a wide range of phenotypic consequences including regulation of **genes** and developmental **toolkit** elements, and (c) several nonrandom features of variation and evolutionary change, including the consequences of **Emergent Evolutionary Potential (EEP)** combined with **Natural Selection**.

The Origin and Evolution of Repetitive DNA Sequences

My doctoral thesis

As a doctoral student at UC Irvine (UCI), I was privileged to play a significant role in the discovery of a fundamental mechanism by which

---

[d]https://en.wikipedia.org/wiki/Paleopolyploidy

repetitive sequences arise and expand genomic DNA, which I called **slipped-strand mispairing (SSM)**.[e] Often called **replication slippage**,[f] the original SSM publication[g] [39] is now widely-cited by the scientific community and well-supported by empirical evidence.

As is so often the case with scientific discoveries, serendipity, and trial-and-error—including pursuit of false leads—played important roles in my discoveries.[h] Throughout my undergraduate studies at UC Berkeley (UCB), and extending into my graduate research at UC Irvine, I was intrigued by the role that **development** plays in the evolution and reproduction of complex organized structures and functions. I began working with **fruitflies**, the powerful model system that ushered in the revolution in **evolutionary developmental biology**. I was working with a DNA sequence found in fruitflies that exhibited **cross-hybridization** with sequences in other species that had been implicated in determining gender (**sex determination**) during development. Since cross-hybridization is often due to **evolutionary conservation** of **homology** (evolutionary descent), I had hoped to learn more about the genetics of sex determination in fruitflies.

As fate would have it, this was a thoroughly false lead. Not only did this fruitfly sequence have nothing to do with sex determination, it was also cross-hybridizing not because of homology, but rather because it contained **simple repetitive DNA sequences**.[i] These simple sequences included a variety of long tracts of short tandem repeats such as GATA GATA GATA GATA…

Although the accidental discovery of this artifact felt like a disaster at the time, the good news is that I became fascinated with the evolutionary origins of simple repetitive sequences, which were starting to turn up in a broad range of organisms—particularly in non**protein-coding** DNA, that is, sequences that do *not* specify amino acid sequences of protein chains. Against the strong objections of my principal investigator, I decided to

---

[e] https://en.wikipedia.org/wiki/Slipped_strand_mispairing

[f] https://en.wikipedia.org/wiki/Replication_slippage

[g] https://drive.google.com/file/d/0B1zIUrf9YzRWa0lSSEJwUWhRcWs/view?usp=sharing

[h] https://owlcation.com/stem/Serendipity-The-Role-of-Chance-in-Making-Scientific-Discoveries

[i] Available online at: https://pdfs.semanticscholar.org/7617/6729c9fb79783b8494045731b8086c6cf141.pdf#page=41

change the direction of my doctoral thesis, and to do so, I needed to find a new mentor. Fortunately, I found George Gutman, who kindly provided both resources and guidance for a thesis that fell outside of the main focus of his own laboratory.

I was astounded to find that not only were simple repetitive sequences extremely common in nonprotein-coding regions of virtually every **eukaryotic** species that I looked at in the database—including a broad range of plants and animals as well as humans—but surprisingly, no one had published a good explanation for how they arise in the first place. Presumably, these repetitive sequences were produced by enzymes that synthesize DNA. But how?

One possibility was that they were produced by the enzymes that routinely replicate DNA, known as **DNA polymerases** (Figure 1).

I hypothesized that short, simple runs of bases that initially occur randomly—for example, a run of GGGG or AAAAA—would be prone to polymerase errors, in which the two strands of replicating DNA are accidently mispaired during replication, which can result in the insertion (or deletion) of repeat units.

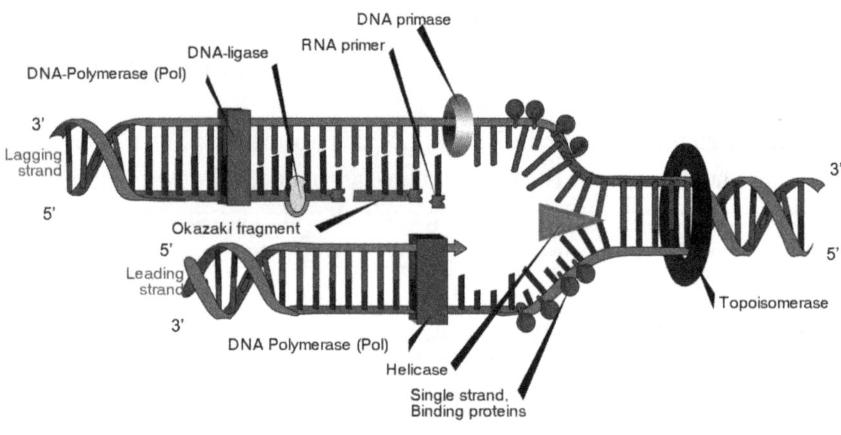

Figure 1. Diagram illustrating the variety of enzymes and other proteins involved in DNA replication.

Source: Page-link: https://commons.wikimedia.org/wiki/File:DNA_replication_en.svg. File-link: https://upload.wikimedia.org/wikipedia/commons/8/8f/DNA_replication_en.svg. Attribution: By LadyofHats Mariana Ruiz [Public domain], via Wikimedia Commons. Rendered in B&W.

Not only that, but the longer the repeat, the greater the chances of mispairing. Therefore, production of simple repeats by slipped-strand mispairing (also known as **replication slippage**) might be a **self-accelerating** process, under conditions that favor (or select for) insertions more than deletions.

I went about testing several of these assumptions by observing the expansion or contraction of synthetic repetitive sequences that I introduced into clones that I introduced into bacteria, using recombinant DNA technology. This provided a model system where I could systematically detect and subsequently sequence newly occurring insertions (or deletions). I found that not only did these simple cloned sequences expand, but they expanded at extraordinarily high rates. These early studies laid the groundwork for numerous subsequent studies by other scientists.

Synergies between replication slippage and point mutations

Point mutations would, I proposed, make simple repeats more complex, in a random way. For example, a sequence of GAGAGAGA might change into a sequence of GAGATAGATAGA if two point mutations occur subsequently. These longer repeating units would also be prone to slipped-strand mispairing and could therefore expand (see [39] for illustrations[j]).

A comprehensive survey of the relevant scientific literature—which is routinely expected for any scientist—led me to an obscure paper that had been published many years earlier. The Nobel Laureate who had played a major role in the discovery of DNA polymerase—Arthur Kornberg—found that purified DNA polymerase would sometimes generate simple repetitive sequences in a template-free manner. Kornberg observed that simple sequences combined with nucleotides and DNA polymerase in a test tube readily generate long repetitive sequences, starting with a short DNA template. In other words, Kornberg observed DNA polymerase slippage events that are similar to those that I hypothesized take place in living cells. In my case, this hypothesis resulted from my analysis of widespread simple repetitive sequences in databases of genomic DNA sequences.

---

[j] https://drive.google.com/file/d/0B1zIUrf9YzRWa0lSSEJwUWhRcWs/view?usp=sharing

Synergy between replication slippage and unequal crossing-over

Unequal crossing-over[k] is a process that occurs between chromosomes during meiosis, when sex cells are produced. This is another mechanism of great importance for DNA sequence evolution—not only for production of repetitive sequences, but also because large segments of chromosomes—often consisting of entire genes, or even many genes—are replicated in this way.

Synergy between tandem repetition and gene duplication

In my doctoral thesis work, I also noted what had been suggested by others—namely, that simple repetitive sequences would be prone to unequal crossing-over, which plays a major role in **gene duplication**. Gene duplication refers to long sequences of DNA containing both protein-coding regions as well as repetitive sequences. The following sections will help to illustrate the major importance of this mechanism for genomic evolution.

Global significance of simple repeats

The hypothesis of simple repeat evolution by SSM combined with point mutations has stood the test of time. Simple repeats often expand at higher rates than the frequency of point mutations—which makes them an important source of variation and expansion in genomic DNA.

The Widespread Significance of Gene Duplication in Genomic Evolution

Darwin's quotation on reuse of repeated elements

Although Darwin's *On the Origin of Species* focused on gradual accumulation of variation as the source of evolutionary change [20], he did clearly envision that similar or repetitive elements would likely be used in a variety of contexts:

---

[k] https://en.wikipedia.org/wiki/Chromosomal_crossover

> We have formerly seen that parts many times repeated are eminently liable to ... it is quite probable that natural selection, during a long-continued course of modification, should have seized on a certain number of the primordially similar elements, many times repeated, and have adapted them to the most diverse purposes.

So, just how important is gene duplication for an updated evolutionary theory? Let's take a closer look at the evolutionary history, extraordinary capabilities, and organizing power of several examples. Today, several relatively new empirical techniques are routinely combined to track the fates of duplicated genes.

For several reasons, duplicated genes represent some of the most compelling illustrations of EEP.

When genes are duplicated, they make it possible for one or more new copies to diverge, and evolve potentially useful functions, while retaining the original structure and functions of a conserved copy.

The new copies already have the features that have proven to be useful in the original. The DNA sequences that have already proven useful are instantly available. Any or all of the structural and functional features of the original—features that have already been tested and refined in the struggle for existence—including the ability to participate in signaling pathways, the ability to catalyze specific chemical reactions, and the ability to bind and interact with other specific molecular targets—are available to be deployed in a variety of useful ways. Sometimes, the protein structure will be conserved, while the flanking regulatory sequences will diverge. This makes it possible for specific copies of the gene to be expressed in different contexts during development or in particular circumstances during the life of the individual.

Sometimes, slight modifications in the shape of the protein copy—which can easily arise as a result of small changes in the DNA sequence—will allow it to fit to other reactants (substrates) and modify them enzymatically.

Since many gene duplications result in tandem copies, these copies will often be subject to self-accelerating gene duplication events which can rapidly and dramatically increase the numbers of individual copies of the gene. This can rapidly give rise to **multigene families** which have the potential to acquire new and useful functions—either individually, or as groups. This helps to explain why duplicated genes are often members of multigene families that carry out diverse functions that were not possible with individual genes.

The widespread significance of gene duplication

In 1967, Susumu Ohno pointed out that[l]:

> When the scope is broadened to consider the evolution of vertebrates as a whole, allelic mutations of already existing genes cannot possibly account for all the genetic changes that occurred during 300 million years. Gene duplication now emerges as the single most important factor in evolution.

Is it really true that gene duplication has become the single most important factor in the evolution of the vertebrates? As it turns out, this is no exaggeration, because most of the protein-coding genes among a broad range of organisms—including the nonprotein-coding DNA sequences that regulate the expression of those genes—have evolved by this mechanism in the past 300 million years and are responsible for the diversity of adaptations among not only vertebrates, but diverse species of plants and animals in general. We now know that the frequency at which entire, functional genes are duplicated is similar to the rate of point mutations.[m]

The history and significance of gene duplication theory

As John Taylor and Jeroen Raes revealed in their fascinating 2004 review [116], the history of scientific ideas concerning gene duplication has several surprises in store. One is that published reports of gene duplication extend as far back as 1911.

Large-scale duplications of large segments of plant chromosomes containing many genes—as well as duplications of entire chromosomes and even entire genomes—led to a number of early discoveries by plant

---

[l]https://link.springer.com/book/10.1007%2F978-3-642-88178-7#about

[m]Duplication and Divergence: The Evolution of New Genes and Old Ideas Annu. Rev. Genet.2004: 38:072902-092831.

cytologists. Such large-scale duplication events are relatively frequent in plant evolution.

Long before DNA sequencing became available, both plant and animal cytologists studied chromosomes that are condensed during cell division. Duplications can be visualized by repeated patterns of banding rendered visible with special chromosomal stains.

The banding patterns of these stained chromosomes make it possible to systematically compare the structure of the chromosomes of related species, and to detect changes in chromosome number and chromosome structure when they occur. When such changes occur in the sex cells, they can contribute to evolution if they prove to be useful in the struggle for existence. The idea of gene duplication became more widely known in the scientific community with Ohno's 1970 publication of *Evolution by Gene Duplication* [28]. But to this day, most of the general public is blissfully unaware of the significance of gene duplication, which has arguably played at least as important a role as point mutations in the evolution of our own species.

Natural Selection of duplicated genes and their protein products or regulatory functions

One of the most important evolutionary consequences of gene duplication is that the duplicated copies are free to diverge and be reused, while preserving the function of the original copy.

During the divergence of duplicated copies, **point mutations** (changes in single DNA nucleotides) can play a synergistic role with gene duplication by leading to either sudden or cumulative incremental changes in the duplicates.

When duplicated pairs of genes take on different functions, this is called subfunctionalization.[n] Alternatively, the new duplicated copy may evolve a new and separate function: this is called neofunctionalization.[o]

---

[n] https://en.wikipedia.org/wiki/Subfunctionalization

[o] https://en.wikipedia.org/wiki/Neofunctionalization

Complex Phenotypes Have Evolved from Multigene Families

Gene duplication plays a major role in the emergence of higher levels of organization via EEP and Natural Selection.

For example, the sense of smell plays a central role in interactions between predators and prey. Several Generative Phenotypes work together to generate the structures and functions that make this possible.

The ability to detect odors depends on sensitive interactions between cell surface receptors known as **olfactory** or **odorant receptors**[p] and the molecules that represent various aromas.

Odorant receptors are members of a large multigene family that arose by numerous gene duplication events. This multigene family consists of a particular class of **transmembrane proteins** known as **G protein-coupled receptors** (GPCR) that play a variety of remarkable functions in specialized cells.[q]

One of the principle ways that keen senses of smell have evolved is by expansion of the olfactory receptor multigene family. Dogs, for example, have up to 40 times greater smell sensitivity than humans, because they have up to 40 times more odorant receptors. Dog owners are quite familiar with the extraordinary ability of their pets to track and distinguish between faint odors.[r]

Of course, the evolution of the phenotypic gestalt of each species—the Ecological Phenotype—involves the coevolution of several Generative Phenotypes. In addition to the expansion of the olfactory receptor family, dogs have also evolved a bony shelf inside the nose that makes it easy for odors to stick to the receptors. Unlike the human brain, in which vision plays a dominant role because of the highly developed visual cortex, the sense of smell plays a dominant role in the dog brain because it has evolved a more highly developed olfactory cortex.

The olfactory receptor multigene family provides a concrete example of the important role played by gene duplication in the evolution of

---

[p] https://en.wikipedia.org/wiki/Olfactory_receptor

[q] https://en.wikipedia.org/wiki/G_protein-coupled_receptor

[r] https://en.wikipedia.org/wiki/Dog_anatomy#Smell

complex adaptations. Since duplicated genes are derived from already functional copies, this type of genetic variation has nonrandom characteristics.

Evolution of the Immunoglobulin Superfamily and the Adaptive Immune System (AIS)

The biology of the **Adaptive Immune System**[s] (**AIS**) is a broad and captivating subject. Readers unfamiliar with the AIS are encouraged to take advantage of the Wikipedia links provided in the footnotes and glossary. Space permits only a brief discussion of some relevant topics here.

Brief overview

The AIS provides a quintessential example of **shape-specific molecular interactions and binding events** (**SSM-IBE**, Chapter 12). Gene duplication has combined with other potent sources of genetic diversity, including **V(D)J recombination**[t] and **somatic hypermutation,**[u] to generate an unrivaled higher-level adaptation that effectively seeks out and destroys potential threats to multicellular individuals, while distinguishing foreign entities from the self. It does this by binding and internalizing **antigens**[v] (which are often fragments of protein chains) and then presenting them to antigen-presenting cells. Then, the antigens stimulate production of clones of individual **T cells**[w] and **B cells**.[x] These clones express specific **T-cell receptors**[y] and **antibodies**[z] that bind to specific antigens with exquisite specificity. Often, B cells secrete soluble forms of the antibodies.

---

[s] https://en.wikipedia.org/wiki/Adaptive_immune_system

[t] https://en.wikipedia.org/wiki/V(D)J_recombination

[u] https://en.wikipedia.org/wiki/Somatic_hypermutation

[v] https://en.wikipedia.org/wiki/Antigen

[w] https://en.wikipedia.org/wiki/T_cell

[x] https://en.wikipedia.org/wiki/B_cell

[y] https://en.wikipedia.org/wiki/T-cell_receptor

[z] https://en.wikipedia.org/wiki/Antibody

Under the aegis of Natural Selection, the AIS has been expanded and fine-tuned so that it can effectively recognize, bind, and neutralize the diverse antigens characteristic of a vast number of different potential threats, including invading bacteria, viruses, toxins, and even cancer cells. As with all products of Natural Selection, these capabilities evolved over many generations from more primitive forebears.

Mammals can produce an astronomical number of different antibodies—probably on the order of a thousand billion different ones in humans.[aa] The AIS also generates a vast range of T-cell receptor specificities. T cells include both helpers and cells that can kill other cells (designated **CD4**[ab] and **CD8,**[ac] respectively). CD4 cells send signals that generate clones of numerous specific T cells and B cells that bind to particular antigens. These are called memory cells, and often circulate in the body for years. When the antigen is encountered after this primary immune response, these cells can rapidly proliferate and effectively eliminate the threat before the threat gains a foothold. The purpose of a **vaccine** is to elicit a primary immune response so that individuals are prepared to effectively fight pathogens that might otherwise cause future infections.

Today, our empirical understanding of the expression, cellular functions, and evolution of the various members of the immunoglobulin gene superfamily is extensive. Several empirical techniques, including DNA sequencing, databases, and computational tools for systematic comparisons of DNA, RNA, and protein sequences, **monoclonal antibodies,**[ad] and **flow cytometry**[ae] are just some of the powerful tools that have been fruitfully applied to the subject.

Evolution of molecular mechanisms that generate diversity in the AIS during T- and B-cell development

How did the diversity of the AIS evolve? During the molecular evolution of the AIS, tandem arrays of literally hundreds of genes—members of a

[aa] https://www.ncbi.nlm.nih.gov/books/NBK26860/

[ab] https://en.wikipedia.org/wiki/CD4

[ac] https://en.wikipedia.org/wiki/CD8

[ad] https://en.wikipedia.org/wiki/Monoclonal_antibody

[ae] https://en.wikipedia.org/wiki/Flow_cytometry

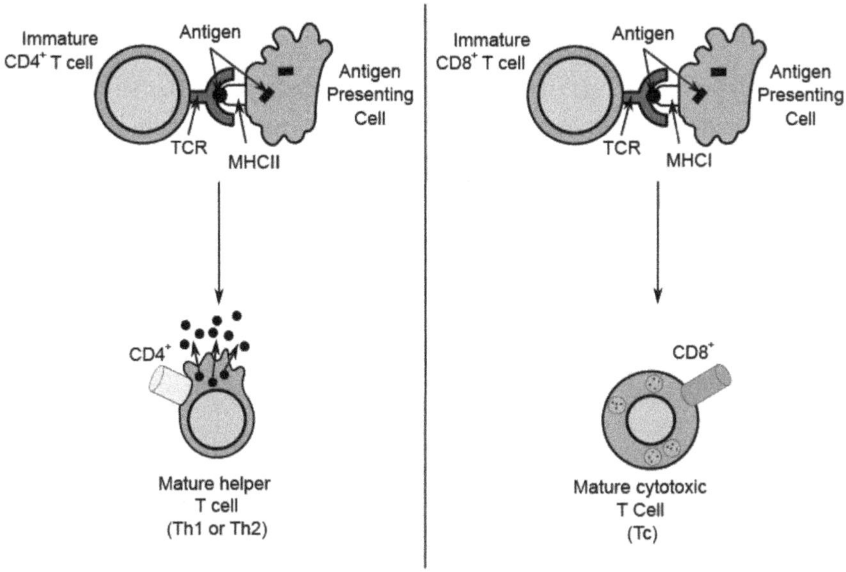

Figure 2. Numerous cell-surface proteins that are members of the immunoglobulin superfamily of duplicated genes play important roles in the functioning of the mammalian immune system. These cell-surface proteins play roles as antigens (genes that elicit an immunological response), as receptors on T cells and B cells, and some are secreted by B cells as antibodies. Various members of these multigene families are involved in recognizing a broad array of potential threats such as pathogens, toxins, and cancer cells. Some bind to other receptors and regulate immune responses in complex ways. Still others are involved in distinguishing between self (cells of the individual that should not be attacked) and foreign antigens (which often should be eliminated).

Source: Page-link: https://commons.wikimedia.org/wiki/File:Antigen_presentation.svg. File-link: https://upload.wikimedia.org/wikipedia/commons/4/4d/Antigen_presentation.svg. Attribution: By user: Sjef [GFDL (http://www.gnu.org/copyleft/fdl.html) or CC-BY-SA-3.0 (http://creativecommons.org/licenses/by-sa/3.0/)], via Wikimedia Commons. Rendered in B&W.

large and diverse multigene family known as the **immunoglobulin superfamily**[af] (**ISF**) evolved via extensive gene duplication. Numerous DNA and protein sequences of ISF family members were identified by Leroy Hood and colleagues in the 1980s [117]. ISF family members play a variety of now well-understood roles in the cellular functions of the AIS (Figure 2).

[af]https://en.wikipedia.org/wiki/Immunoglobulin_superfamily

A second crucial molecular component of the AIS is V(D)J recombination [118], a novel mechanism that is active during the development of T cells and B cells. V(D)J recombination randomly shuffles and splices diverse members of several multigene families, to generate a vast repertoire of random recombinant T-cell receptors and antibodies that can bind to diverse antigens. This random shuffling is facilitated by a major generator of diversity known as **recombination-activating genes**[ag] (RAG).

RAG are thought to have evolved from transposable elements[ah] (TE). Recently, a "living molecular fossil" thought to be the progenitor of RAG was discovered by Shengfeng Huang and colleagues [119, 120].

The widespread activity of TE in molecular evolution was mentioned in a section above.

A third major molecular component of the AIS is a novel process that rapidly generates diversity among the recombined T-cell receptor and antibody sequences during T- and B-cell development, a process known as somatic hypermutation [121].

Conclusion

Increased complexity and efficacy of adaptations take on new meaning in the context of major mechanisms of genomic change, including but not limited to gene duplication and multigene families, transposable elements, V(D)J recombination, somatic hypermutation, and gene regulation mediated by cell–cell communication and antigen presentation.

This is a very different paradigm than the concept of gradual accumulation of random point mutations in protein-coding genes. In other words, paradigms such as the "Modern Synthesis", which were developed in the absence of subsequent empirical discoveries, require a rethink.

The underlying principles of Natural Selection remain the same whether we are discussing point mutations or the synergy of major mechanisms of genomic change, but those major mechanisms have far more organizing power.

[ag] https://en.wikipedia.org/wiki/Recombination-activating_gene
[ah] https://en.wikipedia.org/wiki/Transposable_element

As illustrated by the 2017 publication of the "molecular living fossil" of RAG1 and RAG2, namely proto-RAG, significant pieces of the puzzle of reconstructing past events continue to fall into place. These concrete steps are good to know, because they represent the empirical facts that test the validity of evolutionary hypotheses.

But in the press of urgent specialized research goals, we often fail to step back and reflect on the Big Picture of what we already know and is reflected in the **Updated Evolutionary Synthesis** (**UES**). One does not need to fit all of the pieces of a jigsaw puzzle into place in order to see the whole picture.

Consider a thought experiment with the following sequence of events:

(1) A gene duplication event generates a primitive cell-surface protein (call this protein 1) that fortuitously immobilizes a bacterial toxin or a viral particle, and thereby decreases its pathogenic potential.
(2) Additional gene duplication events and point mutations generate a small family of related proteins (**subfunctionalization**). Call these protein 1a, 1b, 1c, etc. Protein 1b fortuitously binds the toxin or virus more effectively, or bind more effectively to new variants of the virus.
(3) Fortuitous insertion and movement of a transposon resembling proto-RAG results in a recombination event that joins protein 1b transposon to a nearby signal peptide of a **secreted** protein. Now we have a secreted version of protein 1b that binds circulating toxin or virus more effectively.
(4) These fortuitous events lay the groundwork for subsequent gene duplication and transposition events and gradual refinement and fine-tuning of these mechanisms, and the primordial immune system continues to increase in complexity and efficacy by means of Natural Selection.

For purposes of this discussion, the specific sequence of imaginary events is not important—they do not represent actual facts. Other creative readers can undoubtedly bring their own knowledge to bear on the subject and come up with scenarios that are more plausible or consistent with known facts.

Cause and effect become blurred in these scenarios. It is equally valid to say that during the course of evolution of the immune system, gene duplications laid the groundwork for V(D)J recombination, or to say that V(D)J recombination laid the groundwork for additional gene duplications, or to say that V(D)J recombination laid the groundwork for primitive forms of somatic hypermutation, and so on.

At each step in the process, we are witnessing the role played by Natural Selection in molecular evolution. Accumulation of useful variation is taking place.

When combined with the more modern perspectives of emergent evolutionary potential (EEP), shape-specific interactions and molecular binding events (SSM-IBE) and other modern evolutionary perspectives, the sheer organizing power of Natural Selection is clear.

Major mechanisms of genomic change blur the distinction between gradual *quantitative* accumulation of changes and relatively sudden leaps in complexity, organizing power, and *qualitative* innovation.

I would argue that the UES is well-supported by both extensive empirical data and by a rational conceptual framework. Biological evolution by means of natural processes is more plausible and more fascinating than ever before. The metaphorical blind watchmaker continues to thrive.

The effects of EEP on evolution can be seen at every level of complexity, including fascinating Darwinian examples of both **coevolution** and **sexual selection**. Consider, for example, the following excerpt from Sir David Attenborough's narrative of the beautiful film series titled *Our Planet*:[ai]

> Male orchid bees need a rich perfume with which to impress their females. And the orchids provide it. But, the bucket is slippery, and the liquid into which the bee has fallen is slippery. The only way to get out is through a narrow tunnel. As it emerges, the bee is gripped tight, and that gives enough time for the plant to glue pollen sacs on the bee's back.

[ai]This Netflix Original series is described on the web at https://www.netflix.com/title/80049832

As we approach what many have called **the singularity**—a perilous time when the very survival of our species can no longer be taken for granted—there are both spiritual and practical reasons why a deep understanding of the UES among the general public—including voters—is essential.

Fortunately, the UES offers a more plausible explanation for evolution than ever before, thanks largely to an abundance of sound empirical evidence describing the inner-workings of cells.

Global warming is arguably the most urgent and important of the practical reasons for the general public to understand evolutionary principles. Interdependent species—including humans—simply cannot evolve fast enough to offset the life-threatening environmental consequences of global warming and other aspects of climate change.

In the final chapter of this book, I have listed several principles of the UES. This is, of course, not the final word on the subject, and I look forward to constructive feedback from the scientific community, as well as suggestions about how to better bridge the gap between professional researchers and theorists and the general public.

Chapter 17

Biological Principles of the Updated Evolutionary Synthesis

Classical Darwinian Principles That Are Well-Supported by Thousands of Modern Examples

(1) In each generation, individuals of every species must struggle to survive and reproduce their own kind.

(2) In natural environments, resources such as energy, nutrients, and water are generally in limited supply.

(3) Each environment and potential way of life provides both opportunities and perils for survival, including both abiotic factors such as temperature, sunlight, and moisture, and biotic factors such as producers, predators, prey, and mutualism.

(4) In general, among living species, more individuals are born than can possibly survive and reproduce their own kind.

(5) Individual offspring vary among themselves, and some of this variation is heritable.

(6) Some variations are useful to offspring, in that they facilitate survival and reproduction in particular environments and ways of life (*niches*).

(7) Some variations are harmful to offspring and diminish the capacity for survival and reproduction in a particular environment and way of life (*niche*).

(8) Those individuals that inherit factors that are useful will tend to survive longer and reproduce in larger numbers.

(9) Heritable factors that are useful will tend to be preserved and will increase in frequency in future generations.
(10) Slight variations that are preserved will tend to accumulate in natural populations.
(11) Large changes in heritable structures and functions will tend to be preserved over many generations.
(12) Eventually, differences between populations will be sufficiently large to give rise to separate species. This is a natural cause for the origins of species.
(13) New structures and functions, such as various types of legs or wings for example, often arise because they promote survival and production in particular *niches*.
(14) Structures and functions that help adapt species to various ways of life are referred to as adaptations.
(15) The process by which new species arise by means of variation and selection is called Natural Selection.

Enduring Principles Established by the Modern Synthesis

(1) Among sexually reproducing, multicellular diploid organisms such as plants and animals, individuals routinely inherit one set of chromosomes from their mother and one set from their father. Each pair of maternal and paternal chromosomes is referred to as a homologous pair of chromosomes.
(2) Random segregation of homologous chromosomes, independent segregation of nonhomologous chromosomes, and random recombination events, during the production of sperm and egg cells—plus random fertilization of sperm and egg cells, all routinely generate random variation among offspring in each generation.
(3) Units of heredity, referred to as genes, are associated with particular chromosomes.
(4) Genes often exist in multiple versions known as alleles.
(5) For each gene, individuals may inherit identical or different alleles when they inherit pairs of homologous chromosomes, one from each parent.

(6) Genes can undergo spontaneous changes that are called mutations. They can be accelerated by chemical mutagens or by ionizing radiation, such as ultraviolet light or X-rays.
(7) Mutations can generate new alleles.
(8) Genes are determinants that influence the shape and form, structure and function, and capabilities of each individual and each species.
(9) The alleles that are inherited determine the individual genotypes.
(10) The shape and form, structures and functions, and capabilities of the individual, are referred to as the individual phenotypes.
(11) In the absence of specific factors that change the frequencies of alleles, relative frequencies of alleles in natural populations will tend to generate stable and predictable numbers of individuals that are heterozygous (carry two different alleles) or homozygous (carry two of the same alleles. This is called Hardy–Weinberg equilibrium.
(12) Natural Selection is one factor that can change allele frequencies in natural populations. Other factors include random genetic drift due to small population sizes, nonrandom mating, mutations, and migration in or out of a population.
(13) Speciation can take place when gene flow is disrupted in natural populations, leading to separate gene pools. Speciation can depend on both abiotic and biotic factors.

New Insights Provided by Molecular Genetics

(1) Most of the structures and functions of all living cells are carried out by complex molecular binding events that often involve proteins.
(2) Some, but not all, of these proteins function as enzymes in biochemical pathways by catalyzing highly-specific steps in the breakdown and synthesis of organic molecules.
(3) Viruses are not cells but are relatively simple structures consisting of a few proteins, single or double-stranded DNA or RNA, and outer coats that may contain carbohydrate and/or lipid components. They prey on specific host cells and hijack normal cellular functions in order to replicate more virus particles. They often kill their hosts.
(4) DNA is the genetic material of all living cells.

(5) The structure of DNA is a double-helix consisting of two single-stranded polymers of nucleotides. Each subunit of the polymer consists of a sugar, a phosphate group, and a nitrogenous base. Nucleotides are linked into polymeric strands by strong chemical bonds.
(6) The two strands of the double-helix are held together by weak chemical bonding (hydrogen bonding) between complementary bases on each strand. The four bases, designated A, C, G, and T, come in two complementary pairs: A is complementary to T, and C is complementary to G.
(7) In both DNA and RNA, and in DNA–RNA hybrids, hydrogen bonding between complementary bases is called base pairing.
(8) When DNA replicates, the existing strands separate and are base-paired with two new strands consisting of complementary bases.
(9) DNA replication usually results in two identical double-stranded copies of DNA, but sometimes there are copying errors that can introduce heritable mutations into a species if they occur in germ line cells that produce sperm and egg cells.
(10) Many of the genetic determinants that we call alleles and genes specify the amino acid sequences of specific protein chains.
(11) Different alleles often consist of protein chains with different amino acid sequences. Some alleles result in production of defective chains or the absence of a protein chain.
(12) In addition to "protein-coding" genes, many other DNA sequences are involved in various ways in regulating the expression of specific proteins in various cell types and under various circumstances.
(13) When genes are expressed during protein synthesis, the DNA sequence of one strand is typically transcribed into a single-stranded RNA polymer, with bases that are complementary to those in the DNA template.
(14) In eukaryotic cells, RNA transcripts are typically shortened by splicing and then processed in other ways to produce mature molecules known as messenger RNA (mRNA).
(15) Messenger RNA molecules move from the nucleus to the cytoplasm of eukaryotic cells and bind to ribosomes, which can be thought of as metaphorical protein factories.
(16) Relatively short, folded RNA polymers known as transfer RNA molecules (tRNA) are bonded to particular amino acids by enzymes

known as amino-acyl tRNA synthetases (aaRS). The specific linkage of the correct amino acids is determined by the complementary shape and physical properties of each enzyme's active site, each tRNA, and each amino acid.

(17) At the ribosome, a specific three-base triplet called an anticodon base-pairs to its complementary three-base triplet (codon) in the mRNA, and the enzymatic activity of the ribosome breaks the chemical bond linking the amino acid to the tRNA and links the amino acid to a growing amino acid sequence—a new protein chain—by a strong chemical bond.

(18) A table that maps each of 20 possible amino acids to specific codons (including start and stop signals) is referred to as a table of the "genetic code".

(19) Especially in eukaryotes, a variety of regulatory DNA sequences provide genetic control over expression of specific protein chains.

(20) Transcription factors which are often proteins, bind to regulatory DNA sequences, and to each other, providing nuanced and redundant control over gene expression.

(21) A variety of regulatory DNA sequences have been cataloged by the ENCODE project.[a] In eukaryotic chromosomes, DNA is complexed with proteins such as nucleosomes[b], and various types of chemical modifications modify gene expression in a cell-lineage-specific manner.

New Insights into the Origin of Life from Undersea Alkaline Hydrothermal Mounds

(1) The prediction and subsequent discovery of Undersea Alkaline Hydrothermal Mounds (UAHM) provide a more plausible family of hypotheses for the origin of life than the earlier concepts of a "warm little pond" or a "prebiotic soup".

(2) Prior to the evolution of RNA, proteins, DNA, and the genetic apparatus, abundant sources of energy, raw materials such as carbon dioxide and hydrogen, and naturally occurring mineral catalysts at ancient

[a]https://www.encodeproject.org/

[b]https://en.wikipedia.org/wiki/Nucleosome

UAHM drove the synthesis of organic compounds and early metabolic pathways within protected natural mineral compartments.

(3) Proton gradients, which represent an essential component of both prokaryotic and eukaryotic cells, provided natural energy sources at UAHM.

Expansion of the Fossil Record and Molecular Phylogeny

(1) New fossilized remains of both hard-bodied and soft-bodied organisms have filled in many of the previous gaps in the fossil record, such as the ancestry of whales, as well as common ancestors and close relatives of our own species, *Homo sapiens*.

(2) High-throughput genomic sequencing and advanced sequence comparison software provides a more reliable and quantitative method for reconstructing trees of evolutionary ancestry when genomes are passed from parents to offspring in a vertical fashion.

(3) In addition to vertical inheritance, prokaryotic cells frequently exchange genetic material in a lateral or horizontal fashion, which means that the concept of distinct species of prokaryotic organisms needs to be tempered with the understanding that many species share, to varying degrees, common pools of genetic information.

New Evidence Concerning the Deep History and Genomic Evolution of Prokaryotic and Eukaryotic Cells

(1) Several lines of evidence demonstrate that there are three domains of life: (a) bacterial (eubacterial) and (b) archaeal prokaryotes, and (c) eukaryotes.

(2) Eukaryotic cells arose from endosymbiotic events in which archaeal and bacterial forebears were combined into larger cells with emergent properties.

(3) Mitochondria of eukaryotic cells evolved from symbiotic incorporation of bacteria-like prokaryotic ancestors into eukaryotic cells.

(4) Chloroplasts of plant cells evolved from symbiotic incorporation of photosynthetic cyanobacteria-like ancestors into eukaryotic cells.

(5) Multicellularity has independently evolved on several occasions. Multicellular organisms include animals, plants, and fungi.
(6) Animals have gene families that have deep homology with common ancestors of the choanoflagellates.
(7) An "explosion" of diversity in the Cambrian era,[c] approximately 500 million years ago, gave rise to modern animal phyla.

Emergent Evolutionary Potential[d]

(1) It is not possible to fully appreciate how complex organization can evolve without taking two distinct aspects of reality into account: actual biological events, and emergent evolutionary potential (EEP).
(2) Actual events change the range of possibilities for future events.
(3) Actual events can change distant potential—that is, potential that has no direct links to local or recent events.
(4) The relationship of EEP to reality is a logical one that is not constrained by space and time.
(5) Emergence is a scientific principle in which new properties emerge when entities come together and interact. For example, hydrogen and oxygen chemically combine to form water, which has emergent properties.
(6) Emergence also occurs at various levels of complexity within and between living cells and organisms when larger or more complex entities arise from smaller or simpler ones.
(7) EEP shifts the odds of future events in a nonrandom fashion. Previous actual events can (and often do) influence potential future events.
(8) EEP and incremental changes via classical Darwinian Natural Selection are complementary and have greater organizing power when combined.
(9) The diversity and efficacy of biological adaptations are more plausible in the light of the combined effects of both Natural Selection and EEP.

---

[c] https://en.wikipedia.org/wiki/Cambrian_explosion

[d] The concept of EEP was first introduced by the author in *Rethinking Evolution*, and has not been available for vetting by the broader scientific community prior to publication.

Shape-Specific Molecular Interaction and Binding Events (SSM-IBE)

(1) From the earliest carbon-based primitive life forms to cellular organisms, all biological phenomena depend on the emergent potential of organic molecules.
(2) The emergent potential of amino acid and nucleotide polymers to form primary and secondary structures that fold into complex and unique 3D shapes led to the evolution of organization based on binding between complementary shapes and chemical structures.
(3) What we refer to as the "genetic code" is actually determined by the evolution of SSM-IBE. These interactions and binding events link specific amino acids to particular codons (three-base DNA and RNA triplets). This ultimately results in amino acid sequences that correspond to accumulated changes in "protein-coding" sequences in the genome.
(4) The emergent properties of primitive forms of proteins such as enzymes, transcription factors, membrane-bound receptors, and antibodies laid the groundwork for the evolution of higher-level organized structures and functions such as biochemical pathways, gene regulatory networks, signal transduction pathways, and the immune system.

Evolutionary Developmental Biology

(1) Major insights into biological complexity in animals of all kinds were found by studying the Molecular Genetics of newly discovered genes expressed during the development of fruitflies. These include highly conserved elements such as *Hox* genes that define regional segmentation and patterns during development, as well as a variety of transcription factors and morphogens.
(2) A broad range of animals, ranging from worms, flies and butterflies to snakes, birds, and mammals such as human beings, all share a highly conserved set of related gene families often referred to as metaphorical "toolkit" genes. These toolkits play major roles in the formation of segmentation, body plans, and other patterns during embryonic development.

(3) Diversity of shape, form, structure, and function is generated by deploying developmental toolkit elements in a variety of different ways that control sets of genes that have independently evolved in various lines of descent.
(4) Only a tiny fraction of the genome determines the amino acid sequences of protein chains. A larger fraction of the genome has evolved various roles in the regulation of gene expression, and much of the remainder represents a metaphorical "scrapyard" of reusable elements that may evolve new roles in the future.
(5) Genes rarely function in isolation. Most evolutionary adaptations involve complex structures and functions that arise during development. The most interesting phenotypes can be understood by studying molecular and cellular interactions, involving several "protein-coding" as well as regulatory genes, at a variety of levels of complexity.

A Broader Perspective on Organisms, Levels of Organization, and Ecological Relationships

(1) Composite organisms, composite cells, and composite genomes all represent evolutionary innovations where formerly separate cells and genetic elements have been brought together to generate new, emergent forms of life.
(2) Horizontal gene transfer (lateral gene transfer) results in gene pools that transcend individual organisms.
(3) The quest to find new sources of energy and nutrients has driven the evolution of a broad range of ecological relationships such as mutualism, producers and consumers, coevolution, predators and prey, and symbiotic relationships.

Generative and Ecological Phenotypes, and Evolvability

(1) In the classical Darwinian view, adaptations such as wings or legs arise from the gradual accumulation of useful incremental changes. Variants are "useful" when they increase the probabilities for survival and reproduction of an individual and its descendants. Selection takes

place within a specific context, that is, in a particular environment and way of life, i.e. a particular ecological *niche*.

(2) In the light of evolutionary developmental biology, it is clear that (a) multicellular structures and functions must be reproduced by a process of development and (b) flexible, redundant, and reusable "toolkit" genes help determine the molecular and cellular interactions by which development takes place. Therefore, in addition to adaptations that arise in specific ecological contexts, we also must recognize the evolution of developmental toolkit elements.

(3) It is conceptually useful to distinguish between (a) Generative Phenotypes, describing flexible and reusable structures and functions deployed during development, and (b) Ecological Phenotypes, describing the structural and functional gestalts that make specific ecological *niches* possible within particular environmental contexts.

(4) Modularity increases the organizing power of numerous cellular structures and functions and their genetic determinants. The phenotypic meaning or significance of modular structures and functions depends on its interactions with other elements. A variety of higher levels of emergent structure and function arise in various cellular contexts. Modules are often readily modified, fine-tuned and redeployed by Natural Selection, which extends their organizing power in various generative and ecological contexts.

(5) Many, if not most, evolutionary innovations are subsequently derived from structures and functions that first arose in distinct ecological contexts. Exaptation and co-option are the terms used to describe this widespread evolutionary phenomenon.

(6) Evolvability—that is, the flexibility and efficacy of future evolutionary potential—must be recognized as an important aspect of evolution. In other words, the ability to evolve is potentially useful in and of itself. Lineages that may include large numbers of different species are more likely to remain extant and generate new species if they have high degrees of evolvability.

(7) Since genetic elements can be derived, modified and reused, since we can recognize genetic determinants of both Generative Phenotypes and Ecological Phenotypes, and since evolvability is a useful attribute, since genetic elements can be transmitted both vertically and

horizontally, and since group selection is now widely accepted, both Natural Selection and EEP have characteristics that transcend individual organisms and individual species.

(8) Facilitated variation contributes to biological complexity through differential reuse of preexisting components. Regulatory changes in the usage of various conserved core components play important roles in development, cell–cell communication, and the inner-workings of cells.[e]

The Varieties of Genomic Variation

(1) Slipped-strand mispairing (SSM) can rapidly generate and expand simple repetitive sequences in most eukaryotic genomes.
(2) SSM and point mutations act synergistically to generate various lengths of tandem repeats with a variety of simple sequence motifs.
(3) Simple-repetitive sequences become hotspots for unequal crossing-over.
(4) Most eukaryotic genes are members of multigene families that arise via gene duplication.
(5) Gene duplication has laid the groundwork for high-level structures and functions, such as the adaptive immune system, to evolve.

Human Evolution and Evolutionary Psychology

(1) The Baldwin effect[f] helps to explain how learned behaviors can evolve into heritable instincts.
(2) Humans are products of biological evolution.
(3) Evolutionary psychology sheds light on human psychology, behavior, and social history, and can help us to understand and overcome our negative tendencies.

---

[e] https://en.wikipedia.org/wiki/Facilitated_variation
[f] https://en.wikipedia.org/wiki/Baldwin_effect

References

1. Carroll, S.B., Grenier, J.K. and Weatherbee, S.D. (2001). *From DNA to Diversity: Molecular Genetics and the Evolution of Animal Design.* (Blackwell Science, Malden, MA).
2. Gould, S.J. (1994). Evolution as fact and theory. In *Hen's Teeth and Horse's Toes: Further Reflections in Natural History* (Reissue Edition). (W.W. Norton, New York). [Originally published 1983].
3. Gould, S.J. (1981). Evolution as fact and theory. *Discover*, 2 (5), pp. 34–37.
4. Sagan, C. (1977). *The Dragons of Eden. Speculations on the Evolution of Human Intelligence*. (Ballantine Books, New York).
5. Dawkins R. (1986). *The Blind Watchmaker. Why the Evidence of Evolution Reveals a Universe Without Design*. (W.W. Norton, New York).
6. Dawkins, R. (2009). *The Greatest Show on Earth: The Evidence for Evolution.* (Free Press, New York).
7. Gould, S.J. (1977). *Ever Since Darwin. Reflections in Natural History.* (W.W. Norton, New York).
8. Gould, S.J. (1980). *The Panda's Thumb. More Reflections in Natural History*. (W.W. Norton, New York).
9. Mayr, E. (2001). *What Evolution Is*. (Basic Books, New York).
10. Quammen, D. (2018). *The Tangled Tree. A Radical New History of Life.* (Simon and Schuster, New York).
11. Lane, N. (2015). *The Vital Question: Energy, Evolution, and the Origins of Complex Life*. (W.W. Norton, New York).
12. Lawrence, P.A. (1992). *The Making of a Fly: The Genetics of Animal Design.* (Wiley-Blackwell, Hoboken, New Jersey).

13. Kirschner, M.E. and Gerhart, J.C. (2005). *The Plausibility of Life*. (Yale University Press, New Haven).
14. Noble, D. (2008). *The Music of Life: Biology Beyond Genes*. (Oxford University Press, New York).
15. Carroll, S.B. (2006). *Endless Forms Most Beautiful: The New Science of Evo-Devo*. (W.W. Norton, New York).
16. Kastner, R.E., Kauffman, S. and Epperson, M. (2017). *Taking Heisenberg's Potentia Seriously*. *arXiv*:1709.03595v4[quant-ph]
17. Kastner, R.E. (2015). *Understanding Our Unseen Reality*. (Imperial College Press, London).
18. Behe, M.J. (2006). *Darwin's Black Box: The Biochemical Challenge to Evolution*, 2nd Edition. (Free Press, New York).
19. Kuhn, T.S. (1969). *The Structure of Scientific Revolutions,* 2nd Edition. (University of Chicago Press, Chicago).
20. Darwin, C. (1859). *On the Origin of Species by Means of Natural Selection, or the Preservation of Favoured Races in the Struggle for Life,* 1st Edition. (John Murray, London).
21. Carroll, S.M. (2016). *The Big Picture: On the Origins of Life, Meaning, and the Universe Itself.* (Penguin Audio, New York).
22. Morgan, T.H., Sturtevant, A.H., and Muller, H.J. (1915; Revised 1922). *The Mechanism of Mendelian Heredity.* Henry Holt, New York).
23. Huxley, J. (1942). *Evolution: The Modern Synthesis.* (George Allen and Unwin).
24. Avery, O.T., McLeod, M.D., and McCarty, M. (1944). Studies on the chemical nature of the substance inducing transformation of pneumococcal types. Induction of transformation by a desoxyribonucleic acid fraction isolated from Pneumococcus type III. *J. Exp. Med.,* 79 (2), pp. 137–159.
25. Watson, J.D. and Crick, F.H.C. (1953). A structure for deoxyribose nucleic acid. *Nature* 171, pp. 737–738.
26. Miller, S.J. (1953). A production of amino acids under possible primitive Earth conditions. *Science*, 117 (3046), pp. 528–529.
27. Griffiths, A.J.F., Miller, J.H., Suzuki, D.T. *et al.* (2000). *An Introduction to Genetic Analysis,* 7th Edition. (W.H. Freeman, New York).
28. Ohno, S. (1970). *Evolution by Gene Duplication*. (Springer-Verlag, New York).
29. King, M.C. and Wilson, A.C. (1975). Evolution at two levels in humans and chimpanzees. *Science* 188 (4184), pp. 107–116. Also available on the web

at: https://books.google.com/books?id=3ULyAgAAQBAJ&lpg=PA301&ots=j_e6FUgBuo&dq=%E2%80%9CEvolution%20at%20Two%20Levels%20in%20Humans%20and%20Chimpanzees%E2%80%9D&lr&pg=PA267#v=onepage&q=%E2%80%9CEvolution%20at%20Two%20Levels%20in%20Humans%20and%20Chimpanzees%E2%80%9D&f=false

30. Nüsslein-Volhard, C. and Wieschaus, E. (1980). Mutations affecting segment number and polarity in *Drosophila*. *Nature*, 287, pp. 795–803.
31. Russell, M.J., Hall, A.J., Cairns-Smith, A.G., and Braterman, P.S. (1988). Submarine hot springs and the origin of life. *Nature*, 336, p. 117.
32. Wright, R. (1994). *The Moral Animal. Why We Are the Way We Are: The New Science of Evolutionary Psychology.* (Pantheon Books, New York).
33. Wilson, E.O. (1998). *Consilience: The Unity of Science.* (Alfred E. Knopf, New York).
34. Madigan, M.T. and Oren, A. (1999). Thermophilic and halophilic extremophiles. *Curr. Opinion Microbiol.*, 2 (3), pp. 265–269.
35. Martin, W. and Müller, M. (1998). The hydrogen hypothesis for the first eukaryote. *Nature*, 392, pp. 37–41.
36. Diemer G.S. and Stedman K.M. (2012). A novel virus genome discovered in an extreme environment suggests recombination between unrelated groups of RNA and DNA viruses. *Biology Direct*, 7, p. 13.
37. Huxley, J. (1953). *Evolution in Action.* (Harper and Brothers, New York).
38. Makalowski, (2000). Genomic scrap yard: how genomes utilize all that junk. *Gene*, 259 (1–2), pp. 61–67.
39. Levinson, G. and Gutman, G.A. (1987). Slipped-strand mispairing: a major mechanism for DNA sequence evolution. *Molecular Biology and Evolution* 4 (3), pp. 203–221, Also, PDF available on the web at: https://www.researchgate.net/profile/Gene_Levinson/publication/19827130_Slipped-Strand_Mispairing_A_Major_Mechanism_for_DNA_Sequence_Evolution/links/59e64467a6fdcc0e882483c4/Slipped-Strand-Mispairing-A-Major-Mechanism-for-DNA-Sequence-Evolution.pdf
40. Hysi, P.G. *et al.* (2018). Genome-wide association meta-analysis of individuals of European ancestry identifies new loci explaining a substantial fraction of hair color variation and heritability. *Nature Genetics*, 50, pp. 652–656.
41. Dawkins, R. (1978). *The Selfish Gene.* (Oxford University Press, New York).
42. Kelley, D.S. *et al.* (2005). A serpentinite-hosted ecosystem: the lost city hydrothermal field. *Science,* 307 (5714), pp. 1428–1434.
43. Moody, J.B. (1976). Serpentinization: a review. *Lithos,* 9, pp. 125–138.

44. Russell, M.J. and Hall, A.J. (1997). The emergence of life from iron monosulphide bubbles at a submarine hydrothermal redox and pH front. *Journal of the Geological Society,* 154, pp. 377–402.
45. Lane, N., Martin, W.F., Raven, J.A., and Allen, J.F. (2013). Energy, genes and evolution: introduction to an evolutionary synthesis. *Philosophical Transactions of the Royal Society B.*, 368, 20120253, pp. 1–5.
46. Martin, W.F., Garg, S., and Zimorski, V. (2018). Endosymbiotic theories for eukaryote cell origin. *Philosophical Transactions of the Royal Society of London. Series B, Biological Sciences*, 370 (20140330), pp. 1–18.
47. Archibald, J.M. (2015). Endosymbiosis and eukaryotic cell evolution. *Current Biology,* 25 (19), pp. R911–R921.
48. Lane, N. and Martin, W.F. (2012). The origin of membrane bioenergetics. *Cell*, 151, pp. 1406–1416.
49. Martin, W. and Russell, M.J. (2003). On the origins of cells: a hypothesis for the evolutionary transitions from abiotic geochemistry to chemoautotrophic prokaryotes, and from prokaryotes to nucleated cells. *Philosophical Transactions of the Royal Society of London. Series B, Biological Sciences*, 358, pp. 59–85.
50. Kutschera, U. (2003). A comparative analysis of the Darwin-Wallace papers and the development of the concept of natural selection. *Theory in Biosciences*, 122, pp. 343–349.
51. Joyce, G.F. (1989). RNA evolution and the origins of life. *Nature*, 338, pp. 217–224.
52. Kaminski, J. *et al.* (2019). Evolution of facial muscle anatomy in dogs. *Proceedings of the National Academy of Sciences of the United States of America.* DOI: https://doi.org/10.1073/PNAS.1820653116.
53. Sapolsky, R.M. (2017). *Behave: The Biology of Humans at Our Best and Worst.* (Penguin Books, New York).
54. Hom, E.F.Y. and Murray, A.W. (2014). Niche engineering demonstrates a latent capacity for fungal-algal mutualism. *Science*, 345 (6192), pp. 94–98.
55. Martin, W. and Kowallik, K. (1999). Annotated English translation of Mereschkowsky's 1905 paper "Uber Natur und Ursprung der Chromatophoren imPflanzenreiche". *European Journal of Phycology*, 34 (3), pp. 287–295.
56. Sagan, L. (1967). On the origin of mitosing cells. *The Journal of Theoretical Biology*, 14 (3), pp. 225–274.
57. Margulis, L. (1970). *Origin of Eukaryotic Cells.* (Yale University Press, New Haven).

58. W. F. Doolittle, W.F., C. R. Woese, C.R., Sogin, M. L., Bonen, L., and Stahl, D. (1975). Sequence studies on 16S ribosomal RNA from a blue-green alga. *Journal of Molecular Evolution*, 4 (4), pp. 307–315.
59. Woese, C.R. and Fox, G.E. (1977). Phylogenetic structure of the prokaryotic domain: The primary kingdoms. *Proceedings of the National Academy of Sciences of the United States of America*, 74 (11), pp. 5088–5090.
60. Lane, N. and Martin, W. (2010). The energetics of genome complexity. *Nature*, 467, pp. 929–934.
61. Cox, C., Foster, P.G., Hirt, R.P., Harris, S.R., and Embley, T.M. (2008). The archaebacterial origin of eukaryotes. *Proceedings of the National Academy of Sciences of the United States of America*, 105 (51), pp. 20356–20361.
62. Darwin, C.R. (1851). [Note on a Galapagos lichen]. In Hooker, J.D., An enumeration of the plants of the Galapagos Archipelago; with descriptions of those which are new. [Read 4 March, 6 May, and 16 December 1845.] *Transactions of the Linnean Society of London*, 20, p. 164.
63. Wilson, E.O. (1992). *The Diversity of Life*. (W.W. Norton, New York).
64. Gould, S.J. (1977). *Ever Since Darwin. Reflections in Natural History #1*. (W.W. Norton, New York).
65. Carson, R., Pratt, C., and Kelsh, N. (1998). *The Sense of Wonder*. (Harper, New York. First published in 1956).
66. Papenfort, K. and Bassler, B.L. (2016). Quorum sensing signal–response systems in Gram-negative bacteria. *Nature Reviews Microbiology*, 14, pp. 576–588.
67. White, G.F. *et al.* (2016). Mechanisms of bacterial extracellular electron exchange. *Advances in Microbial Physiology*, 68, pp. 87–138.
68. Gould, S.J. and Lewontin, R.C. (1979). The spandrels of San Marco and the panglossian paradigm: a critique of the adaptationist program. *Proceedings of the National Academy of Sciences of the United States of America*, 205 (1161), pp. 581–598.
69. Priest G. (2017). Framing causal questions about the past: The Cambrian explosion as case study. *Studies in History and Philosophy of Biological and Biomedical Sciences*, 63, pp. 55–63.
70. Noble, D. (2015). Evolution beyond neo-Darwinism: a new conceptual framework. *Journal of Experimental Biology*, 218, pp. 7–13. Corrected version available on the web as a pdf: http://jeb.biologists.org/content/jexbio/218/1/7.full.pdf 14 March 2018.

71. Paley, W. (1802). *Natural Theology, Or, Evidences of the Existence and Attributes of the Deity. Collected from the Appearances of Nature.* (John Morgan, Philadelphia).
72. Barry, J.M. (2006). Narrated by Brick, S. *The Great Influenza: The Epic Story of the Deadliest Plague in History, audiobook version.* (Penguin Audio, New York).
73. Mayr, E. (1942). *Systematics and the Origin of Species, from the Viewpoint of a Zoologist.* (Harvard University Press, Cambridge Massachusetts).
74. Fisher, R.A. (1930). *The Genetical Theory of Natural Selection.* (Clarendon Press, Oxford).
75. Dobzhansky, T. (1937). *Genetics and the Origin of Species.* (Columbia University Press, New York).
76. Simpson, G.G. (1944). *Tempo and Mode in Evolution.* (Columbia University Press, New York).
77. Beadle, G.W. and Tatum, E.L. (1941). Genetic control of biochemical reactions in Neurospora. *Proceedings of the National Academy of Sciences of the United States of America*, 27, pp. 499–506.
78. Crick, F.H.C. (1958). On Protein Synthesis. In F.K. Sanders. *Symposium of the Society for Experimental Biology, Number XII: The Biological Replication of Macromolecules*, pp. 138–163. (Cambridge University Press, Cambridge United Kingdom).
79. Doudna, J.A. and Sternberg, S.H. (2017). *A Crack in the Creation: Gene Editing and the Unthinkable Power to Control Evolution.* (Houghton-Mifflin Harcourt, Boston Massachusetts).
80. Britten, R.J. and Davidson, E.H. (1969). Gene regulation for higher cells: a theory. *Science,*165, pp. 349–357.
81. Maslov, S. *et al.* (2009). Toolbox model of evolution of prokaryotic metabolic networks and their regulation. *Proc. Natl. Acad. Sci. USA*, 106 (24), pp. 9743–9748.
82. Jiggins, C.D., Wallbank, R.W.R., and Hanly, J.J. (2017). Waiting in the wings: what we can learn about gene co-option from the diversification of butterfly wing patterns. *Philosophical Transactions of the Royal Society B*, 372, 20150485.
83. Bar-On, Y.M., Phillips, R., and Milo, R. (2018). The biomass distribution on Earth. *Proceedings of the National Academy of Sciences of the United States of America*, 115 (25), pp. 6506–6511.
84. Brunet, T. and King, N. (2017). The origin of animal multicellularity and cell differentiation. *Dev. Cell,* 43, pp. 124–140.

85. Richter, D.J., Fouzouni, P., Eisen, M.B., and King, N. (2018). Gene family innovation, conservation and loss on the animal stem lineage. *eLife,* 7, e342267, pp. 1–43.
86. Stephen R., Fairclough, S.R., Dayel, M.J., and King, N. (2010). Multicellular development in a choanoflagellate. *Current Biology*, 20(20), pp. R875–R876.
87. Biernaskie, J.M. and West, S.A. (2015). Cooperation, clumping and the evolution of multicellularity. *Proceedings of the Royal Society B*, 282, 2015075, pp. 1–9.
88. Escalante, R. *et al.* (2004). The MADS-box transcription factor SrfA is required for actin cytoskeleton organization and spore coat stability during Dictyostelium sporulation. *Mechanisms of Development*, 121 (1), pp. 51–56.
89. Purugganan, M.D. *et al.* (1995). Molecular evolution of flower development: diversification of the plant MADs-box regulatory gene family. *Genetics*, 140 (1), pp. 345–356.
90. Becker, A. and Theissen, G. (2003). The major clades of MADS-box genes and their role in the development and evolution of flowering plants. *Molecular Phylogenetics and Evolution,* 29 (3), pp. 464–489.
91. Gramzow, L. and Theissen, G. (2010). A hitchhiker's guide to the MADS world of plants. *Genome Biology*, 11 (6), p. 214.
92. Shore, P. and Sharrocks, A.D. (1995). The MADS-box family of transcription factors. *European Journal of Biochemistry*, 229 (1), pp. 1–13.
93. Rensing, S.A. *et al.* (2007). The *Physcomitrella* genome reveals evolutionary insights into the conquest of land by plants. *Science,* 319, pp. 58–59 (available on the web via scienceexpress.org).
94. Hagemann, M. *et al.* (2016). Evolution of photorespiration from cyanobacteria to land plants, considering protein phylogenies and acquisition of carbon concentrating mechanisms. *Journal of Experimental Botany*, 67 (10), pp. 2963–2976.
95. Harrison, C.J. (2017). Development and genetics in the evolution of land plant body plans. *Philosophical Transactions of the Royal Society B*, 372, 20150490, pp. 1–12.
96. Lawrence, P.A. and Levine, M. (2006). Mosaic and regulative development: two faces of one coin. *Current Biology*, 16 (7), pp. R236–R239.
97. Wilkins, A.S. (1993). *Genetic Analysis of Animal Development,* 2nd Edition. (Wiley-Liss, New York).
98. Pontius, J.U. *et al.* (2007). Initial sequence and comparative analysis of the cat genome. *Genome Research,* 17, pp. 1675–1689.

99. Sagan, C. and Druyan, A. (2013). *Cosmos.* (Ballantine Books, New York).
100. Osaka, S. *et al.* (1992). Recent evidence for evolution of the genetic code. *Microbiology Reviews*, 56 (1), pp. 229–264.
101. Forrest, S. (1993). Genetic algorithms: principles of natural selection applied to computation. *Science*, 261 (5123), pp. 872–878.
102. Kahneman, D. (2011). *Thinking, Fast and Slow.* (Farrar, Straus and Giroux, New York).
103. Holland, J. (1992). *Adaptation in Natural and Artificial Systems: An Introductory Analysis with Applications to Biology, Control, and Artificial Intelligence.* (MIT Press, Cambridge, Massachusetts).
104. Levinson, G. (1994). Crossovers generate nonrandom recombinants under Darwinian selection. In *Artificial Life IV* (Cambridge MA, M.I.T. Press).
105. Levinson, G. (1995). Rethinking Evolution. The World & I magazine, January 1995 issue, pp. 195–201. Also, PDF available on the web at: https://www.researchgate.net/publication/320471802_Rethinking_Evolution
106. Bohm, D. and Peat, F.D. (1987). *Science, Order, and Creativity: A Dramatic New Look at the Creative Roots of Science and Life* (Reissue Edition). (Bantam, New York).
107. Woodworth, M.B. *et al.* (2017). Building a lineage from single cells: genetic techniques for lineage tracking. *Nature Reviews Genetics*, 18, pp. 230–244.
108. Cochard, L.R. (2012). *Netter's Atlas of Human Embryology, Updated Edition.* (Saunders, Philadelphia PA).
109. Shubin, N., Tabin, C. and Carroll, S. (2009). Deep homology and the origins of evolutionary novelty. *Nature*, 457, pp. 818–823.
110. Tschopp, P. and Tabin, C.J. (2017). Deep homology in the age of next-generation sequencing. *Philosophical Transactions of the Royal Society B*, 372, 20150475, pp. 1–8.
111. Held, L.I. (2010). The evolutionary geometry of human anatomy: discovering our inner fly. *Evolutionary Anthropology*, 19, pp. 227–235.
112. Martin, A. and Reed, R.D. (2014). Wnt signaling underlies evolution and development of the butterfly wing pattern symmetry systems. *Developmental Biology*, 395, pp. 367–378.
113. Reed, R.D. *et al.* (2011). *optix* drives the repeated convergent evolution of butterfly wing mimicry. *Science,* 333 (6046), pp. 1137–1141.
114. Zhang, L. *et al.* (2017). Single master regulatory gene coordinates the evolution and development of butterfly color and iridescence. *Proceedings of the National Academy of Sciences of the United States of America*, 114 (40), pp. 10707–10712.

115. Fortey, R. (2010). *Dry Store Room No. 1: The Secret Life of the Natural History Museum*. UK edition. (Harper Press, London).
116. Taylor, J.S. and Raes, J. (2004). Duplication and divergence: the evolution of new genes and old ideas. *Annual Review of Genetics*, 38, pp. 615–643.
117. Hunkapiller, T. and Hood, L. (1989). Diversity of the immunoglobulin gene superfamily. *Advances in Immunology*, 44, pp. 1–63.
118. Markert, E. and Papavasillou, F.N. (2003). *PLoS Biology*, 1 (1), pp. 24–27.
119. Flajnik, M.F. (2016). Evidence of G.O.D.'s miracle: unearthing a RAG transposon. *Cell*, 166 (1), pp. 11–12.
120. Huang *et al*. (2016). Discovery of an active RAG transposon illuminates the origins of V(D)J recombination. *Cell*, 166 (1), pp. 102–114.
121. Diaz, M *et al*. (2001). Evolution and the molecular basis of somatic hypermutation of antigen receptor genes. *Philosophical Transactions of the Royal Society of London. Series B, Biological Sciences*, 356, pp. 67–72.
122. Gerhart, J. and Kirschner, M. (2007). The theory of facilitated variation. *Proceedings of the National Academy of Sciences of the United States of America*, 104(Suppl 1), pp. 8582–8589.
123. Gould, S. J. (1989). *Wonderful Life. The Burgess Shape and the Nature of History*. (W.W. Norton, New York).
124. Stent G.S. (1977). Explicit and implicit semantic content of the genetic information. In: Butts R.E., Hintikka J. (eds.), *Foundational Problems in the Special Sciences*. The University of Western Ontario Series in Philosophy of Science, vol 10. (Springer, Dordrecht).
125. Lane, N. Narrated by Patterson, N. (2005). *Power, Sex, Suicide: Mitochondria and the Meaning of Life, Unabridged*. (Tantor Audio, Old Saybrook, CT.).
126. Krebs, J.E., Goldstein, E.S. and Kilpatrick, S.T. (2017). Lewin's *Genes XII*. 12th edition. (Jones & Bartlett Learning, Burlington MA).
127. Alberts, B. *et al*. (2014). Molecular Biology of the Cell, 6th edition. (Garland Publishing, Oxford U.K.).
128. Chen, H. *et al*. (2012). A system of repressor gradients spatially organizes the boundaries of bicoid-dependent target genes. *Cell*, 149 (3), pp. 618–629.
129. Tuschman, A. (2013). *Our Political Nature: The Evolutionary Origins of What Divides Us*. (Prometheus, Amherst, New York).

Glossary

Glossary Term	Descriptions or Definitions, Including Modified Wikipedia Excerpts	Relevant Link (Usually Wikipedia)
A, C, G, T	Abbreviations for each of the four possible nitrogenous bases in each nucleotide of a DNA sequence, including adenine (A), cytosine (C), guanine (G), and thymine (T).	(not applicable)
A, C, G, U	Abbreviations for each of the four possible nitrogenous bases in each nucleotide of an RNA sequence, including adenine (A), cytosine (C), guanine (G), and uracil (U).	(not applicable)
aaRS (aminoacyl-tRNA synthetase)	Abbreviation for amino-acyl RNA synthetase enzymes. These enzymes link specific amino acids to specific transfer RNA (tRNA) molecules. In this way, amino acids that correspond to particular codons in DNA and messenger RNA (mRNA) are brought to the ribosomes, where they are chemically bonded, in the correct sequence, in each growing protein chain, during protein synthesis.	https://en.wikipedia.org/wiki/Aminoacyl_tRNA_synthetase

(Continued)

Glossary Term	Descriptions or Definitions, Including Modified Wikipedia Excerpts	Relevant Link (Usually Wikipedia)
Abiogenesis	Informally known as the origin of life, it is the natural process by which life arises from nonliving matter, such as simple organic compounds. The transition from nonliving to living entities was a gradual process of increasing complexity.	https://en.wikipedia.org/wiki/Abiogenesis
Abstract	A succinct formal summary of the contents of a published scientific journal article, frequently included at the beginning of the article. Publishers often provide free access to Abstracts.	(not applicable)
Accessory pigments	Light-absorbing compounds, found in photosynthetic organisms, that work in conjunction with chlorophyll a. They include other forms of this pigment, such as chlorophyll b, c, or d.	https://en.wikipedia.org/wiki/Accessory_pigment
Acetogen	A microorganism that generates acetate as an end product of respiration without oxygen. There are also what are thought to be genuine acetogens, known as "homoacetogens". These can produce biochemical reactions from two molecules of carbon dioxide and four molecules of molecular hydrogen.	https://en.wikipedia.org/wiki/Acetogen
Acetogenesis	A process through which acetate is produced from carbon dioxide and an electron source, with no requirement for oxygen. This takes place via biochemical reactions.	https://en.wikipedia.org/wiki/Acetogenesis
Actin	Subunit containing alpha and/or beta actin protein chains. Actin subunits are often found as polymerized actin microfilaments that make up part of the cytoskeleton of a cell.	https://en.wikipedia.org/wiki/Actin

(*Continued*)

Glossary Term	Descriptions or Definitions, Including Modified Wikipedia Excerpts	Relevant Link (Usually Wikipedia)
Adaptation	In biology, there are three related meanings. Firstly, it is the dynamic evolutionary process that fits organisms to their environment, enhancing their evolutionary fitness. Secondly, it is a state reached by the population during that process. Thirdly, it is an observable trait, with a functional role in each individual organism, that is maintained and has been evolved by Natural Selection.	https://en.wikipedia.org/wiki/Adaptation
Adaptations	Observable traits, with functional roles in each individual organism, that are maintained and have been evolved by Natural Selection.	https://en.wikipedia.org/wiki/Adaptation
Adaptive immune system	Specialized cells such as T cells, B cells and accessory cells, and members of the immunoglobulin superfamily of genes, that interact and bind to specific antigens and proactively increase the numbers of circulating T cells and B cells that bind to particular antigens, thereby optimizing secondary immune responses. Also responds to self-antigens by removing or suppressing T cell and B cells that respond to those self-antigens (immune tolerance).	https://en.wikipedia.org/wiki/Adaptive_immune_system
Adenosine triphosphate (ATP)	A "high-energy" compound widely used as a readily accessible energy source for active cellular processes. Most energy is captured by breaking the covalent bond between the second and third phosphates attached to the sugar. These bonds represent high potential energy because repulsion of the electron clouds make them highly unstable.	https://en.wikipedia.org/wiki/Adenosine_triphosphate

(*Continued*)

Glossary Term	Descriptions or Definitions, Including Modified Wikipedia Excerpts	Relevant Link (Usually Wikipedia)
Algae (algal)	An informal term for a large, diverse group of photosynthetic eukaryotic organisms that are not necessarily closely related. Included organisms range from unicellular microalgae genera, such as Chlorella and the diatoms, to multicellular forms, such as the giant kelp, a large brown alga which may grow up to 50 m in length.	https://en.wikipedia.org/wiki/Algae
Alkaline	An aqueous solution that has a pH above 7.0 due to a lower concentration of protons (hydrogen ions) than found in pure water.	https://en.wikipedia.org/wiki/Alkali
Alkaline hydrothermal vent (Lost City Hydrothermal Field)	The first alkaline deep-sea hydrothermal vent, known as the Lost City, was discovered on the mid-Atlantic sea floor in 2,000. Distinguished from high-temperature hydrothermal vents known as black smokers, alkaline vents are formed by serpentinization, in which olivine (magnesium iron silicate) and similar rock on the seafloor reacts with water. This produces large volumes of hydrogen. Precipitation reactions between the warm (45–90°C) alkaline (pH 9–11) vent fluids and the colder, more acidic seawater create tall, porous calcium carbonate chimneys containing embedded metals and minerals. It is thought that proton gradients generated by similar, ancient vents provided the energy required for the metabolic reactions that led to the origin of life.	https://en.wikipedia.org/wiki/Hydrothermal_vent

(Continued)

Glossary Term	Descriptions or Definitions, Including Modified Wikipedia Excerpts	Relevant Link (Usually Wikipedia)
Allele	A variant form of a given gene, resulting from DNA sequence differences. In protein-coding genes, may result in production of a variant amino acid sequence, or failure to produce a functional protein.	https://en.wikipedia.org/wiki/Allele
Allele frequency	The relative frequency of an allele (a variant) of a specific gene in a population. Alleles are represented by specific DNA sequences at particular chromosomal positions. The allele frequency can be expressed as a fraction or percentage.	https://en.wikipedia.org/wiki/Allele_frequency
Alpha-helix	A common motif in the secondary structure of proteins. Also, a helix in which the N–H backbone donates a hydrogen bond to the C=O groups of the amino acid.	https://en.wikipedia.org/wiki/Alpha_helix
Alpha-proteobacteria	A class of bacteria in the Proteobacteria phylum. Its members are highly diverse and possess few commonalities, but share a common ancestor.	https://en.wikipedia.org/wiki/Alphaproteobacteria
Amino acid	Organic compounds containing amine and carboxyl functional groups, along with a side chain specific to each amino acid. The key elements are carbon, hydrogen, oxygen, and nitrogen.	https://en.wikipedia.org/wiki/Amino_acid
Amino acid sequence	The sequence of amino acids in a protein chain, or the sequence that can be predicted from analysis of a specific protein-coding DNA or RNA segment.	https://en.wikipedia.org/wiki/Protein_primary_structure
Amino group	Functional group consisting of a nitrogen atom attached to carbon or hydrogen atoms.	https://en.wikipedia.org/wiki/Amino_acid

(*Continued*)

Glossary Term	Descriptions or Definitions, Including Modified Wikipedia Excerpts	Relevant Link (Usually Wikipedia)
Amplify (PCR)	To amplify is to use PCR to generate a large number of identical copies of nucleic acid target sequences, starting with a small number of DNA or RNA template molecules. PCR is a powerful, rapid, and widely-used technique in biotechnology.	https://en.wikipedia.org/wiki/Polymerase_chain_reaction
Animal	A member of the animal kingdom. Animals represent one of the three kingdoms of multicellular eukaryotic organisms, the others being plants and fungi.	https://en.wikipedia.org/wiki/Animal
Anthropomorphic	The attribution of human traits, emotions, or intentions to nonhuman entities. Anthropomorphism is considered to be an innate tendency of human psychology.	https://en.wikipedia.org/wiki/Anthropomorphism
Antibiotic resistance	An evolved insensitivity to one or more antibiotics, which takes place by a variety of known mechanisms, and represents a serious threat to the global human population.	https://en.wikipedia.org/wiki/Antimicrobial_resistance
Antibody	One of an extremely diverse set of proteins produced by B cells. The variable regions of these proteins can bind to a very large number of distinct antigens in a highly selective manner.	https://en.wikipedia.org/wiki/Antibody
Anticodon	Three-base triplet on transfer RNA molecule that base-pairs with messenger RNA codon during synthesis of a protein chain (translation). Assures that the amino acid will be incorporated into the growing chain in the correct sequence, at the ribosome.	https://en.wikipedia.org/wiki/Transfer_RNA#Anticodon

(*Continued*)

Glossary Term	Descriptions or Definitions, Including Modified Wikipedia Excerpts	Relevant Link (Usually Wikipedia)
Antigen (immunology)	A molecule capable of inducing an immune response in the host organism.	https://en.wikipedia.org/wiki/Antigen
Antigen-presenting cell	Also known as an accessory cell, it is a cell that displays a molecule which will induce an immune response. It is complexed with cell surface proteins essential for the immune system to recognize foreign molecules.	https://en.wikipedia.org/wiki/Antigen-presenting_cell
Antiporter	A transporter, which in modern cells consists of a transport protein embedded in a biological membrane. Antiporters move two different types of molecules in opposite directions. This may take advantage of the energy flux of one type of molecule to move the other type in the opposite direction.	https://en.wikipedia.org/wiki/Antiporter
Apoptosis	A type of programmed-cell death that represents a normal part of development and also provides a way of eliminating cells that are dangerous because they are infected or cancerous.	https://en.wikipedia.org/wiki/Apoptosis
Aqueous solution	A solution in which the solvent that dissolves a solute is water. Solutes can be liquid, solid or gas. The cytoplasm of all cells mostly consists of an aqueous solution that is packed with large numbers of macromolecules and other substances.	https://en.wikipedia.org/wiki/Aqueous_solution
Archaea (archaebacteria)	One of the two domains of prokaryotic organisms, which can be distinguished from bacteria by comparison of their genome, metabolism, and cell wall structure. Reported in 1977 by Carl Woese and colleagues.	https://en.wikipedia.org/wiki/Archaea

(*Continued*)

Glossary Term	Descriptions or Definitions, Including Modified Wikipedia Excerpts	Relevant Link (Usually Wikipedia)
Astrobiology	Searches for empirical evidence and principles related to past or present extraterrestrial life forms.	https://en.wikipedia.org/wiki/Astrobiology
Atom	The smallest constituent unit of ordinary matter that has the properties of a chemical element. Every solid, liquid, gas, and plasma is composed of atoms.	https://en.wikipedia.org/wiki/Atom
Atomic nuclei	Relatively small dense core of atoms containing positively charged protons and neutrally charged neutrons.	https://en.wikipedia.org/wiki/Atomic_nucleus
ATP	A complex organic chemical that provides energy to drive many processes in living cells. It is also a precursor to DNA and RNA, and is used as a coenzyme.	https://en.wikipedia.org/wiki/Adenosine_triphosphate
ATP synthase	An enzyme that synthesizes ATP (adenosine triphosphate) by transforming the energy of protons that flow through the enzyme to equalize a proton gradient.	https://en.wikipedia.org/wiki/ATP_synthase
ATP synthesis	Synthesis of adenosine triphosphate, a widespread energy currency of all living cells.	https://en.wikipedia.org/wiki/ATP_synthase
Autonomous specification (determinate cleavage)	A type of embryonic cleavage in which cytoplasmic determinants combined with cell lineage define patterns of cell differentiation, rather than intercellular communication between neighboring embryonic cells (blastomeres).	https://en.wikipedia.org/wiki/Cleavage_(embryo)
Autosome	A chromosome that is not a sex chromosome. Diploid organisms have pairs of autosomes that are similar in form and DNA sequence. Each gene in a pair of autosomes may contain the same alleles or different alleles.	https://en.wikipedia.org/wiki/Autosome

(*Continued*)

Glossary Term	Descriptions or Definitions, Including Modified Wikipedia Excerpts	Relevant Link (Usually Wikipedia)
Autotroph	An organism that captures energy from inorganic chemicals or sunlight and use raw materials such as carbon dioxide and minerals to build their own organic compounds.	https://en.wikipedia.org/wiki/Autotroph
B cell	A specialized white blood cell, the part of the adaptive immune system that produces specific antibodies.	https://en.wikipedia.org/wiki/B_cell
Bacteria (eubacteria)	Single-celled organisms from one of the two domains of prokaryotes. The other domain is archaea.	https://en.wikipedia.org/wiki/Bacteria
Bacteriophage	A virus that infects and replicates within bacteria and archaea. Bacteriophages are composed of proteins that encapsulate a DNA or RNA genome.	https://en.wikipedia.org/wiki/Bacteriophage
Baldwin effect	The effect of a learned behavior on evolution. James Mark Baldwin hypothesized an organism's ability to learn new behaviors could affect its reproductive success. This would increase the probability that half of its genome would be passed on to the next generation. Any genes that contributed to that ability to learn could be passed on. This could increase the learning capabilities in the gene pool. The Baldwin effect was consistent with other empirical data and was considered part of the Modern Synthesis.	https://en.wikipedia.org/wiki/Baldwin_effect
Base (nitrogenous)	One of four different nitrogen-containing organic ring structures found in nucleotides of DNA or RNA (see A, C, G, T and A, C, G, U, respectively). Nitrogenous bases pair with complementary bases by hydrogen bonding; A pairs with T or U and G pairs with C.	https://en.wikipedia.org/wiki/Nitrogenous_base

(*Continued*)

Glossary Term	Descriptions or Definitions, Including Modified Wikipedia Excerpts	Relevant Link (Usually Wikipedia)
Base-pair (base-pairing)	As a verb, refers to hydrogen bond formation between complementary pairs of nitrogenous bases in strands of DNA and/or RNA. As a noun, referred to pair of hydrogen-bonded nitrogenous bases. In DNA, A pairs with T and G with C; in RNA, A pairs with U and G pairs with C.	https://en.wikipedia.org/wiki/Base_pair
Basic research (fundamental research)	Research intended to expand and improve knowledge concerning natural phenomena, usually without a specific or immediate application in mind.	https://en.wikipedia.org/wiki/Basic_research
Beta-pleated sheet	A common motif of regular secondary structure in proteins. They consist of beta strands connected laterally by two or three backbone hydrogen bonds. This forms the twisted, pleated sheet.	https://en.wikipedia.org/wiki/Beta_sheet
Bicoid	A maternal effect gene whose protein concentration gradient patterns the anterior–posterior (A–P) axis during *Drosophila* (fruitfly) embryogenesis.	https://en.wikipedia.org/wiki/Bicoid_(gene)
Binary bits	Units of digital information used in computer data. Has two alternative states, designated as 0 or 1, representing electrical states that switch between "off" or "on". Sequences of bits can encode larger units of information, such as ASCII characters.	https://en.wikipedia.org/wiki/Bit
Biofilm	Any consortium of microorganisms exchanging nutrients, in which cells stick to each other and often also to a surface. These adherent cells become embedded within a slimy extracellular matrix.	https://en.wikipedia.org/wiki/Biofilm

(*Continued*)

Glossary Term	Descriptions or Definitions, Including Modified Wikipedia Excerpts	Relevant Link (Usually Wikipedia)
Bioinformatics	An interdisciplinary field that develops methods and software tools for understanding biological data. As an interdisciplinary field of science, bioinformatics combines biology, computer science, information engineering, mathematics and statistics to analyze and interpret biological data.	https://en.wikipedia.org/wiki/Bioinformatics
Biological evolution	The empirically proven natural process of change in all living things on Earth. Natural Selection is one of the most fundamental mechanisms that account for biological evolution, but it is not the only relevant mechanism involved. Biological evolution includes the precellular molecular changes that led to the origin of life on Earth, as well as subsequent changes in the complexity of structures and functions in all living cells and organisms. Evolution is also the fundamental process responsible for the origin of species, as pointed out by Charles Darwin in 1859, in the context of Natural Selection.	https://en.wikipedia.org/wiki/Category:Biological_evolution
Biological membrane	An enclosing or separating membrane consisting of a lipid bilayer that acts as a selectively permeable barrier within or surrounding a living cell.	https://en.wikipedia.org/wiki/Biological_membrane
Biological organization	The hierarchy of complex biological structures, functions, and systems, that helps to define the complexity of life in terms of simpler ideas. This is a fundamental premise for numerous areas of scientific research.	https://en.wikipedia.org/wiki/Biological_organisation

(Continued)

Glossary Term	Descriptions or Definitions, Including Modified Wikipedia Excerpts	Relevant Link (Usually Wikipedia)
Biomass (ecology)	The mass of living biological organisms in a given area or ecosystem at a given time. Biomass can refer to species biomass, which is the mass of one or more species, or to community biomass, which is the mass of all species in the community. It can include microorganisms, plants, or animals. The mass can be expressed as the average mass per unit area, or as the total mass in the community.	https://en.wikipedia.org/wiki/Biomass
Bithorax complex	A group of homeotic genes in fruitflies (*Drosophila melanogaster*) which control the differentiation of the abdominal and posterior thoracic segments, located on chromosome III. The name is derived from the fact that when these genes are mutated, the third thoracic segment becomes a repeat of the second thoracic segment, creating what is essentially a second thorax.	https://en.wikipedia.org/wiki/Bithorax_complex
Body plan	A body plan is a set of morphological features common to many members of a phylum of animals. Refers to diverse vertebrate and invertebrate phyla.	https://en.wikipedia.org/wiki/Body_plan
Boveri–Sutton chromosomal theory of inheritance	A now well-supported unifying theory of genetics that identifies chromosomes as the carriers of genetic material.	https://en.wikipedia.org/wiki/Boveri%E2%80%93Sutton_chromosome_theory
Buoyant density centrifugation (biology)	A separation technique in which a dense solution is centrifuged at high speed so that substances form bands equivalent to their density. Was commonly used in recombinant DNA technology to separate cloned plasmid DNA from bacterial DNA.	https://en.wikipedia.org/wiki/Buoyant_density_centrifugation

(*Continued*)

Glossary Term	Descriptions or Definitions, Including Modified Wikipedia Excerpts	Relevant Link (Usually Wikipedia)
Cambrian explosion	A relatively short period starting approximately 541 million years ago and lasting about 20–25 million years, when most major animal phyla appeared in the fossil record. It resulted in the divergence of most modern metazoan phyla and diversification of other organisms.	https://en.wikipedia.org/wiki/Cambrian_explosion
Capillary action	The ability of a liquid to flow in narrow spaces without external forces like gravity. It is the result of intermolecular forces between the liquid and surrounding solid surfaces.	https://en.wikipedia.org/wiki/Capillary_action
Carbohydrate	Macromolecule composed of carbon, hydrogen, and oxygen atoms. Includes simple sugars or polymers of sugars such as cellulose, glycogen, starch, and chitin.	https://en.wikipedia.org/wiki/Carbohydrate
Carbon fixation	Capture of carbon dioxide by photosynthetic organisms, from which organic compounds are synthesized.	https://en.wikipedia.org/wiki/Carbon_fixation
Catalyst	An inorganic substance, a co-factor or an enzyme that lowers the activation energy for a chemical reaction, thereby increasing its speed. Catalysts are typically not consumed or destroyed by chemical reactions, so they are reusable.	https://en.wikipedia.org/wiki/Catalysis
Catalytic (catalyst)	An entity or process involving a catalyst. A catalyst is an inorganic substance, a co-factor or an enzyme that lowers the activation energy for a chemical reaction, thereby increasing its speed. Catalysts are typically not consumed or destroyed by chemical reactions, so they are reusable.	https://en.wikipedia.org/wiki/Catalysis

(*Continued*)

Glossary Term	Descriptions or Definitions, Including Modified Wikipedia Excerpts	Relevant Link (Usually Wikipedia)
CD4	"Helper" T cells that interact with other T cells, B cells, and antigen presenting cells in the adaptive immune system to regulate immune responses.	https://en.wikipedia.org/wiki/CD4
CD8	"Cytotoxic" T cells that can eliminate infected or cancerous cells. May also be associated with suppression of immune responses.	https://en.wikipedia.org/wiki/CD8
cDNA (complementary DNA)	Complementary DNA strand synthesized from an RNA template by reverse transcriptase enzyme. Associated with retroviruses and widely used in biotechnology.	https://en.wikipedia.org/wiki/Complementary_DNA
Cell	The fundamental semi-autonomous unit found in all forms of unicellular or multicellular life, including the three domains of bacteria, archaea, and eukaryotes. All cells contain a cell membrane, chromosomes containing DNA, and ribosomes that are the sites of protein synthesis.	https://en.wikipedia.org/wiki/Cell_(biology)
Cell adhesion	Binding of two or more cells, often in multicellular organisms, which usually involves specific specialized proteins and cell junctions.	https://en.wikipedia.org/wiki/Cell_adhesion
Cell biology	The broad interdisciplinary fields of biology that study various aspects of cell structure and function.	https://en.wikipedia.org/wiki/Cell_biology
Cell cycle	A regulated process by which eukaryotic cells grow, replicate their genetic material and divide in an orderly fashion, in response to internal and external signals. The cell cycle includes stages of growth and DNA replication designated G1, S, and G2, mitosis in which chromosomes	https://en.wikipedia.org/wiki/Cell_cycle

(*Continued*)

Glossary Term	Descriptions or Definitions, Including Modified Wikipedia Excerpts	Relevant Link (Usually Wikipedia)
	segregate into newly forming daughter cells, and cytokinesis in which the cytoplasmic division is completed. Specialized cells often exit the cell cycle and enter a stage known as G0. In multicellular organisms, cancer cells generally lose control of the cell cycle and become a danger to the rest of the individual.	
Cell division	The division of parent cells into daughter cells. Cell division in prokaryotes is called binary fission. Cell division in eukaryotes includes mitotic division, which produces genetically identical daughter cells, and meiosis in which diploid germ line cells divide into haploid cells that develop into sperm or egg cells, known as sex cells, in which the daughter cells are genetically diverse.	https://en.wikipedia.org/wiki/Cell_division
Cell fractionation	The process used to separate cellular components while preserving individual functions of each component.	https://en.wikipedia.org/wiki/Cell_fractionation
Cell membrane (plasma membrane)	Biological membrane that separates the interior of all types of cells from the outside environment. Consists of two lipid layers with embedded proteins and other molecules. Cell membrane designation refers to outer membrane, which distinguishes that biological membrane from internal lipid membranes which are also widespread.	https://en.wikipedia.org/wiki/Cell_membrane

(*Continued*)

Glossary Term	Descriptions or Definitions, Including Modified Wikipedia Excerpts	Relevant Link (Usually Wikipedia)
Cell motility	A general term referring to various types of cell movement, migration, and shape changes, often mediated by the internal cytoskeleton and/or by cilia or flagella which can propel a cell through an external fluid medium. Cell migration often involves signals from extracellular materials as well as changes in cell adhesion.	https://en.wikipedia.org/wiki/Cell_migration
Cell signaling	A general term referring to local and long-distance signals passed between cells that coordinate their activities. In multicellular organisms, local signals may involve physical contact between cell surface proteins, including receptors, or short-range diffusion of signals known as paracrine signals. Long-range signaling often involves hormones secreted by glands, such as endocrine glands.	https://en.wikipedia.org/wiki/Cell_signaling
Cell-free preparation	A laboratory preparation that includes specific isolated or fractionated materials from cells, which often continue to carry out particular functions that can be isolated, tagged, and studied by observations and experiments under controlled conditions.	https://en.wikipedia.org/wiki/Cell-free_system
Cell-surface receptor	A protein on the cell surface that receives and responds to signals from other cells or from the environment. Often, eukaryotic cell surface receptors extend through the cell membrane and respond to signals by triggering internal signal transduction pathways involving events in the cytoplasm and nucleus of the cell.	https://en.wikipedia.org/wiki/Cell_surface_receptor

(*Continued*)

Glossary Term	Descriptions or Definitions, Including Modified Wikipedia Excerpts	Relevant Link (Usually Wikipedia)
Cellular differentiation	The process whereby initially flexible embryonic stem cells become specialized and develop into specific cell types with specialized patterns of gene expression.	https://en.wikipedia.org/wiki/Cellular_differentiation
Cellular metabolism	The set of life-sustaining chemical transformations within the cells of organisms. Its main purposes are in the breakdown (catabolism) and synthesis (anabolism) of macromolecules, and in energy transformations. Additionally, metabolism eliminates nitrogenous wastes.	https://en.wikipedia.org/wiki/Metabolism
Cellular slime mold	A type of fungus that has both solitary and aggregated multicellular stages in its life cycle.	https://en.wikipedia.org/wiki/Slime_mold
Central dogma of molecular genetics	A tongue-in-cheek expression attributed to Francis Crick, which hypothesizes that genetic information can flow from DNA to RNA to proteins but not the other way around. Today, we know that these macromolecules are parts of a complex adaptive system that interact in complex ways. For example, transcription factors are proteins that control the expression of genes by binding to DNA, which means that genetic information is flowing from proteins back to DNA.	https://en.wikipedia.org/wiki/Central_dogma_of_molecular_biology
Chaperonin	Proteins that provide favorable conditions for the correct folding of other proteins. This prevents the formation of clusters and misfolding, which in turn prevents diseases.	https://en.wikipedia.org/wiki/Chaperonin

(*Continued*)

Glossary Term	Descriptions or Definitions, Including Modified Wikipedia Excerpts	Relevant Link (Usually Wikipedia)
Charged (electric)	An entity that has a static positive or negative charge, usually associated with gains or losses of electrons.	https://en.wikipedia.org/wiki/Electric_charge
Charles Darwin	The famous 19th century biologist who published *On the Origin of Species* in 1859 which correctly hypothesized that biological evolution of adaptations and of species depend on Natural Selection.	https://en.wikipedia.org/wiki/Charles_Darwin
Chemical mutagens	Chemicals that speed up the rate of changes in DNA sequences over the spontaneous rate.	https://en.wikipedia.org/wiki/Mutagen
Chemoautotrophic	Organisms that use chemical energy from the environment to fix carbon and build macromolecules.	https://en.wikipedia.org/wiki/Chemotroph
Chloroplast	Organelles, or specialized compartments, in plant and algal cells. The main role is to conduct photosynthesis.	https://en.wikipedia.org/wiki/Chloroplast
Choanoflagellate	A group of free-living unicellular and colonial flagellate eukaryotes considered to be the closest living relatives of the animals.	https://en.wikipedia.org/wiki/Choanoflagellate
Chromatin	A complex of DNA, RNA, and protein found in eukaryotic cells. It plays various roles in the structural integrity of the DNA, in cell division, in preventing DNA damage, and in regulating gene expression and DNA replication.	https://en.wikipedia.org/wiki/Chromatin
Chromosome	A long DNA molecule often complexed with proteins and RNA sequences that represents part or all of the genetic material (genome) of an organism.	https://en.wikipedia.org/wiki/Chromosome

(*Continued*)

Glossary Term	Descriptions or Definitions, Including Modified Wikipedia Excerpts	Relevant Link (Usually Wikipedia)
Classes of organic compounds	Organic compounds are classified on the basis of the presence or absence of specific structural formulas and functional groups.	https://en.wikipedia.org/wiki/Organic_compound
Classical Darwinism	A theory of biological evolution developed by the English naturalist Charles Darwin (1809–1882) and others, stating that all species of organisms arise and develop through the Natural Selection of small, inherited variations that increase the individual's ability to compete, survive, and reproduce. Also called Darwinian theory.	https://en.wikipedia.org/wiki/Darwinism
Cleavage (embryo)	Division of cells during development of an embryo. Usually begins with a fertilized egg (zygote) in sexually reproducing organisms.	https://en.wikipedia.org/wiki/Cleavage_(embryo)
Cloning (cellular)	The process of producing genetically identical individuals of an organism from identical cells, either naturally or artificially.	https://en.wikipedia.org/wiki/Cloning
Cloning (molecular)	Production of large numbers of identical copies of DNA or RNA. Often refers to recombinant DNA technology.	https://en.wikipedia.org/wiki/Molecular_cloning
Co-dominance	Alleles that simultaneously produce distinct visible phenotypes in organisms. For example, A, B, and AB blood types are caused by the presence of co-dominant alleles, and represent proteins that act as distinct cell-surface antigens on red blood cells.	https://en.wikipedia.org/wiki/Dominance_(genetics)
Codon	A nucleotide triplet in DNA or RNA that is associated with the addition of a particular amino acid (or a start or stop signal) during translation of new protein chains.	https://en.wikipedia.org/wiki/Genetic_code

(*Continued*)

Glossary Term	Descriptions or Definitions, Including Modified Wikipedia Excerpts	Relevant Link (Usually Wikipedia)
Coevolution	Occurs when two or more species reciprocally affect each other's evolution.	https://en.wikipedia.org/wiki/Coevolution
Colonial (colony, biology)	Two or more individuals of the same species living in close proximity to each other.	https://en.wikipedia.org/wiki/Colony_(biology)
Commensalism	A long-term biological interaction in which members of one species gain benefits while members of another species are neither benefited nor harmed.	https://en.wikipedia.org/wiki/Commensalism
Complementary DNA (cDNA)	DNA synthesized from a single-stranded RNA template. This reaction is catalyzed by an enzyme called reverse transcriptase which generates a single DNA strand that is complementary to an RNA strand used as the template.	https://en.wikipedia.org/wiki/Complementary_DNA
Complex adaptive system	A system in which a perfect understanding of individual parts does not convey a perfect understanding of the whole system's behavior. Complex adaptive systems often change or evolve over time.	https://en.wikipedia.org/wiki/Complex_adaptive_system
Complex organization (biological)	The hierarchy of complex biological structures and systems that define life using a reductionist approach. Each level in the hierarchy represents an increase in organizational complexity, with each "object" being primarily composed of the previous level's basic unit. The basic principle behind the organization is the concept of emergence—the properties and functions found at a hierarchical level are not present and irrelevant at the lower levels.	https://en.wikipedia.org/wiki/Biological_organisation

(*Continued*)

Glossary Term	Descriptions or Definitions, Including Modified Wikipedia Excerpts	Relevant Link (Usually Wikipedia)
Complexity (complex adaptive system)	Biological complexity can be described as a special type of complex adaptive system, that is, a system in which a perfect understanding of the individual parts does not automatically convey a perfect understanding of the whole system's behavior. Biological systems are complex in that they are dynamic networks of interactions. They are adaptive in that the individual and collective behavior mutate, self-organize, and evolve under Natural Selection.	https://en.wikipedia.org/wiki/Complex_adaptive_system
Composite biological organization	Refers to composite cells, genomes, or organisms. Examples of composite cells are eukaryotic cells which have evolved by endosymbiosis. Examples of composite genomes are bacteria and archaea that frequently exchange genetic material by horizontal gene transfer. Examples of composite organisms are lichens which are composed of both algal and fungal components.	https://en.wikipedia.org/wiki/Biological_organisation
Composite cell (symbiogenesis)	Eukaryotic cells are composite cells because they evolve by symbiogenesis (endosymbiosis) from prokaryotic forebears. This is well-supported by a consilience of empirical evidence. The composition of the genome and presence of organelles such as mitochondria and/or chloroplasts that retain their own separate chromosomes represent some of these lines of evidence.	https://en.wikipedia.org/wiki/Symbiogenesis
Composite genome	Genome composed of genetic material from more than one species, usually arising in evolution by lateral gene transfer or endosymbiosis.	(not applicable)

(*Continued*)

Glossary Term	Descriptions or Definitions, Including Modified Wikipedia Excerpts	Relevant Link (Usually Wikipedia)
Composite organism	Organisms such as lichens that have cellular structures composed of two or more species of organisms.	https://en.wikipedia.org/wiki/Organism
Concentrate	A form of substance which has had the majority of its base component removed.	https://en.wikipedia.org/wiki/Concentrate
Conditional specification	A type of embryonic cleavage in which the differentiation of cells remains flexible and depends on intercellular communication between neighboring embryonic cells (blastomeres).	https://en.wikipedia.org/wiki/Cleavage_(embryo)
Consciousness	Consciousness is the state or quality of awareness or of being aware of an external object or something within oneself. It has been defined variously in terms of sentience, awareness, qualia, subjectivity, the ability to experience or to feel, wakefulness, having a sense of selfhood or soul, the fact that there is something "that it is like" to "have" or "be" it, and the executive control system of the mind. Despite the difficulty in definition, many philosophers believe that there is a broadly shared underlying intuition about what consciousness is.	https://en.wikipedia.org/wiki/Consciousness
Consilience	The principle that evidence from independent sources can converge on strong conclusions. When multiple sources of evidence are in agreement, the conclusion can be very strong.	https://en.wikipedia.org/wiki/Consilience
Consumer (heterotroph)	An organism that must consume (internalize) and metabolize organic compounds from other organisms to obtain energy and nutrients for synthesizing macromolecules.	https://en.wikipedia.org/wiki/Heterotroph

(*Continued*)

Glossary Term	Descriptions or Definitions, Including Modified Wikipedia Excerpts	Relevant Link (Usually Wikipedia)
Control of gene expression	Regulation of production of gene products, which sometimes, but not always, involves control of transcription by transcription factors.	https://en.wikipedia.org/wiki/Regulation_of_gene_expression
Covalent bond	A strong chemical bond between atoms that involves sharing of electron pairs between atoms. Electron sharing adds to stability with hybrid orbitals that have equivalent of full outer shell. Separation of the atoms often requires chemical reaction that breaks the bond.	https://en.wikipedia.org/wiki/Covalent_bond
Creationism	A religious belief that the universe and life originated from specific acts of divine creation. Usually includes human origins. This opposes the scientific conclusion that the universe and life arose through natural processes such as the Big Bang and Natural Selection, respectively. Often masquerades as a scientific theory.	https://en.wikipedia.org/wiki/Creationism
CRISPR	An abbreviation of Clustered Regularly Interspaced Short Palindromic Repeats. A family of DNA sequences in bacteria and archaea.	https://en.wikipedia.org/wiki/CRISPR
Cross-hybridization (DNA or RNA)	Formation of a double-stranded hybrid nucleic acid segment or molecule in which one strand of DNA and/or RNA is derived from one species (or individual) and the other strand is derived from another. May indicate homology (conservation and common origin) of sequences or may occur between repetitive sequences.	https://en.wikipedia.org/wiki/Hybridisation

(*Continued*)

Glossary Term	Descriptions or Definitions, Including Modified Wikipedia Excerpts	Relevant Link (Usually Wikipedia)
Crossing-over (crossovers)	Exchanges of genetic material between two homologous chromosomes that results in recombinant chromosomes during sexual reproduction.	https://en.wikipedia.org/wiki/Chromosomal_crossover
Cyanobacteria	An ancient phylum of bacteria that performs photosynthesis to fix carbon from carbon dioxide and generate energy through photosynthesis. Oxygen is produced as a byproduct.	https://en.wikipedia.org/wiki/Cyanobacteria
Cysteine	One of the 20 amino acids found in all living things. Contains a sulfur atom in its side chain, and is often found covalently bonded to other cysteine residues, forming S–S which is called cystine.	https://en.wikipedia.org/wiki/Cysteine
Cytoplasm	All of the internal material within a living cell (except the nucleus), which is surrounded by the cell membrane. In eukaryotes, the cytoplasm excludes the nucleus. In prokaryotes, the cytoplasm includes all of the internal material of the cell.	https://en.wikipedia.org/wiki/Cytoplasm
Cytoplasmic determinant	Materials such as RNA and proteins that are of maternal origin and are usually asymmetrically distributed in fertilized eggs (zygotes). Play important roles in early embryonic development.	https://en.wikipedia.org/wiki/Cytoplasmic_determinant#Mosaic_development
Cytoskeleton	Cytoplasmic elements such as actin microfilaments, microtubules, and intermediate filaments that are involved in cellular transport, shape, and motility.	https://en.wikipedia.org/wiki/Cytoskeleton

(*Continued*)

Glossary Term	Descriptions or Definitions, Including Modified Wikipedia Excerpts	Relevant Link (Usually Wikipedia)
Dark matter	A hypothetical form of matter that is thought to account for approximately 85% of the matter in the universe. Metaphorically, may also refer to as yet uncharacterized genomic elements that contribute to development.	https://en.wikipedia.org/wiki/Dark_matter
Deep homology	Characteristic of highly-conserved genes and structures found in distant organisms, which often play important roles in development.	https://en.wikipedia.org/wiki/Deep_homology
Deletion	A mutation in which part of a chromosome or a sequence of DNA is lost, often during DNA replication. Deletions can range in length from a single nucleotide up to large portions of an entire chromosome.	https://en.wikipedia.org/wiki/Deletion_(genetics)
Denaturation	Unfolding and/or inactivation of a macromolecule such as a protein. This can occur by extremes of temperature, pH, or presence or absence of other chemicals such as ions, desiccation, or interfaces between gaseous and aqueous phases.	https://en.wikipedia.org/wiki/Denaturation_(biochemistry)
Desiccation	A general term that refers to removal or absence of water. Most living cells require environmental conditions that prevent desiccation, although some spores, seeds, and some specialized species have evolved so that they can remain dormant but viable following desiccation.	https://en.wikipedia.org/wiki/Desiccation

(*Continued*)

Glossary Term	Descriptions or Definitions, Including Modified Wikipedia Excerpts	Relevant Link (Usually Wikipedia)
Development (embryogenesis)	A term describing a broad range of processes that take place throughout all stages of the life cycle, primarily in multicellular organisms. Often refers to the process whereby biological organization is reproduced from a fertilized egg which undergoes cleavage, gastrulation, embryogenesis, and organogenesis. During this process the body plan is formed and cells cooperate to generate integrated patterns and specialized cells, tissues, and organs. Development also refers to growth that takes place after embryogenesis, differentiation of stem cells, pattern formation, metamorphosis, and more.	https://en.wikipedia.org/wiki/Embryogenesis
Developmental biology	A broad interdisciplinary field in biology that studies animal or plant growth and development. In addition to development of embryos, it also encompasses the biology of normal growth, regeneration, asexual reproduction, metamorphosis, stem cells, and cancer.	https://en.wikipedia.org/wiki/Developmental_biology
Developmental toolkit	A metaphorical description of modular structures, functions, and genetic determinants at various levels of complexity that contribute to development. Toolkit elements are often reused and modified in a variety of ways by natural processes, during biological evolution.	

(Continued)

Glossary Term	Descriptions or Definitions, Including Modified Wikipedia Excerpts	Relevant Link (Usually Wikipedia)
Differentiation (cellular)	Specialization of embryonic cells or stem cells during development, resulting from factors such as cytoplasmic determinants, cellular communication, epigenetic changes in specific cell lineages, and transcription factors. Examples include muscle cells, lymphocytes, and sex cells. May occur during embryonic development or during growth or regeneration in the juvenile or adult.	https://en.wikipedia.org/wiki/Cellular_differentiation
Digital physics	A collection of theoretical perspectives based on the premise that the universe is describable by information. It is a form of digital ontology about the physical reality. According to this theory, the universe can be conceived of as either the output of a deterministic or probabilistic computer program, a vast, digital computation device, or mathematically isomorphic to such a device.	https://en.wikipedia.org/wiki/Digital_physics
Diploid	Eukaryotic cells that have two sets of homologous chromosomes, commonly found among sexually reproducing plants and animals.	https://en.wikipedia.org/wiki/Ploidy
Dipole (molecular)	An asymmetrical directional charge distribution in polar molecules with a positive end and a negative end, which may exert attractive or repulsive forces on other nearby molecules. Transient dipoles can also be formed by random interactions between electron clouds of adjacent molecules.	https://en.wikipedia.org/wiki/Dipole_(disambiguation)

(*Continued*)

Glossary Term	Descriptions or Definitions, Including Modified Wikipedia Excerpts	Relevant Link (Usually Wikipedia)
Dipole–dipole attractions	Attractions between dipoles of two polar molecules, such as water. One type of intermolecular force responsible for phase changes from gas to liquid. In water and in base-pairs of nucleic acids, relatively strong dipole–dipole attractions called hydrogen bonds play important roles in all living cells.	https://en.wikipedia.org/wiki/Intermolecular_force
Disequilibrium (chemistry)	A dynamic state in which chemical reactions between reactants and products are not taking place in both directions at equal rates. This does not imply that amounts of reactants or products are unequal. Another meaning of the term refers to gradients in the distribution of particles, such as across a semi-permeable membrane. In equilibrium, the distribution of the particles is greater on one side of the membrane.	emergence (properties and processes)
Disorder	The absence of order or the lack of organization. Also, may refer to entropy.	https://en.wikipedia.org/wiki/Disorder
Dispersal vector	An agent transporting seeds or other dispersal units. They can be living or nonliving parts of the environment.	https://en.wikipedia.org/wiki/Dispersal_vector
DNA	A polymer of nucleotides, each containing the sugar deoxyribose, a phosphate group, and a nitrogenous base. In the genome, DNA usually consists of two base-paired chains that coil around each other to form a double helix. Strands can be transcribed into RNA molecules and can be replicated to form two identical copies. DNA carries the genetic instructions used in the growth, development, function, and reproduction of all living organisms.	https://en.wikipedia.org/wiki/DNA

(*Continued*)

Glossary Term	Descriptions or Definitions, Including Modified Wikipedia Excerpts	Relevant Link (Usually Wikipedia)
DNA polymerase	Enzyme that synthesizes complementary strands of DNA from DNA template strands. Required for DNA replication, and may also perform DNA repair functions.	https://en.wikipedia.org/wiki/DNA_polymerase
DNA replication	The biological process of producing two identical replicas of DNA from the original DNA molecule. With double-stranded DNA, two identical copies are produced when the strands separate and two new complementary strands are synthesized by DNA polymerase enzymes. DNA replication occurs in all living organisms and is the basis for biological inheritance.	https://en.wikipedia.org/wiki/DNA_replication
DNA sequence variation	Usually refers to heritable differences in genomic DNA. DNA sequence variation plays a critical role in Natural Selection.	https://en.wikipedia.org/wiki/Genetic_variation
DNA sequencing	The process of determining the precise order of organic molecules within a DNA. It includes any method used to determine the order of the four bases. These are adenine, guanine, cytosine, and thymine.	https://en.wikipedia.org/wiki/DNA_sequencing
DNA synthesis	The natural or artificial creation of DNA molecules, usually starting with a template strand.	https://en.wikipedia.org/wiki/DNA_synthesis
DNA/RNA templates	During DNA replication or transcription, a single strand of DNA serves as a template for the synthesis of a complementary (base-paired) strand of DNA or RNA by DNA or RNA polymerase enzymes, respectively. During reverse transcription, a strand of RNA serves as a template for synthesis of a complementary DNA strand by reverse transcriptase.	(not applicable)

(Continued)

Glossary Term	Descriptions or Definitions, Including Modified Wikipedia Excerpts	Relevant Link (Usually Wikipedia)
Dominance (genetics)	A relationship between alleles of one gene in which the effect on phenotype of one allele masks the contribution of the second allele. The first is dominant over the second.	https://en.wikipedia.org/wiki/Dominance_(genetics)
Double-stranded (base-paired)	Refers to DNA, RNA, or DNA/RNA hybrids in which nitrogenous bases are paired and held together by hydrogen bonding. May occur within single stranded molecules such as tRNA which fold into a cloverleaf shape, or may occur between two formerly separate strands, or may arise during DNA or RNA synthesis.	https://en.wikipedia.org/wiki/Base_pair
Downstream (DNA or RNA)	Relative position in DNA, RNA, or protein molecules. In nucleic acids, usually means towards the 3′ end of a strand of DNA or RNA, relative to the direction of transcription or translation. In protein chains, usually means towards the carboxyl terminus of the chain.	https://en.wikipedia.org/wiki/Downstream
Drosophila melanogaster (fruitfly)	A species of fly that has been productively used as a model organism. Extensively used in Molecular Genetics. Common name is "fruitfly". Still used extensively in biology and medicine.	https://en.wikipedia.org/wiki/Drosophila_melanogaster
Ecological Phenotype	Term introduced in *Rethinking Evolution* referring to the fully-developed structures, functions, and organization of the individual that adapt the individual to a particular environmental *niche* or way of life.	https://en.wikipedia.org/wiki/Phenotype

(*Continued*)

Glossary Term	Descriptions or Definitions, Including Modified Wikipedia Excerpts	Relevant Link (Usually Wikipedia)
Ecological relationships	Relationships between organisms and abiotic (nonliving) features of shared environments. All species depend on effective ecological relationships to survive and reproduce their own kind. Relatively sudden changes in abiotic factors (such as temperature) play major roles in extinction—particularly when other species depend on a producer at the bottom of the food web, or a keystone species, that is affected by the change.	https://en.wikipedia.org/wiki/Ecology
Ecosystem	A community made up of living organisms and nonliving components, such as air, water, minerals, and/or soil. Living and nonliving components interact through nutrient cycles and energy transformation and exchange.	https://en.wikipedia.org/wiki/Ecosystem
Ectosymbiotic	Symbiosis in which the symbiont lives on the body surface of the host. This can be internal surfaces like digestive tubes or ducts of glands.	https://en.wikipedia.org/wiki/Ectosymbiosis
Efficacy (wiktionary)	The effectiveness of an entity in performing specific tasks.	https://en.wiktionary.org/wiki/efficacy
Egg cell	The female reproductive cell (sex cell) in sexually reproducing species. A zygote is formed when the egg is fertilized by a sperm cell, which can lead to development of an embryo.	https://en.wikipedia.org/wiki/Egg_cell
Electron	A particle smaller than an atom with a negative charge. Electrons play essential roles in both weak and strong chemical bonding, ionization, electricity, magnetism, thermal conductivity, and more.	https://en.wikipedia.org/wiki/Electron

(*Continued*)

Glossary Term	Descriptions or Definitions, Including Modified Wikipedia Excerpts	Relevant Link (Usually Wikipedia)
Electron bifurcation (from published abstract)	The coupling of exergonic and endergonic redox reactions to simultaneously generate (or utilize) low- and high-potential electrons. It is the third recognized form of energy conservation in biology and was recently described for select electron-transferring flavoproteins.	https://jb.asm.org/content/199/21/e00440-17
Electron cloud	The 3D space containing electron orbitals of a molecule or atom.	https://en.wikipedia.org/wiki/Electron-cloud_effect
Electron microscope	A microscope that uses a focused beam of accelerated electrons rather than light to form an image. They are used to investigate the structure or microorganisms, large molecules, cell samples, metals, and crystals.	https://en.wikipedia.org/wiki/Electron_microscope
Electron pair	Two electrons that occupy the same molecular orbital. Often, this may be a hybrid orbital representing a covalent bond between two atoms.	https://en.wikipedia.org/wiki/Electron_pair
Electron transport chain	A series of complexes that transfer electrons from electron donors to electron acceptors via redox (both reduction and oxidation occurring simultaneously) reactions, and couples this electron transfer with the transfer of protons (H^+ ions) across a membrane. This creates an electrochemical proton gradient that drives the synthesis of adenosine triphosphate (ATP), a molecule that stores energy chemically in the form of highly strained bonds.	https://en.wikipedia.org/wiki/Electron_transport_chain

(*Continued*)

Glossary Term	Descriptions or Definitions, Including Modified Wikipedia Excerpts	Relevant Link (Usually Wikipedia)
Embryo (multicellular eukaryote)	An embryo is an early stage of development of a multicellular diploid eukaryotic organism. In general, in organisms that reproduce sexually, an embryo develops from a zygote, the single cell resulting from the fertilization of the female egg cell by the male sperm cell. The zygote possesses half the DNA from each of its two parents. In plants, animals, and some protists, the zygote will begin to divide by mitosis to produce a multicellular organism. The result of this process is an embryo.	https://en.wikipedia.org/wiki/Embryo
Embryonic development (embryogenesis)	Embryogenesis is the process by which the embryo forms and develops. In mammals, the term refers chiefly to early stages of prenatal development, whereas the terms fetus and fetal development describe later stages. Embryogenesis starts with the fertilization of the egg cell (ovum) by a sperm cell, (spermatozoon). Once fertilized, the ovum is referred to as a zygote, a single diploid cell. The zygote undergoes mitotic divisions with no significant growth (a process known as cleavage) and cellular differentiation, leading to development of a multicellular embryo.	https://en.wikipedia.org/wiki/Embryogenesis
Emergence (properties and processes)	Resulting from a process whereby larger entities arise through interactions among smaller or simpler entities such that the larger entities exhibit properties the smaller/simpler entities do not exhibit.	https://en.wikipedia.org/wiki/Emergence#Emergence_and_evolution

(*Continued*)

Glossary Term	Descriptions or Definitions, Including Modified Wikipedia Excerpts	Relevant Link (Usually Wikipedia)
Emergent evolutionary potential (EEP)	Term introduced in *Rethinking Evolution*. Refers to the changing potential for higher-level organization to arise when new entities interact in ways that prove useful in survival and reproduction.	https://en.wikipedia.org/wiki/Emergent_evolution
Emergentism (system)	A property of a system is said to be emergent if it is a new outcome of some other properties of the system and their interaction, while it is itself different from them.	https://en.wikipedia.org/wiki/Emergentism
Empirical data	Data gathered from the real world through observations and experiments. The scientific method demands that such data should be based on true objective facts that are reproducible, and that are precisely described in an unambiguous way.	https://en.wikipedia.org/wiki/Empirical_evidence
Empirical evidence	Evidence gathered from the real world through observations and experiments. The scientific method demands that such data should be based on true objective facts that are reproducible, and that are precisely described in an unambiguous way. Empirical evidence is used to test the validity of hypotheses, claims and ideas about the real world.	https://en.wikipedia.org/wiki/Empirical_evidence
Empirical research	Research using observable and documented evidence. This is accomplished through direct or indirect observation or experience.	https://en.wikipedia.org/wiki/Empirical_research
Endergonic reaction	A chemical reaction that absorbs energy from the surroundings. Since the energy is gained, the energy change has a positive sign.	https://en.wikipedia.org/wiki/Endergonic_reaction

(*Continued*)

Glossary Term	Descriptions or Definitions, Including Modified Wikipedia Excerpts	Relevant Link (Usually Wikipedia)
Endosymbiosis	A well-supported theory for the origin of eukaryotic cells from prokaryotic organisms. Organelles such as mitochondria and chloroplasts evolved from bacteria that were engulfed by other cells.	https://en.wikipedia.org/wiki/Symbiogenesis
Energy	A quantifiable property of matter. Energy can often be harnessed to do work. Energy can be converted into other forms, but cannot be created nor destroyed.	https://en.wikipedia.org/wiki/Energy#Biology
Energy dissipation	The natural spontaneous tendency of energy to become distributed. Energy spontaneously moves from regions or objects of higher energy to regions of lower energy if there is nothing blocking that movement. Energy dissipation takes place over time, and is affected by the nature of the materials involved. This is an area of study in thermodynamics.	https://en.wikipedia.org/wiki/Dissipation
Entity	A general term for any type of discrete identifiable object, unit or subunit, whether simple or complex.	https://en.wikipedia.org/wiki/Entity
Entropy (thermodynamics)	An extensive property of a thermodynamic system. Representing the amount of disorder in a system, entropy can be measured by determining the number of microstates of particles, temperature, or energy of a system.	https://en.wikipedia.org/wiki/Entropy
Enzymatic activity	Binding of reactants (substrates) to the active site of enzymes, followed by catalytic facilitation of the chemical reaction and release of the products.	https://en.wikipedia.org/wiki/Enzyme

(*Continued*)

Glossary Term	Descriptions or Definitions, Including Modified Wikipedia Excerpts	Relevant Link (Usually Wikipedia)
Enzyme	Macromolecular biological catalysts. Most enzymes are proteins, but some RNA sequences also have catalytic activity. Enzymes facilitate chemical reactions by binding to specific reactant molecules and lowering the activation energy required for chemical reactions. Each step in a metabolic pathway is usually carried out by a specific enzyme, and enzymes may also change the activity of other molecules by modifying them in various ways. For example, protein kinases are enzymes that regulate many cellular processes by adding phosphate groups to protein chains.	https://en.wikipedia.org/wiki/Enzyme
Epigenetic	An added genetic affect that is determined by factors other than the DNA sequence *per se*. May involve chemical modifications of the DNA, condensation of chromosomes, and other factors.	https://en.wikipedia.org/wiki/Epigenetics
Epistasis	The phenomenon where the effect of one gene is dependent on other genes in the genome.	https://en.wikipedia.org/wiki/Epistasis
Epistemology	The branch of philosophy concerned with the theory of knowledge. It involves studying the nature of knowledge, justification, and the rationality of belief.	https://en.wikipedia.org/wiki/Epistemology
Equilibrium (chemical)	A dynamic state in which chemical reactions between reactants and products are taking place in both directions at equal rates. This does not imply that amounts of reactants or products are equal. Another meaning of the term refers to gradients in the distribution of	https://en.wikipedia.org/wiki/Chemical_equilibrium

(*Continued*)

Glossary Term	Descriptions or Definitions, Including Modified Wikipedia Excerpts	Relevant Link (Usually Wikipedia)
	particles, such as across a semi-permeable membrane. At equilibrium, the distribution of the particles is equal on both sides of the membrane.	
Escherichia coli	Bacterial commonly found in the lower intestine of warm-blooded organisms, including humans. Most strains are harmless, but some can cause serious food poisoning in their hosts. Commonly used as model organism and recombinant DNA technology.	https://en.wikipedia.org/wiki/Escherichia_coli
Eubacteria (bacteria)	Single-celled organisms from one of the two domains of prokaryotes. The other domain is archaea.	https://en.wikipedia.org/wiki/Bacteria
Euglenid	One of the best-known groups of flagellates, commonly found in freshwater, especially when it is rich in organic materials. Most euglenids are unicellular. Many euglenids have chloroplasts and produce their own food through photosynthesis, but others feed by phagocytosis, or strictly by diffusion.	https://en.wikipedia.org/wiki/Euglenid
Eukaryotic	Cells or individuals that are classified in the domain of organisms designated Eukaryota. Eukaryotic cells have a nucleus enclosed within membranes as well as membrane-bound organelles. Distinguished from prokaryotic cells.	https://en.wikipedia.org/wiki/Eukaryote
Evo-devo (evolutionary developmental biology)	Abbreviation for evolutionary developmental biology. A field of biological research that compares the developmental processes of organisms. This allows inference of ancestral relationships and evolutionary origins of developmental processes.	https://en.wikipedia.org/wiki/Evolutionary_developmental_biology

(*Continued*)

Glossary Term	Descriptions or Definitions, Including Modified Wikipedia Excerpts	Relevant Link (Usually Wikipedia)
Evo-psych (evolutionary psychology)	Abbreviation for evolutionary psychology. A theoretical approach that draws on both social and natural sciences. Psychological structure is examined from a modern evolutionary perspective. Usually requires interdisciplinary knowledge in evolutionary biology, psychology, and social sciences.	https://en.wikipedia.org/wiki/Evolutionary_psychology
Evolution (biology)	The empirically proven natural process of change in all living things on Earth. Natural Selection is one of the most fundamental mechanisms that account for biological evolution, but it is not the only relevant mechanism involved. Biological evolution includes the precellular molecular changes that led to the origin of life on Earth, as well as subsequent changes in the complexity of structures and functions in all living cells and organisms. Evolution is also the fundamental process responsible for the origin of species, as pointed out by Charles Darwin in 1859, in the context of Natural Selection.	https://en.wikipedia.org/wiki/Evolution
Evolutionary conservation	Conservation (preservation and hereditary transmission) of genetic determinants in the genome, as well as the higher-level developmental processes and developed structures and functions that arise from such conserved genetic elements. For example, conservation of the widespread uniformity of transcription and translation as well as ATP synthases that convert the energy of proton gradients are conserved	https://en.wikipedia.org/wiki/Conserved_sequence

(*Continued*)

Glossary Term	Descriptions or Definitions, Including Modified Wikipedia Excerpts	Relevant Link (Usually Wikipedia)
	elements that require conservation of the DNA sequences that determine the amino acid sequences of the proteins involved. Similarly, deep conservation of the body plans of vertebrates depends on conservation of genetic elements such as Hox genes.	
Evolutionary developmental biology (evo-devo)	A field of biological research that compares the developmental processes of organisms. This allows inference of ancestral relationships and evolutionary origins of developmental processes.	https://en.wikipedia.org/wiki/Evolutionary_developmental_biology
Evolutionary psychology (evo-psych)	A theoretical approach that draws on both social and natural sciences. Psychological structure is examined from a modern evolutionary perspective. Usually requires interdisciplinary knowledge in evolutionary biology, psychology, and social sciences.	https://en.wikipedia.org/wiki/Evolutionary_psychology
Evolvability	The capacity of a system for adaptive evolution. It is the ability of a population of organisms to generate adaptive genetic diversity.	https://en.wikipedia.org/wiki/Evolvability
Exaptation	The natural reuse, with or without modification, of genetic determinants for new purposes, in biological evolution.	https://en.wikipedia.org/wiki/Exaptation
Exergonic	A chemical reaction that loses energy to the surroundings. Since the energy is lost, the energy change has a negative sign.	https://en.wikipedia.org/wiki/Exergonic_reaction
Exon	The protein-coding portions of eukaryotic genes that are spliced together during processing of RNA to generate messenger RNA sequences.	https://en.wikipedia.org/wiki/Exon

(*Continued*)

Glossary Term	Descriptions or Definitions, Including Modified Wikipedia Excerpts	Relevant Link (Usually Wikipedia)
Expanded evolutionary synthesis	Although there is debate among evolutionary biologists, the term is generally used to describe an evolutionary synthesis that retains certain core principles of classical Darwinism and the Modern Synthesis while expanding and modifying it with newly acquired concepts that are well-supported by empirical evidence.	http://courses.washington.edu/biol354/niklas.pdf
Experimental embryology	Experiments primarily involving disaggregation and reaggregation of cells, transplantation of cells, or mixing of cells, that provided phenomenological insights into the mechanisms of development. This was an active area of investigation in the 20th century in several countries throughout the world.	(not applicable)
Extended evolutionary synthesis	A set of theoretical concepts more comprehensive than the earlier modern synthesis of evolutionary biology that took place between 1918 and 1942. The extended evolutionary synthesis was called for in the 1950s by C. H. Waddington, argued for on the basis of punctuated equilibrium by Stephen Jay Gould and Niles Eldredge in the 1980s, and was reconceptualized in 2007 by Massimo Pigliucci and Gerd B. Müller.	https://en.wikipedia.org/wiki/Extended_evolutionary_synthesis
Extinction	The termination of evolutionary lineages, including one or more species, during the evolutionary process. Most of the species that have ever lived on planet Earth have gone extinct, and only a small fraction continues to live (these	

(*Continued*)

Glossary Term	Descriptions or Definitions, Including Modified Wikipedia Excerpts	Relevant Link (Usually Wikipedia)
	are extant species). Extinction occurs when populations of a species no longer leave descendent offspring, and when the last living individual dies.	https://en.wikipedia.org/wiki/Extinction
Extracellular matrix	Macromolecular structures that surround cells, outside of the cell membrane, consisting of a variety of complex molecules such as interacting carbohydrates and proteins. The extracellular matrix is involved in a broad range of functions both during development and also in mature tissues and organs.	https://en.wikipedia.org/wiki/Extracellular_matrix
Extremophile	An organism or species that thrives under conditions that are usually lethal to most other species, such as extremes of temperature, ionizing radiation, salinity, desiccation, and other extreme environmental conditions.	https://en.wikipedia.org/wiki/Extremophile
Eyespot (mimicry)	A pigmented structure, for example on a butterfly wing or a caterpillar, that resembles an eye but does not actually function at all in vision. Eyespots presumably evolve because they fool would-be predators, because they play roles in mate preferences in sexual selection, or because in general they make an individual look larger or more threatening to other individuals.	https://en.wikipedia.org/wiki/Eyespot_(mimicry)
Facilitate	To help bring about or to make easier.	https://en.wiktionary.org/wiki/facilitate
Facilitated variation	Demonstrates how seemingly complex biological systems can arise through a limited number of regulatory genetic changes. This occurs through differential re-use of preexisting developmental components.	https://en.wikipedia.org/wiki/Facilitated_variation

(*Continued*)

Glossary Term	Descriptions or Definitions, Including Modified Wikipedia Excerpts	Relevant Link (Usually Wikipedia)
Fate map (fate mapping)	A mapping of embryonic cells that demonstrates how particular cells will ultimately migrate, differentiate, specialize, and form specific mature tissues and organs.	https://en.wikipedia.org/wiki/Fate_mapping
Fatty acid	A carboxyl functional group covalently bonded to a long hydrocarbon chain. Found as components of biological membranes and function as stored energy sources.	https://en.wikipedia.org/wiki/Fatty_acid
Fertilization	In sexual reproduction, the fusion of a sperm cell and an egg cell to form a zygote or fertilized egg, which may then go on to undergo subsequent cleavage, gastrulation, and embryonic development. Usually, one set of maternal genes and one set of paternal genes are combined to form a diploid genome with two sets of chromosomes.	https://en.wikipedia.org/wiki/Fertilisation
Fertilized egg (zygote)	An egg that has combined with the genome of a sperm cell, the prerequisite to embryonic development.	https://en.wikipedia.org/wiki/Zygote
Flagellum	A lash-like appendage that protrudes from the cell body of certain bacterial and eukaryotic cells and whose primary function is locomotion, but it also often has function as a sensory organelle, being sensitive to chemicals and temperatures outside the cell. The similar structure in the archaea functions in the same way but is structurally different.	https://en.wikipedia.org/wiki/Flagellum
Flavins	A group of organic compounds based on pteridine, a chemical compound. Flavins are capable of undergoing redox reactions.	https://en.wikipedia.org/wiki/Flavin_group

(*Continued*)

Glossary Term	Descriptions or Definitions, Including Modified Wikipedia Excerpts	Relevant Link (Usually Wikipedia)
Flow cytometry	A technique in which individual cells are sorted into separate groups, usually by means of cell-surface proteins that have been attached to fluorescent antibodies. Flow cytometry may be performed for analytical and/or preparative purposes.	https://en.wikipedia.org/wiki/Flow_cytometry
Food web	A natural interconnection of food chains, in which organisms consume other organisms to obtain energy and nutrients. A major feature of ecosystems.	https://en.wikipedia.org/wiki/Food_web
Frameshift mutation	A genetic mutation caused by insertion or deletion of nucleotides in a protein-coding DNA sequence. This can alter downstream codons and result in changes in the amino acid sequence and/or length of a protein chain.	https://en.wikipedia.org/wiki/Frameshift_mutation
Fruitfly (*Drosophila melanogaster*)	A fly of the species *Drosophila melanogaster* that has played crucial roles in Mendelian Genetics, Molecular Genetics, evolutionary biology, and evo-devo as a model organism.	https://en.wikipedia.org/wiki/Drosophila_melanogaster
Functional group	A chemical group that is usually attached to one or more carbon atoms in organic compounds. Examples include amino groups and carboxyl groups found in all amino acids.	https://en.wikipedia.org/wiki/Functional_group
Fungus (fungal)	Any member of the group of eukaryotic organisms in the Fungi kingdom, such as yeasts, molds, and mushrooms. Fungi are usually decomposers in ecological systems.	https://en.wikipedia.org/wiki/Fungus

(*Continued*)

Glossary Term	Descriptions or Definitions, Including Modified Wikipedia Excerpts	Relevant Link (Usually Wikipedia)
G protein-coupled receptor	Protein products of a large gene family that function as transmembrane cell-surface receptors that activate G-proteins in the cytoplasm, that play various roles in signal transduction pathways. Examples include photoreceptors involved in vision and olfactory receptors involved in the sense of smell (odor detection).	https://en.wikipedia.org/wiki/G_protein-coupled_receptor
Gap gene	A type of gene involved in the development of the segmented embryos of some arthropods. Gap genes are defined by the effect of a mutation in that gene, which causes the loss of contiguous body segments, resembling a gap in the normal body plan. Each gap gene, therefore, is necessary for the development of a section of the organism. *In situ* hybridization against mRNA for some of the gap genes in the *Drosophila* early embryo. Gap genes were first described by Christiane Nüsslein-Volhard and Eric Wieschaus in 1980.	https://en.wikipedia.org/wiki/Gap_gene
Gastrulation	A widespread developmental process in a variety of animals that establishes cell layers in the embryo by means of cellular motility and migration. In general terms, cells change shape and turn inward to form new layers beneath other cells. Essential to the establishment of the overall shape and form of the body, and to organ development.	https://en.wikipedia.org/wiki/Gastrulation

(*Continued*)

Glossary Term	Descriptions or Definitions, Including Modified Wikipedia Excerpts	Relevant Link (Usually Wikipedia)
Gene	A sequence of DNA or RNA that "codes" for a gene product such as a protein, or exerts a regulatory effect on gene expression. Most biological traits are determined by the effects of multiple interacting genes and gene products.	https://en.wikipedia.org/wiki/Gene
Gene duplication	A major mechanism through which new genetic material is generated during molecular evolution. It involves duplication of a region of DNA that contains one or more genes.	https://en.wikipedia.org/wiki/Gene_duplication
Gene expression	The regulated process by which genes produce their gene product or products, such as protein chains or various types of functional RNA sequences. Usually begins with control of transcription.	https://en.wikipedia.org/wiki/Gene_expression
Gene pool	A set of genes in a population of individual organisms. A gene pool implies relatively free flow of genetic material between individuals, usually referring to sexual reproduction within a single eukaryotic species, but also possibly by horizontal gene transfer, for example between prokaryotes.	https://en.wikipedia.org/wiki/Gene_pool
Gene product	The biochemical material resulting from expression of a gene. Usually, genes are expressed as protein chains or as RNA molecules.	https://en.wikipedia.org/wiki/Gene_product
Generalist (natural sciences)	A researcher or theorist with interdisciplinary interests. Often, generalists try to understand the metaphorical "Big Picture" that represents a broad understanding of reality. In the biological sciences and biomedical research, most funded research is carried out by specialists rather than generalists.	https://en.wikipedia.org/wiki/Generalist

(*Continued*)

Glossary Term	Descriptions or Definitions, Including Modified Wikipedia Excerpts	Relevant Link (Usually Wikipedia)
Generative phenotype	A term introduced in *Rethinking Evolution* to help clarify the difference between developmental toolkit elements and the overall phenotype of the developed individual. Refers to the shape and form and capabilities of structures and functions at various levels of complexity that function as reusable modules in development. These modular elements play major roles during the development of organisms, and are themselves subject to both evolutionary conservation and evolutionary modification and refinement. Distinguished from the Ecological Phenotype.	https://en.wikipedia.org/wiki/DNA_phenotyping
Genetic algorithm	A software technique inspired by Natural Selection, in which strings of computer code are recombined in a random fashion and then subjected to artificial selection according to some predetermined criteria. Selected strings are replicated, artificially varied, and then subjected to subsequent recombination and selection. A powerful technique for solving multivariate problems in which the best ways to achieve the desired ends are difficult for the human mind to conceive.	https://en.wikipedia.org/wiki/Genetic_algorithm
Genetic apparatus (genetics, gene expression)	Modular structures and functions including chromosomes (the full genome), transcription factors, enzymes, and proteins involved in transcription, translation, mitosis, and meiosis, and various RNA molecules, which all work together to store, transmit, diversify, shuffle, and express genetic determinants to reproduce biological structures and functions via the development of each individual.	https://en.wikipedia.org/wiki/Genetics#Gene_expression

(*Continued*)

Glossary Term	Descriptions or Definitions, Including Modified Wikipedia Excerpts	Relevant Link (Usually Wikipedia)
Genetic blueprint (metaphorical)	A misleading metaphor that arose from a teleological view of the ways that genes and their regulated expression determine biological structure and function.	https://en.wikipedia.org/wiki/Blueprint_(disambiguation)#Other_uses
Genetic code	A table that links specific triplets (codons) in DNA or RNA to specific amino acids (or start or stop signals). These amino acids are incorporated into a growing protein chain during protein synthesis.	https://en.wikipedia.org/wiki/Genetic_code
Genetic determinants	Transcribed and translated RNA molecules and protein chains, as well as a variety of regulatory DNA sequences that interact in complex ways to generate and reproduce the phenotypes characteristic of individuals and species.	https://en.wikipedia.org/wiki/Genetics
Genetic material	DNA is the universal genetic material of all types of cells. It is located in the chromosomes of prokaryotic and eukaryotic cells, as well as in the smaller genomic elements of organelles such as mitochondria and chloroplasts.	https://en.wikipedia.org/wiki/Genetics
Genetic transformation	The genetic alteration of a cell resulting from direct uptake and incorporation of DNA originating outside the organism. This occurs from its surroundings through the cell membranes. It can occur naturally or can be facilitated by specialized techniques in biotechnology.	https://en.wikipedia.org/wiki/Transformation_(genetics)
Genetic variation	A general term for heritable variation which can be traced to specific sequences of genomic DNA.	https://en.wikipedia.org/wiki/Genetic_variation

(*Continued*)

Glossary Term	Descriptions or Definitions, Including Modified Wikipedia Excerpts	Relevant Link (Usually Wikipedia)
Genetics	The study of the nature, storage, transmission, and expression of the heritable factors found in genomic DNA. Some of the major interdisciplinary fields in genetics are Mendelian Genetics, Molecular Genetics, and Developmental Genetics.	https://en.wikipedia.org/wiki/Genetics
Genome	The complete set of genetic material of an organism of a given species. May refer to an individual genome or the collective gene pool of the species as a whole. It consists of DNA.	https://en.wikipedia.org/wiki/Genome
Genome evolution	Accumulated changes in the genetic material of various populations and species over time, resulting from changes in the frequencies of various genetic determinants.	https://en.wikipedia.org/wiki/Genome_evolution
Genomic DNA	The complete set of DNA sequences contained in the cells of, and passed from generation to generation, during the reproduction of a species of organism.	https://en.wikipedia.org/wiki/Genomic_DNA
Genomic toolkit	A metaphorical description of modular, reusable genomic elements, and processes such as gene duplication and conserved regulatory or protein-coding sequences. Toolkit elements play major roles in development and general cellular functions and are often modified and reused in various ways by Natural Selection.	(not applicable)
Genotype	The specific alleles usually represented by the DNA sequences of one or two copies of a specific gene in an individual, which contributes to the genetic makeup of a cell, and may contribute to the phenotype.	https://en.wikipedia.org/wiki/Genotype

(Continued)

Glossary Term	Descriptions or Definitions, Including Modified Wikipedia Excerpts	Relevant Link (Usually Wikipedia)
Geology	An earth science concerned with the solid Earth made of rock and the processes by which it changes over time.	https://en.wikipedia.org/wiki/Geology
Germ layers	The layers of developing embryonic cells that interact to form tissues and organs. The three major germ layers are ectoderm, mesoderm, and endoderm.	https://en.wikipedia.org/wiki/Germ_layer
Germ line (cells or DNA)	Usually refers to diploid cells that produce haploid sex cells during sexual reproduction. Germ line DNA may refer to the genome of immature lymphocytes that have not undergone V(D)J recombination.	https://en.wikipedia.org/wiki/Germline
Gestalt (biology)	A general term for overall, interconnected wholes that have modules that interact to generate higher levels of emergent structure and function.	https://en.wikipedia.org/wiki/Gestalt
Gradient (chemical or morphogenetic)	An unequal distribution or concentration of a substance, which may or may not be separated by a semi-permeable membrane. Gradients can consist of ions, small molecules, or DNA or RNA.	(not applicable)
Green algae	A large, informal grouping of algae. The land plants, or embryophytes, are thought to have emerged from the charophytes.	https://en.wikipedia.org/wiki/Green_algae
Group selection	A proposed mechanism of evolution in which Natural Selection acts at the level of the group, instead of at the more conventional level of the individual.	https://en.wikipedia.org/wiki/Group_selection

(*Continued*)

Glossary Term	Descriptions or Definitions, Including Modified Wikipedia Excerpts	Relevant Link (Usually Wikipedia)
H^+	Hydrogen ions, which are positively charged protons from hydrogen atoms that have been ionized and lost their electrons. Hydrogen ions play important roles in biological proton gradients and their concentration determines the acidity or alkalinity (pH) of aqueous solutions.	https://en.wikipedia.org/wiki/Hydrogen_ion
Heat	Energy transferred from one system to another as a result of thermal interactions.	https://en.wikipedia.org/wiki/Heat
Hedgehog (signaling pathway)	A signaling pathway that transmits information to embryonic cells required for proper cell differentiation. Different parts of the embryo have different concentrations of hedgehog signaling proteins. The pathway also has roles in the adult. Diseases associated with the malfunction of this pathway include basal cell carcinoma.	https://en.wikipedia.org/wiki/Hedgehog_signaling_pathway
Hereditary factor	A genetic determinant. Includes DNA sequences that are transcribed and translated into RNA molecules and amino acid sequences, as well as DNA sequences involved in control of gene expression.	https://en.wikipedia.org/wiki/Heredity
Heredity (genetics)	The nature, storage, transmission, and expression of genetic determinants that are passed from generation to generation in all species.	https://en.wikipedia.org/wiki/Heredity
Heterogeneous	Refers to a mixture containing components that are diverse and/or unevenly distributed.	https://en.wikipedia.org/wiki/Homogeneity_and_heterogeneity

(*Continued*)

Glossary Term	Descriptions or Definitions, Including Modified Wikipedia Excerpts	Relevant Link (Usually Wikipedia)
Heterotroph	An organism that must consume (internalize) and metabolize organic compounds from other organisms to obtain energy and nutrients for synthesizing macromolecules.	https://en.wikipedia.org/wiki/Heterotroph
Heterozygous	If two alleles of a gene carried on homologous chromosomes of a diploid organism are distinct, the organism is heterozygous at that genetic locus.	https://en.wikipedia.org/wiki/Zygosity
High-energy compounds (biology)	In biology, usually refers to molecules such as ATP (as well as other nucleotide triphosphates), NADH, NADPH, and FADH2. Energy is usually released by enzymes that break unstable bonds that are the source of energy in these compounds. They are widely used sources of energy that drive cellular, metabolic processes requiring an external energy source, including photosynthesis, cellular respiration, and in general anabolic reactions that build up larger molecules from smaller ones.	https://en.wikipedia.org/wiki/High-energy_phosphate
Holon	Something that is part of a whole but with some degree of autonomy. They have a degree of autonomy but are subject to control from higher authorities.	https://en.wikipedia.org/wiki/Holon_(philosophy)
Homeodomain (homeobox)	A DNA sequence, around 180 base pairs long, found within genes that are involved in the regulation of patterns of anatomical development (morphogenesis) in animals, fungi and plants. These genes encode homeodomain protein products that are transcription factors sharing a characteristic protein fold structure that binds DNA.	https://en.wikipedia.org/wiki/Homeobox

(Continued)

Glossary Term	Descriptions or Definitions, Including Modified Wikipedia Excerpts	Relevant Link (Usually Wikipedia)
Homologous pair	A set of one maternal and one paternal chromosome that pair up with each other inside a cell during meiosis.	https://en.wikipedia.org/wiki/Homologous_chromosome
Homology (biology)	A characteristic of conserved sequences in DNA (or RNA or protein) sequences, or phenotypic structures and functions determined by genetic elements, that are derived from common ancestral genes or organisms. Homologous DNA sequences are found either by sequence comparisons or by experimental methods in which similar single-stranded sequences from two genes or two organisms will base-pair (hybridize) to form double-stranded hybrid sequences.	https://en.wikipedia.org/wiki/Homology_(biology)
Homozygous	If both alleles of a gene carried on homologous chromosomes of a diploid organism are the same, the organism is homozygous at that genetic locus.	https://en.wikipedia.org/wiki/Zygosity
Horizontal gene transfer (lateral gene transfer)	The movement of genetic material between organisms other than by the transmission of DNA from parent to offspring. It is an important factor in the evolution of many organisms.	https://en.wikipedia.org/wiki/Horizontal_gene_transfer
Hox gene	A group of related genes that control the body plan of a broad range of animal embryos along the head-tail axis. A subset of the homeotic gene family.	https://en.wikipedia.org/wiki/Hox_gene
Hybrid orbital (chemistry)	An orbital occupied by a pair of electrons that extends between two atoms in a covalent bond. Hybrid orbitals have different shapes than orbitals of isolated atoms and bind atoms together by a strong electromagnetic force that often requires a chemical reaction to separate the bonded atoms.	https://en.wikipedia.org/wiki/Orbital_hybridisation

(*Continued*)

Glossary Term	Descriptions or Definitions, Including Modified Wikipedia Excerpts	Relevant Link (Usually Wikipedia)
Hybridization (nucleic acid)	Formation of a double-stranded hybrid sequence from DNA and RNA strands, or from two different genes or organisms.	https://en.wikipedia.org/wiki/Hybridisation
Hydrogen atom (H)	An atom of the simplest chemical element, which in neutral form contains one proton and one electron. Hydrogen atoms constitute about 75% of the baryonic mass of the universe.	https://en.wikipedia.org/wiki/Hydrogen_atom
Hydrogen bonding	A partially electrostatic attraction between molecules, where two relatively strong dipoles formed by hydrogen and oxygen, hydrogen and nitrogen, or hydrogen and fluorine interact. Hydrogen bonding plays critical roles in base-pairing of nucleotides and in the structure of macromolecules.	https://en.wikipedia.org/wiki/Hydrogen_bond
Hydroxide ion (OH^-)	An oxygen atom covalently bonded to a hydrogen atom which has gained an extra electron with its negative charge. Contributes to alkalinity in aqueous solutions.	https://en.wikipedia.org/wiki/Hydroxide#Hydroxide_ion
Hypothesis	A general feature of scientific inquiry and research and the scientific method. Hypotheses should be precisely stated and testable by experiments, observations, and principles that are objective and reproducible.	https://en.wikipedia.org/wiki/Hypothesis
Immune system	A host defense system comprising many biological structures and processes within an organism. Protects an organism from diseases and threats caused by pathogens, toxins, and/or cancer cells. Immune systems may include both innate and adaptive components (see adaptive immune system).	https://en.wikipedia.org/wiki/Immune_system

(*Continued*)

Glossary Term	Descriptions or Definitions, Including Modified Wikipedia Excerpts	Relevant Link (Usually Wikipedia)
Immunoglobulin superfamily	A large and diverse gene family (and its protein products) that includes large subfamilies of segments of T-cell receptor and antibody proteins, as well as other cell-surface antigens and functional proteins found on body cells (such as the major histocompatibility complex) and cells of the innate and adaptive immune systems (such as CD4 and CD8 cells that distinguish functional subsets of T cells). The immunoglobulin superfamily is named after classes of antibodies called immunoglobulins. The superfamily is involved in complex molecular, cellular, and developmental interactions and communications, especially involving the adaptive immune system.	https://en.wikipedia.org/wiki/Immunoglobulin_superfamily
Impermeable	A barrier that does not let other substances or objects pass through.	https://en.wikipedia.org/wiki/Permeability_(earth_sciences)
Implicit information content	A concept in developmental and Molecular Genetics advanced by Gunther Stent that recognizes the importance of higher-level emergent properties and interactions—and not just explicit DNA sequences—of genomic DNA, in the reproduction of species-specific biological organization in each generation.	https://en.wikipedia.org/wiki/Implicit
In situ hybridization	A powerful laboratory technique involving nucleic acid hybridization of fluorescent or otherwise tagged probes. This technique is used to	https://en.wikipedia.org/wiki/In_situ_hybridization

(*Continued*)

Glossary Term	Descriptions or Definitions, Including Modified Wikipedia Excerpts	Relevant Link (Usually Wikipedia)
	render patterns of gene expression visible within intact biological specimens, such as the embryos of fruitflies. A similar technique, using fluorescent antibodies, can be used to render specific proteins visible.	
Incomplete dominance	A genetic relationship between alleles in which expression of both alleles in a heterozygous individual determines intermediate phenotypes. For example, red and white alleles that together generate pink flower color represent incomplete dominance.	https://en.wikipedia.org/wiki/Dominance_(genetics)
Induced fit	Shape change in the active site of an enzyme when it binds to reactants that improves the snugness of the fit between the active site and the reactants.	https://en.wikipedia.org/wiki/Enzyme_catalysis#Induced_fit
Infinitesimal variations (Natural Selection)	Refers to the slight variations that Darwin hypothesized are subjected to selection. When the variations are useful in the context of the struggle for existence, they tend to accumulate and lead to larger changes.	https://en.wikipedia.org/wiki/Infinitesimal
Information (genetic)	Refers to the determinants transmitted from generation to generation in the DNA sequences of the genome. Since many types of explicit and implicit information can be found in the genome, this term tends to be an oversimplification when used to describe the relationship between genotype and phenotype.	https://en.wikipedia.org/wiki/Genome

(Continued)

Glossary Term	Descriptions or Definitions, Including Modified Wikipedia Excerpts	Relevant Link (Usually Wikipedia)
Information content (biology)	In biology, attempts to describe DNA sequences in reductionist terms of probability or Shannon entropy. This approach fails to recognize the implicit information and emergent properties of complex molecular, cellular, and developmental interactions.	https://en.wikipedia.org/wiki/Entropy_(information_theory)
Innate immune system	One of the two main immunity strategies found in vertebrates (the other being the adaptive immune system). The innate immune system is an older evolutionary defense strategy, relatively speaking, and it is the dominant immune system response found in plants, fungi, insects, and primitive multicellular organisms.	https://en.wikipedia.org/wiki/Innate_immune_system
Inner-workings of cells	A general term describing all of the complex interactions between molecular structures and functions at various levels of complexity.	https://en.wikipedia.org/wiki/Cell_(biology)
Innovation (evolutionary biology)	In the context of *Rethinking Evolution*, this term is frequently used to denote the evolution of useful new structures and functions at various levels of complexity, by natural processes. Does not imply conscious or supernatural design or intelligence of any kind.	https://en.wikipedia.org/wiki/Key_innovation
Inorganic	Substances that do not contain carbon atoms (excluding allotropes of carbon and small molecules such as carbon dioxide).	https://en.wikipedia.org/wiki/Inorganic_compound
Insertion (DNA)	The addition of one or more nucleotide bases (or base-pairs) into a DNA sequence.	https://en.wikipedia.org/wiki/Insertion_(genetics)

(*Continued*)

Glossary Term	Descriptions or Definitions, Including Modified Wikipedia Excerpts	Relevant Link (Usually Wikipedia)
Integral membrane protein	A protein partially or completely embedded in the lipid bilayer of the cell membrane. Includes transmembrane cell-surface receptors, channels, active pumps, and other functional proteins. Signal sequences on portions of the protein chains lead to transport and integration of integral membrane proteins. This involves some degree of self-assembly that is determined by the chemical and physical properties of both lipids and the amino acid side-chains of the proteins.	https://en.wikipedia.org/wiki/Integral_membrane_protein
Integrated evolutionary synthesis	Although there is debate among evolutionary biologists, the term is generally used to describe an evolutionary synthesis that retains certain core principles of classical Darwinism and the Modern Synthesis while expanding and modifying it with newly acquired concepts that are well-supported by empirical evidence.	http://jeb.biologists.org/content/jexbio/218/1/7.full.pdf
Intelligent design	A religious argument for the existence of God. Masquerades as an evidence-based scientific theory about life's origins, but has been discredited as pseudoscience.	https://en.wikipedia.org/wiki/Intelligent_design
Intermediate products (biosynthetic pathways)	Products from earlier steps that serve as reactants in later steps of biosynthetic pathways.	(not applicable)
Intermolecular binding forces	Weak interactions such as dipole–dipole attractions and transient dipoles that play major roles in phase transitions in chemistry as well as biological processes such as base-pairing and protein-folding.	https://en.wikipedia.org/wiki/Intermolecular_force

(Continued)

Glossary Term	Descriptions or Definitions, Including Modified Wikipedia Excerpts	Relevant Link (Usually Wikipedia)
Intron	Segments of eukaryotic genes that are transcribed but that are spliced out during processing of messenger RNA.	https://en.wikipedia.org/wiki/Intron
Ion	An atom or molecule that has a nonzero net electrical charge. Ions can be negatively or positively charged.	https://en.wikipedia.org/wiki/Ion
Ionic compound	A chemical compound composed of ions held together by ionic bonding.	https://en.wikipedia.org/wiki/Ionic_compound
Ionization	The process whereby an atom or a molecule acquires a negative or positive charge. This occurs when electrons are gained or lost.	https://en.wikipedia.org/wiki/Ionization
Ionizing radiation	High-energy radiation (alpha, beta, or gamma) capable of breaking molecules such as DNA.	https://en.wikipedia.org/wiki/Ionizing_radiation
Irreducible complexity	A pseudoscientific argument that holds that certain complex biological systems cannot evolve by Natural Selection. Irreducible Complexity assumes that evolution by Natural Selection always involves accumulation of infinitesimal variations, which represents only one aspect of 21st century evolutionary theory. Irreducible complexity is central to the creationist concept of Intelligent Design, but has been rejected by the scientific community.	https://en.wikipedia.org/wiki/Irreducible_complexity
Junk DNA (noncoding DNA)	A persistent but misleading 20th century term popularized by Susumu Ohno that refers to what are now more correctly referred to as noncoding DNA sequences. This, too is misleading, because many noncoding	https://en.wikipedia.org/wiki/Non-coding_DNA#Junk_DNA

(*Continued*)

Glossary Term	Descriptions or Definitions, Including Modified Wikipedia Excerpts	Relevant Link (Usually Wikipedia)
	sequences play important roles in a variety of functions, including control of gene expression, chromosome condensation, and more. A more interesting recent term for such sequences is "the genomic scrapyard", because this emphasizes that modular components are often modified and reused in evolution.	
Lateral gene transfer (horizontal gene transfer)	The movement of genetic material between organisms other than by the transmission of DNA from parent to offspring. It is an important factor in the evolution of many organisms.	https://en.wikipedia.org/wiki/Horizontal_gene_transfer
Lichen	A composite organism that arises from algae or bacteria living among multiple fungi in a symbiotic relationship.	https://en.wikipedia.org/wiki/Lichen
Life-history strategy (life-history theory)	The ways that patterns of development, behavior, and other adaptations contribute to the long-term survival of a species. Often, life-history strategies involve major changes that adapt individuals to more than one ecological *niche* or way of life, such as metamorphosis of caterpillars to butterflies. Also includes reproductive strategies that may favor large numbers of offspring or fewer numbers with greater parental care.	https://en.wikipedia.org/wiki/Life_history_theory
Ligase	An enzyme that can catalyze the joining of two molecules by forming a new covalent chemical bond.	https://en.wikipedia.org/wiki/Ligase

(*Continued*)

Glossary Term	Descriptions or Definitions, Including Modified Wikipedia Excerpts	Relevant Link (Usually Wikipedia)
Linkage group	Refers to genetic maps of groups of genes known to be located on a particular homologous pair of autosomes or on a specific sex chromosome. For example, genes on human chromosome 21 are members of a linkage group.	https://en.wikipedia.org/wiki/Genetic_linkage
Lipid	Macromolecule that is soluble in nonpolar solvents. They can store energy, act as signals or represent structural components of cell membranes.	https://en.wikipedia.org/wiki/Lipid
Lipid bilayer	A thin polar membrane made of two layers of lipid molecules. Found in cell membranes and internal membranes of all living cells.	https://en.wikipedia.org/wiki/Lipid_bilayer
Liter	A derived unit of volume in the metric system defined by 1,000 cubic centimeters. Equivalent to 1.06 US quarts.	https://en.wikipedia.org/wiki/Litre
Lock and key (model)	A metaphor for the specific 3D fit of reactants at the active sites of enzymes. The reality is that induced fits as well as shapes, plus chemical and physical properties, are involved in the binding process. Induced fit involves changes in the shape of the enzyme active site initiated by the binding process, leading to tighter and closer binding.	https://en.wikipedia.org/wiki/Enzyme#%22Lock_and_key%22_model
Lone pair (electrons)	A pair of nonbonded valence electrons of one of the atoms of a molecule. Lone pairs tend to have larger electron clouds and cause VSEPR effects (valence shell electron pair repulsion) which can cause molecular asymmetry.	https://en.wikipedia.org/wiki/Lone_pair

(*Continued*)

Glossary Term	Descriptions or Definitions, Including Modified Wikipedia Excerpts	Relevant Link (Usually Wikipedia)
Lost city hydrothermal field	The first alkaline deep-sea hydrothermal vent, known as the Lost City, was discovered on the mid-Atlantic sea floor in 2,000. Distinguished from high-temperature hydrothermal vents known as black smokers, alkaline vents are formed by serpentinization, in which olivine (magnesium iron silicate) and similar rock on the seafloor reacts with water. This produces large volumes of hydrogen. Precipitation reactions between the warm (45–90 degrees Celsius) alkaline (pH 9–11) vent fluids and the colder, more acidic seawater create tall, porous calcium carbonate chimneys containing embedded metals and minerals. It is thought that proton gradients generated by similar, ancient vents provided the energy required for the metabolic reactions that led to the origin of life. Currently, the Lost City provides a unique protected environment supplying both energy and nutrients to thriving miniature local ecosystems of organisms.	https://en.wikipedia.org/wiki/Lost_City_Hydrothermal_Field
Lymphocyte	A white blood cell that develops into a T cell or B cell—part of the adaptive immune system.	https://en.wikipedia.org/wiki/Lymphocyte
Macroevolution	Refers to large, visible adaptations that accumulate over evolutionary time and distinguish organisms. For example, the evolution of limbs of tetrapods is an example of macroevolution. The distinction between macroevolution and microevolution is an artificial distinction, because both are part of a natural continuum of change involving both genotypes and phenotypes.	https://en.wikipedia.org/wiki/Macroevolution

(*Continued*)

Glossary Term	Descriptions or Definitions, Including Modified Wikipedia Excerpts	Relevant Link (Usually Wikipedia)
Macromolecule	A very large molecule commonly created by the polymerization of smaller subunits. They are typically composed of thousands of atoms or more. Classes of macromolecules found in all living cells include carbohydrates, lipids, nucleic acids, and proteins.	https://en.wikipedia.org/wiki/Macromolecule
Macroscopic realm	The realm in which relativistic or classical physics, rather than quantum mechanics, can predict the behavior and outcomes of events and interactions.	https://en.wikipedia.org/wiki/Quantum_realm
Maternal genes (maternal effect)	Usually refers to gene products deposited in the egg, known as cytoplasmic determinants, produced by the mother, that play important roles in early development.	https://en.wikipedia.org/wiki/Maternal_effect
Meiosis	A specialized type of cell division that reduces the chromosome number of a diploid cell by half, usually during the production of sex cells in sexual reproduction. This creates four haploid cells (or discarded nuclei) which are each genetically distinct from the diploid parent cell that gave rise to them.	https://en.wikipedia.org/wiki/Meiosis
Membrane (lipid bilayer)	A selective barrier which allows some things to pass through but stops others. Biological membranes usually consist of lipid bilayers, and may include cell membranes, nuclear membranes, or membranes of organelles in eukaryotic cells, or internal membranes in prokaryotic cells.	https://en.wikipedia.org/wiki/Membrane

(*Continued*)

Glossary Term	Descriptions or Definitions, Including Modified Wikipedia Excerpts	Relevant Link (Usually Wikipedia)
Mendelian genetics (Mendelian inheritance)	The study of a type of biological inheritance that follows the laws originally proposed by Gregor Mendel in 1865 and 1866 and re-discovered in 1900. When Mendel's theories were integrated with the Boveri–Sutton chromosome theory of inheritance by Thomas Hunt Morgan in 1915, they became the core of classical genetics.	https://en.wikipedia.org/wiki/Simple_Mendelian_genetics_in_humans
Meristematic (meristem)	Pertaining to the tissue in most plants containing undifferentiated cells (meristematic cells), found in zones of the plant where growth can take place. Meristematic cells give rise to various organs of a plant and are responsible for growth.	https://en.wikipedia.org/wiki/Meristem
Messenger RNA (mRNA)	A large family of RNA molecules that convey genetic information from DNA to the ribosome. Messenger RNA is transcribed from DNA sequences and then processed to produce mature messenger RNA molecules. At the ribosome, they specify the amino acid sequence of the protein products of gene expression.	https://en.wikipedia.org/wiki/Messenger_RNA
Metabolic pathway	A linked series of chemical reaction occurring within a cell.	https://en.wikipedia.org/wiki/Metabolic_pathway
Metabolic scrapyard	A metaphor describing the modular reuse of specific enzymes and biochemical pathways in different species and in different ecological contexts.	https://en.wikipedia.org/wiki/Metabolism

(*Continued*)

Glossary Term	Descriptions or Definitions, Including Modified Wikipedia Excerpts	Relevant Link (Usually Wikipedia)
Metabolism	A general term for the breakdown (catabolism) and synthesis (anabolism) of organic molecules by living organisms. Includes highly conserved core pathways involved with energy capture and conversion, as well as biosynthesis.	https://en.wikipedia.org/wiki/Metabolism
Metabolomics	The scientific study of chemical processes involving metabolites, the small molecule intermediates and products of metabolism. Specifically, metabolomics is the "systematic study of the unique chemical fingerprints that specific cellular processes leave behind", the study of their small-molecule metabolite profiles. The metabolome represents the complete set of metabolites in a biological cell, tissue, organ, or organism, which are the end products of cellular processes.	https://en.wikipedia.org/wiki/Metabolomics
Metal	A material that has a lustrous appearance and conducts electricity and heat. Metals are typically malleable or ductile. Metals may contain one or more elements. For example, iron is single element, but steel usually contains a mixture of iron plus other elements.	https://en.wikipedia.org/wiki/Metal
Methanogenesis	The formation of methane by methanogens. This is responsible for significant amounts of natural gas accumulations, including atmospheric methane.	https://en.wikipedia.org/wiki/Methanogenesis

(*Continued*)

Glossary Term	Descriptions or Definitions, Including Modified Wikipedia Excerpts	Relevant Link (Usually Wikipedia)
Methanogens	Microorganisms that produce methane as a metabolic byproduct in low oxygen conditions.	https://en.wikipedia.org/wiki/Methanogen
Micro-machine	A metaphorical description of the complex and precise structural and functional roles played by macromolecular interactions, especially proteins.	https://en.wikipedia.org/wiki/Micromachinery
Microbial consortium	Two or more microbial groups living symbiotically. Consortiums can be endosymbiotic or ectosymbiotic.	https://en.wikipedia.org/wiki/Microbial_consortium
Microevolution	Refers primarily to changes in the frequencies of alleles in the gene pool recognized by Population Genetics. These include the same adaptations that accumulate over evolutionary time and distinguish organisms. The distinction between macroevolution and microevolution is an artificial distinction, because both are part of a natural continuum of change involving both genotypes and phenotypes.	https://en.wikipedia.org/wiki/Microevolution
Mimicry	Evolutionary adaptations in which organisms resemble other organisms, usually to mislead potential predators or prey. For example, certain caterpillars have a striking appearance resembling venomous snakes, and are called snake mimics.	https://en.wikipedia.org/wiki/Mimicry
Mitochondrion	A double-membrane-bound organelle found in most eukaryotic cells. Metabolizes glucose or other sugars and generates most of the cell's supply of ATP during aerobic respiration.	https://en.wikipedia.org/wiki/Mitochondrion

(Continued)

Glossary Term	Descriptions or Definitions, Including Modified Wikipedia Excerpts	Relevant Link (Usually Wikipedia)
Mitosis	The part of the cell division cycle of eukaryotes. Usually, identical copies of replicated chromosomes are separated into two new nuclei of daughter cells.	https://en.wikipedia.org/wiki/Mitosis
Modern synthesis	Early 20th century evolutionary synthesis reconciling Charles Darwin's classical theory of Natural Selection with Mendelian Genetics.	https://en.wikipedia.org/wiki/Modern_synthesis_(20th_century)
Module (biology)	A general term describing an entity that serves as both a unit of structure and function and a subunit that contributes to more complex emergent structures and functions. Evolutionary modification and reuse of modular genomic elements and their gene products is widely observed in natural history, and is an important theme in evolutionary developmental biology (evo-devo).	https://en.wikipedia.org/wiki/Module
Molarity	A measure of the concentration of a chemical species in terms of amount of substance per unit volume of a solution.	https://en.wikipedia.org/wiki/Molar_concentration
Mole	The unit of measurement for amount of a substance. Specifically, a mole contains 6.02×10^{23} particles.	https://en.wikipedia.org/wiki/Mole_(unit)
Molecular cloning technology (recombinant DNA)	A set of experimental methods in molecular biology that are used to assemble recombinant DNA molecules. Can also refer to replication of DNA within host organisms.	https://en.wikipedia.org/wiki/Molecular_cloning
Molecular evolution	A general term describing all kinds of changes in the DNA sequences of genomes of various species over time.	https://en.wikipedia.org/wiki/Molecular_evolution

(*Continued*)

Glossary Term	Descriptions or Definitions, Including Modified Wikipedia Excerpts	Relevant Link (Usually Wikipedia)
Molecular genetics	The field of biology that studies the structure and function of genes at a molecular level. Employs methods of molecular biology, including recombinant DNA technology, biochemistry, and genetics.	https://en.wikipedia.org/wiki/Molecular_genetics
Molecular recognition (biology)	In biology, a metaphorical description of shape-specific binding events, such as binding of enzymes and their reactants (substrates), binding of antibodies to specific epitopes of antigens, or binding of signals to cell-surface receptors. The metaphor has teleological overtones, however. *Rethinking Evolution* introduces a more objective and descriptive term: shape-specific molecular interaction and binding events, abbreviated as SSM-IBE.	https://en.wikipedia.org/wiki/Molecular_recognition
Molecule	An electrically neutral group of two or more atoms held together by covalent chemical bonds.	https://en.wikipedia.org/wiki/Molecule
Monoclonal antibody	An artificially produced, highly specific antibody produced by fusion of a B cell clone with a tumor cell to create a hybridoma. Monoclonal antibodies are extremely valuable reagents in biomedical research, because they recognize highly-specific epitopes (binding portions) of highly-specific antigens, can be produced in very large quantities, and can be tagged with molecules such as fluorescent proteins. They also can sometimes be used to deliver toxins to tumor cells with reduced toxicity for normal cells.	https://en.wikipedia.org/wiki/Molecule

(*Continued*)

Glossary Term	Descriptions or Definitions, Including Modified Wikipedia Excerpts	Relevant Link (Usually Wikipedia)
Monomer (biology)	A subunit molecule that can undergo polymerization. In biological polymers, monomers of DNA or RNA are nucleotides, monomers of carbohydrates are sugars, and monomers of protein chains are amino acids.	https://en.wikipedia.org/wiki/Monomer
Morphogen	Usually, a substance such as a diffusible protein or a specific RNA sequence, that forms a gradient in a developing embryo, leading to particular developmental switches, segmentation patterns, or other aspects of pattern formation that help determine the morphology (shape, structure, and appearance) of organisms during development. Some morphogens are transcription factors.	https://en.wikipedia.org/wiki/Morphogen
Morphogenetic field	A region in which a coordinated pattern is established during development. Can refer to various aspects of morphology, such as shape, segmentation, or other aspects structure and appearance of an organism. Morphogenic fields are often influenced by morphogens.	https://en.wikipedia.org/wiki/Morphogenetic_field
Morphology (biology)	A general term describing shape and form and structure and function in biology. Usually refers to multicellular structures involving tissues and organs.	https://en.wikipedia.org/wiki/Morphology_(biology)
Mosaic embryo (determinate cleavage)	A type of embryonic cleavage in which cytoplasmic determinants combined with cell lineage define patterns of cell differentiation, rather than intercellular communication between neighboring embryonic cells (blastomeres).	https://en.wikipedia.org/wiki/Cleavage_(embryo)

(*Continued*)

Glossary Term	Descriptions or Definitions, Including Modified Wikipedia Excerpts	Relevant Link (Usually Wikipedia)
mRNA (messenger RNA)	Processed RNA transcripts that carry sequences of codons (triplets in the nucleotide sequence) to ribosomes, where they determine the amino acid sequences of newly synthesized protein chains.	https://en.wikipedia.org/wiki/Messenger_RNA
Multicellular (organism)	Organisms (generally eukaryotic) composed of multitudes of cells. Includes animals, plants, and fungi. In sexual reproduction of plants and animals, specialized cells arise by cleavage and subsequent development of an embryo derived from a fertilized ovule or egg (zygote), respectively, during development.	https://en.wikipedia.org/wiki/Multicellular_organism
Multigene family (gene family)	Genes that arise by gene duplication. Most of the genes of multicellular organisms are members of multigene families.	https://en.wikipedia.org/wiki/Gene_family
Mutagen	A physical or chemical agent that changes the genetic material of an organism. Mutagens increase the frequency of mutations above the natural background level.	https://en.wikipedia.org/wiki/Mutagen
Mutation (genetics)	The permanent alteration of the nucleotide sequence of a genome. Can occur in an organism, virus, extrachromosomal DNA, or other genetic elements.	https://en.wikipedia.org/wiki/Mutation
Mutualism	The way organisms of different species exist in a relationship in which each individual benefits from the activity of the other.	https://en.wikipedia.org/wiki/Mutualism_(biology)
Myelin sheath (myelin)	Specialized cells containing lipid-rich myelin substances that surround neurons (cells of nervous systems that	https://en.wikipedia.org/wiki/Myelin

(Continued)

Glossary Term	Descriptions or Definitions, Including Modified Wikipedia Excerpts	Relevant Link (Usually Wikipedia)
	transmit electrical signals) and provide structural support, as well as electrical insulation.	
Myosin	A motor protein that uses ATP as an energy source and interacts with actin to generate cell motility.	https://en.wikipedia.org/wiki/Myosin
Natural science	Branch of science concerned with the description, prediction, and understanding of natural phenomena. Based on empirical evidence from observation and experimentation. Examples include biology, chemistry, and physics.	https://en.wikipedia.org/wiki/Natural_science
Natural Selection	The differential survival and reproduction of individuals due to differences in phenotype. It is a key mechanism of evolution, the change due to heritable traits characteristic of a population over generations.	https://en.wikipedia.org/wiki/Natural_selection
Natural theology	A type of theology that provides arguments for the existence of God based on reason combined with observations from nature.	https://en.wikipedia.org/wiki/Natural_theology
Naturalism	The scientific belief, supported by empirical observations and experiments, that only natural laws and forces operate in the universe.	https://en.wikipedia.org/wiki/Naturalism_(philosophy)
Negative charge	The physical force exerted by electrons. An equal and opposite positive force is exerted by protons. Negative charges often refers to static electrical forces generated by electrons. In atoms, ions, and molecules, negative charges arise from a relative excess of orbiting electrons in a particular region.	https://en.wikipedia.org/wiki/Electric_charge

(*Continued*)

Glossary Term	Descriptions or Definitions, Including Modified Wikipedia Excerpts	Relevant Link (Usually Wikipedia)
Negative logarithm	In mathematics, the negative of the number obtained with the inverse function to exponentiation. That means the negative logarithm of a given number x is the negative of the exponent to which another fixed number, the base b, must be raised, to produce that number x.	https://en.wikipedia.org/wiki/Logarithm
Neo-Darwinism	The interpretation of Darwinian evolution through Natural Selection, as it had been modified since it was first proposed. The term "neo-Darwinism" may refer to the Modern Synthesis, or more recent evolutionary syntheses.	https://en.wikipedia.org/wiki/Neo-Darwinism
Neofunctionalization	One of the possible outcomes of functional divergence, occurs when one gene copy, or paralog, takes on a totally new function after a gene duplication event. Neofunctionalization is an adaptive mutation process; meaning one of the gene copies must mutate to develop a function that was not present in the ancestral gene. In other words, one of the duplicates retains its original function, while the other accumulates molecular changes such that, in time, it can perform a different task.	https://en.wikipedia.org/wiki/Neofunctionalization
Niche	The way of life of an organism—how it survives and reproduces in the context of specific environments.	https://en.wikipedia.org/wiki/Niche
Nitrogenous base	An organic molecule with a nitrogen atom that has the chemical properties of a base. Involved in the base-pairing (hydrogen bonding of complementary bases) of DNA and/or RNA strands.	https://en.wikipedia.org/wiki/Nitrogenous_base

(Continued)

Glossary Term	Descriptions or Definitions, Including Modified Wikipedia Excerpts	Relevant Link (Usually Wikipedia)
Nonrepetitive DNA sequences	DNA segments with complex sequences, that is, that lack simple runs of nucleotides or tandem repeats. Usually, protein-coding sequences of DNA (e.g. eukaryotic exons) are relatively nonrepetitive.	https://en.wikipedia.org/wiki/Repeated_sequence_(DNA)
Nonspontaneous (chemistry)	In chemistry, a nonspontaneous reaction or process requires an external source of energy to drive it. Refers to endergonic chemical reactions. May also refer to the movement of heat or other energy sources, or particles in solutions, from regions where they are lower in amount or concentration to regions that are higher in amount or concentration.	https://en.wikipedia.org/wiki/Endergonic_reaction
Noncoding DNA	Components of an organism's DNA that are not transcribed into RNA molecules and/or translated into amino acid sequences of protein chains.	https://en.wikipedia.org/wiki/Non-coding_DNA
Nonmetal	A chemical element that mostly lacks metallic attributes. Tends to have low melting and boiling points and usually is poor conductor of heat and electricity.	https://en.wikipedia.org/wiki/Nonmetal
Nonrandom (variation or events)	Variation or events that are influenced in some way by other factors such that various possible outcomes do not have equal probabilities.	https://en.wikipedia.org/wiki/Randomness#In_biology
Nuance	Subtle, fine, or minor distinction, usually made clear by the precise use of language.	https://en.wiktionary.org/wiki/nuance
Nucleated (cell)	A typical eukaryotic cell, which has a nucleus surrounded by membranes (nuclear envelope). Some specialized eukaryotic cells, such as red blood cells in human adults, lose their nuclei during development and are no longer nucleated cells.	https://en.wikipedia.org/wiki/Cell_nucleus

(*Continued*)

Glossary Term	Descriptions or Definitions, Including Modified Wikipedia Excerpts	Relevant Link (Usually Wikipedia)
Nucleic acid (DNA, RNA)	Macromolecules of DNA or RNA consisting of polymers of nucleotides. May consist of single or double strands. DNA is the genetic material of all types of cells.	https://en.wikipedia.org/wiki/Nucleic_acid
Nucleic acid hybridization	A phenomenon in which single-stranded DNA and/or RNA molecules are base-paired to form hybrid double-stranded structures. An essential tool for finding or tagging related (homologous) sequences in biotechnology.	https://en.wikipedia.org/wiki/Nucleic_acid_hybridization
Nucleic acid structure (DNA and RNA)	Refers to the covalent bonding between nucleotides in strands of DNA or RNA, as well as the 3D structures formed by hydrogen bonds during base-pairing. Both RNA and DNA strands have "backbones" of covalently bonded sugar and phosphate groups, which are covalently bonded to nitrogenous bases.	https://en.wikipedia.org/wiki/Nucleic_acid_structure
Nucleotide	Organic molecules that serve as the monomer units for forming DNA and RNA. These are the building blocks of nucleic acids and play a central role in metabolism.	https://en.wikipedia.org/wiki/Nucleotide
Nucleotide sequence	A sequence of nucleotides within a strand of DNA or RNA. Nucleotide sequence is determined by the sequence of nitrogenous bases in the nucleotides. Letters are used to represent the different nucleotides. (See A, C, G, T for DNA or A, C, G, U for RNA).	https://en.wikipedia.org/wiki/Nucleic_acid_sequence

(*Continued*)

Glossary Term	Descriptions or Definitions, Including Modified Wikipedia Excerpts	Relevant Link (Usually Wikipedia)
Nucleus (cell)	A structure in eukaryotic cells surrounded by a nuclear envelope consisting of a double lipid bilayer that contains nuclear pores through which molecules are transported to and from the cytoplasm in a regulated way. The nucleus contains chromosomes containing genomic DNA, histone proteins, and other proteins and RNA molecules.	https://en.wikipedia.org/wiki/Cell_nucleus
Nutrients	Substances used by an organism to survive, grow, and reproduce. Nutrients are used as raw materials and enzymatic co-factors in metabolism.	https://en.wikipedia.org/wiki/Nutrient
Odorant receptors (olfactory receptors)	Members of the G protein-coupled receptor gene family involved in signal transduction pathways that detect specific odors. Larger numbers of these receptors are associated with species (such as dogs) that have a keen or highly sensitive/selective sense of smell.	https://en.wikipedia.org/wiki/Olfactory_receptor
One gene, one enzyme hypothesis	The hypothesis that each gene acts through the production of an enzyme. Each gene is considered to be responsible for producing a single enzyme that in turn affects a single step in a metabolic pathway.	https://en.wikipedia.org/wiki/One_gene%E2%80%93one_enzyme_hypothesis
One gene, one protein chain	The hypothesis that each gene acts through the production of a protein, or more specifically, a particular chain of amino acids (polypeptide). This broadens the "one gene, one enzyme" hypothesis to include any polypeptide. Each gene is considered to be	https://en.wikipedia.org/wiki/One_gene%E2%80%93one_enzyme_hypothesis

(*Continued*)

Glossary Term	Descriptions or Definitions, Including Modified Wikipedia Excerpts	Relevant Link (Usually Wikipedia)
	responsible for producing a single protein. True for many genes, but some eukaryotic genes may produce multiple gene products per gene, through alternative splicing during messenger RNA production.	
Operator	In prokaryotes, a sequence of DNA that a repressor binds to. When the repressor is bound, nearby genes that are downstream cannot be transcribed. A classic example of regulation of gene expression and regulatory DNA sequences.	https://en.wikipedia.org/wiki/Operon#Operator
Operon	A set of genes containing an operator, involved in the control of expression (transcription) of those genes in response to conditions, such as the availability of a particular sugar, and the need to metabolize that sugar when present.	https://en.wikipedia.org/wiki/Operon
Order (biological organization)	In biology, the organized hierarchy of complex biological structures, functions, and systems, that helps to define the complexity of life in terms of simpler ideas. This is a fundamental premise for numerous areas of scientific research.	(not applicable)
Organ (anatomy)	Structural and functional arrangement of tissues to perform useful tasks for a multicellular organism.	https://en.wikipedia.org/wiki/Organ_(anatomy)
Organelle	A specialized subunit within a cell that has a specific function vital to the cell. Usually contains one or more biological membranes. Examples of eukaryotic organelles include mitochondria and chloroplasts.	https://en.wikipedia.org/wiki/Organelle

(*Continued*)

Glossary Term	Descriptions or Definitions, Including Modified Wikipedia Excerpts	Relevant Link (Usually Wikipedia)
Organic (chemistry)	Pertaining to molecules containing carbon, other than simple molecules such as carbon dioxide or allotropes of carbon, e.g. graphite.	https://en.wikipedia.org/wiki/Organic
Organic compound	Any chemical compound (other than carbon dioxide or pure carbon) that contains carbon atoms.	https://en.wikipedia.org/wiki/Organic_compound
Organic synthesis	Synthesis of organic, i.e. carbon-containing, compounds.	https://en.wikipedia.org/wiki/Organic_synthesis
Organismic systems approach (evo-devo)	In evolutionary developmental biology (evo-devo), a supplement to the Modern Synthesis discussed in a published paper by Callebaut, Müller and Newman.	https://www.researchgate.net/publication/258236066_The_organismic_systems_approach_EvoDevo_and_the_streamlining_of_the_naturalistic_agenda
Organization (biology)	The hierarchy of complex biological structures and systems that define life using a reductionist approach. Each level in the hierarchy represents an increase in organizational complexity, with each "object" being primarily composed of the previous level's basic unit. The basic principle behind the organization is the concept of emergence—the properties and functions found at a hierarchical level are not present and irrelevant at the lower levels.	(not applicable)
Origin of life (biology)	Field of biological research and theory concerned with the first appearance of life forms from nonliving metabolic	https://en.wikipedia.org/wiki/Abiogenesis

(*Continued*)

Glossary Term	Descriptions or Definitions, Including Modified Wikipedia Excerpts	Relevant Link (Usually Wikipedia)
	processes, primary on Earth. Also called abiogenesis. When extended to other planets in the universe, this is generally called astrobiology.	
Oxidation	A chemical process that removes electrons (or removes hydrogen atoms and electrons) from atoms, molecules, or ions, thereby increasing the positive charge or reducing the negative charge and/or increasing the oxidation state of that substance.	https://en.wikipedia.org/wiki/Redox
Pair-rule gene	A type of gene involved in the development of the segmented embryos of insects. Pair-rule genes are expressed as a result of differing concentrations of gap gene proteins, which encode transcription factors controlling pair-rule gene expression. Pair-rule genes are defined by the effect of a mutation in that gene, which causes the loss of the normal developmental pattern in alternating segments. Pair-rule genes were first described by Christiane Nüsslein-Volhard and Eric Wieschaus in 1980.	https://en.wikipedia.org/wiki/Pair-rule_gene
Paradigm shift	A fundamental change in the basic concepts and experimental practices of a scientific discipline.	https://en.wikipedia.org/wiki/Paradigm_shift
Parasitism (parasitic)	A relationship between species where one organism benefits at the expense of another by living on it or within it. This parasite causes its host some harm and will adapt to its way of life.	https://en.wikipedia.org/wiki/Parasitism
Parson	The priest of an independent parish church. This church is not under the control of another organization.	https://en.wikipedia.org/wiki/Parson

(*Continued*)

Glossary Term	Descriptions or Definitions, Including Modified Wikipedia Excerpts	Relevant Link (Usually Wikipedia)
Pathogen	An organism, virus or prion that causes harm or disease to another organism.	https://en.wikipedia.org/wiki/Pathogen
Pattern formation (developmental biology)	The generation of complex organizations of cell fates in space and time. Pattern formation is controlled by genes. The role of genes in pattern formation is an aspect of morphogenesis, the creation of diverse anatomies from similar genes, now being explored in the science of evolutionary developmental biology or evo-devo.	https://en.wikipedia.org/wiki/Pattern_formation
Pax (genes, evo-devo)	Paired box (Pax) genes are a family of genes coding for tissue specific transcription factors containing a paired domain and usually a partial, or in the case of four family members (PAX3, PAX4, PAX6, and PAX7), [1] a complete homeodomain. An octapeptide may also be present. Pax proteins are important in early animal development for the specification of specific tissues, as well as during epimorphic limb regeneration in animals capable of such.	https://en.wikipedia.org/wiki/Pax_genes
Potential biological evolution (PBE)	A term introduced in *Rethinking Evolution* to describe potential change in the heritable characteristics of biological populations over successive generations. As is the case with potential energy, PBE is considered real, not imagined, and dependent on the relative positions and properties of interacting objects.	(not applicable)

(*Continued*)

Glossary Term	Descriptions or Definitions, Including Modified Wikipedia Excerpts	Relevant Link (Usually Wikipedia)
Prebiotic or primordial soup hypothesis (PBS)	Abbreviation for a hypothetical condition of the Earth's atmosphere and oceans before the emergence of life. According to this hypothesis, life arose from biological molecules that were formed by spontaneous natural sources of energy such as lightning, heat, or UV irradiation.	https://en.wikipedia.org/wiki/Primordial_soup
Polymerase chain reaction (PCR)	A widely used, rapid, and sensitive technique in biotechnology involving amplification (production of large numbers of identical copies) of DNA or RNA sequences by repeated cycles of DNA replication in vitro.	https://en.wikipedia.org/wiki/Polymerase_chain_reaction
Peer-review (scientific)	The formal evaluation of work by one or more people of similar competence to the producers of the work. Methods are employed to maintain standards of quality and credibility. A standard practice for most formal publications in scientific journals, and an essential tool for success of the scientific method.	https://en.wikipedia.org/wiki/Peer_review
Persistent conceptual networks (semantic networks)	Refers to semantic relations between concepts that tend to continue for long periods of time. Persistent conceptual networks often influence thoughts and ideas in subtle and subjective ways.	https://en.wikipedia.org/wiki/Semantic_network
pH	A logarithmic scale used to specify the acidity or basicity of an aqueous (water-based) solution. Neutral solutions have a pH of 7.0, whereas acidic solutions and basic solutions have lower and higher pH, respectively.	https://en.wikipedia.org/wiki/PH

(*Continued*)

Glossary Term	Descriptions or Definitions, Including Modified Wikipedia Excerpts	Relevant Link (Usually Wikipedia)
Phenotype	The composite of an organism's observable characteristics or traits. Results from the expression of an organism's genetic code, and usually involves complex interactions between multiple genes and gene products.	https://en.wikipedia.org/wiki/Phenotype
Phenotypic gestalt	A general term referring to the overall phenotype of an organism that results from numerous genetic determinants that arise as a unified whole from complex molecular, cellular, and developmental interactions.	https://en.wikipedia.org/wiki/Phenotype
Phenylalanine	An amino acid that contains a side-chain with a benzyl functional group.	https://en.wikipedia.org/wiki/Phenylalanine
Phosphate backbone (sugar-phosphate backbone, DNA or RNA)	The covalently bonded phosphate and sugar residues that form an alternating "backbone" in DNA and RNA strands.	https://en.wikipedia.org/wiki/Nucleic_acid_structure
Phospholipids	A class of lipids containing a phosphate functional group that are major components of all cell membranes. They are usually found in lipid bilayers.	https://en.wikipedia.org/wiki/Phospholipid
Photoreceptor protein	Light-sensitive transmembrane G-protein coupled receptor proteins involved in light perception in tissues such as the retina.	https://en.wikipedia.org/wiki/Photoreceptor_protein
Photosynthesis	Metabolic process that uses energy from light plus carbon dioxide and water to produce glucose and other sugars. Converts energy into excited electrons, photon gradients, ATP, and NADPH. Takes place in the chloroplasts of plants and in cyanobacteria.	https://en.wikipedia.org/wiki/Photosynthesis

(*Continued*)

Glossary Term	Descriptions or Definitions, Including Modified Wikipedia Excerpts	Relevant Link (Usually Wikipedia)
Phylogenomics	The intersection of the fields of evolution and genomics. Four major areas fall under phylogenomics: (1) prediction of gene function, (2) establishment and clarification of evolutionary relationships, (3) gene family evolution, and (4) prediction and retracing lateral gene transfer.	https://en.wikipedia.org/wiki/Phylogenomics
Pigmentation (biological pigment)	A light-absorbing substance produced by a living cell. The colors of pigments are derived from wavelengths of light that are not absorbed by the pigment.	https://en.wikipedia.org/wiki/Biological_pigment
Plant	One of the three kingdoms of multicellular organisms. Plant cells are distinguished from animal cells by cell walls and must add to the cell wall with the cell plate to divide. Plant cells usually perform photosynthesis which takes place in chloroplasts, and often have large vacuoles in the cytoplasm that can store fluids and contribute to the turgidity of the cell.	https://en.wikipedia.org/wiki/Plant
Plasmid	A small circular DNA molecule within a cell. Physically separated from chromosomal DNA and can replicate independently.	https://en.wikipedia.org/wiki/Plasmid
Plastid	A membrane-bound organelle found in the cells of plants, algae, and some other eukaryotic organisms. Plastids are the site of manufacture and storage of important chemical compounds used by the cells of autotrophic eukaryotes. They often contain pigments used in photosynthesis, and the types of pigments in a plastid	https://en.wikipedia.org/wiki/Plastid

(*Continued*)

Glossary Term	Descriptions or Definitions, Including Modified Wikipedia Excerpts	Relevant Link (Usually Wikipedia)
	determine the cell's color. They have a common evolutionary origin and possess a double-stranded DNA molecule that is circular, like that of prokaryotic cells.	
Pleiotropic	Effects caused by single genes that influence two or more seemingly unrelated observable traits.	https://en.wikipedia.org/wiki/Pleiotropy
Point mutation	A genetic mutation that changes the identity of a single nitrogenous base in a DNA sequence. May result in a change in one codon, which may change a single amino acid or add or eliminate a start or stop codon in a protein-coding sequence.	https://en.wikipedia.org/wiki/Point_mutation
Polar (bond)	An electric dipole resulting from differential attraction of the electron cloud of a hybrid orbital toward two atomic nuclei that have different effective positive nuclear charges in a hybrid orbital. The differential attraction results from differences in the effective positive nuclear charge of the two atomic nuclei.	https://en.wikipedia.org/wiki/Chemical_polarity
Polar (molecule)	A molecule with one or more polar bonds with an asymmetrical surrounding charge distribution. Often, this results from polar bonds combined with VSEPR from lone pairs of electrons (VSEPR = valence shell electron pair repulsion).	(not applicable)
Polar bonds	Asymmetrical positive and negative charge distribution resulting from asymmetrical distribution of electrons in hybrid orbitals of covalent bond. Arise when	https://en.wikipedia.org/wiki/Chemical_polarity

(*Continued*)

Glossary Term	Descriptions or Definitions, Including Modified Wikipedia Excerpts	Relevant Link (Usually Wikipedia)
	elements with relatively large and small effective nuclear charges, such as hydrogen and oxygen, form covalent bonds. Results from greater pull on electron cloud by nucleus with larger effective nuclear charge.	
Polar molecules	Molecules with local asymmetrical positive and negative charge distributions that result from the combination of polar bonds and other sources of asymmetry in the overall molecular shape, such as lone pairs and VSEPR (valence shell electron pair repulsion). Symmetrical molecules with polar bonds are not polar molecules.	https://en.wikipedia.org/wiki/Chemical_polarity
Pollination vector	Usually refers to an animal that transfers pollen from a male part of a plant to a female part of a plant. This enables fertilization and seed production by the plant.	https://en.wikipedia.org/wiki/Pollination
Polymer (biology)	A large molecule composed of many repeated subunits which are called monomers. Major classes of biological polymers include nucleic acids (DNA and RNA, protein chains, and glucose polymers such as cellulose, starch, glycogen, and chitin.	https://en.wikipedia.org/wiki/Polymer
Polymerase chain reaction (PCR)	A technique used in molecular biology to artificially amplify a specific segment of a DNA or RNA template. Starting with one or more template molecules, amplification rapidly and routinely generates hundreds of thousands or millions of identical	https://en.wikipedia.org/wiki/Polymerase_chain_reaction

(*Continued*)

Glossary Term	Descriptions or Definitions, Including Modified Wikipedia Excerpts	Relevant Link (Usually Wikipedia)
	copies. RNA amplification often involves starting with a complementary DNA sequence generated from the RNA.	
Polytene chromosome	Large chromosome that has thousands of identical DNA strands that have not separated by mitosis. They provide a high level of function in certain tissues.	https://en.wikipedia.org/wiki/Polytene_chromosome
Population genetics	A subfield of genetics that deals with genetic differences (especially, allele frequencies) within and between populations. It is a part of evolutionary biology.	https://en.wikipedia.org/wiki/Population_genetics
Porosity	A measure of the openings (void spaces) in a material. This determines the ability of other substances to pass through a porous material. Porosity is expressed as a fraction of the volume of voids over the total volume.	https://en.wikipedia.org/wiki/Porosity
Porous	Containing spaces that allow substances to pass through.	https://en.wikipedia.org/wiki/Porosity
Position-effect variegation	Local effects in gene expression caused by nearby genetic elements. Often caused by transposable elements, which have randomly inserted themselves in the genome in different cells. A visible phenotypic effect is the diverse and random color pattern found in the seeds of maize. Reported by Nobel Laureate Barbara McClintock.	https://en.wikipedia.org/wiki/Position-effect_variegation
Positive charge	A static electric charge associated with the subatomic proton particle. In atoms, molecules, or ions, can be caused by a deficit of negative charges from electrons.	https://en.wikipedia.org/wiki/Electric_charge

(*Continued*)

Glossary Term	Descriptions or Definitions, Including Modified Wikipedia Excerpts	Relevant Link (Usually Wikipedia)
Potential (possibility of specific changes or events)	In general terms, potential describes the possibilities and probabilities of subsequent events. Potential can be changed by actual events.	https://en.wikipedia.org/wiki/Potential
Potential biological evolution	A term introduced in *Rethinking Evolution* to describe potential change in the heritable characteristics of biological populations over successive generations. As is the case with potential energy, PBE is considered real, not imagined, and dependent on the relative positions and properties of interacting objects.	(not applicable)
Potential energy	In physics, potential energy represents energy that results from relative positions, such as a pencil held over a table. In chemistry, potential energy is found in covalent bonds and in differential concentrations of particles, such as proton gradients.	https://en.wikipedia.org/wiki/Potential_energy
Prebiotic soup (primordial soup)	A hypothetical condition of the Earth's atmosphere and oceans before the emergence of life. According to this hypothesis, life arose from biological molecules that were formed by spontaneous natural sources of energy such as lightning, heat, or UV irradiation.	https://en.wikipedia.org/wiki/Primordial_soup
Precipitate	A verb describing the separation of a dissolved substance from a solution, or a noun describing the state of that substance. Precipitates include solids, such as salts for example, or liquids, such as droplets of rain.	https://en.wikipedia.org/wiki/Precipitation_(chemistry)
Precipitation	The formation of a precipitate, such as solids crystallizing out of solution or the formation of rain droplets (see precipitate).	https://en.wikipedia.org/wiki/Precipitation_(chemistry)

(*Continued*)

Glossary Term	Descriptions or Definitions, Including Modified Wikipedia Excerpts	Relevant Link (Usually Wikipedia)
Preformationism (sperm)	The false belief that tiny individuals, such as miniature human individuals, can be found, preformed, in sperm cells.	https://en.wikipedia.org/wiki/Preformationism
Pressure	The force applied perpendicular to the surface of an object.	https://en.wikipedia.org/wiki/Pressure
Primary producer (autotroph)	Uses energy in light or in inorganic chemical compounds to build organic molecules.	https://en.wikipedia.org/wiki/Primary_producers
Primary structure	The shape and characteristics determined by the linear sequence of amino acids in a protein chain, which also helps determine higher-level structure.	https://en.wikipedia.org/wiki/Protein_primary_structure
Producer (autotroph)	An organism such as a plant that can build macromolecules from simple elements such as carbon dioxide and water. Photoautotrophs use the energy of sunlight to do this; chemoautotrophs use energy of inorganic compounds from the environment. In ecosystems, producers such as plants generate excess carbohydrates and other materials that can be used as energy and nutrient sources by consumers (heterotrophs).	https://en.wikipedia.org/wiki/Autotroph
Products (chemistry)	The substances resulting from a chemical reaction.	https://en.wikipedia.org/wiki/Product_(chemistry)
Programmed cell death	Programmed cell death (or PCD) is the death of a cell in any form, mediated by an intracellular program. PCD is carried out in a biological process, which usually confers advantage during an organism's life-cycle. PCD serves fundamental functions during both plant and animal tissue development as well as the removal of infected, genetically damaged, or cancerous cells.	https://en.wikipedia.org/wiki/Programmed_cell_death

(*Continued*)

Glossary Term	Descriptions or Definitions, Including Modified Wikipedia Excerpts	Relevant Link (Usually Wikipedia)
Prokaryotic (prokaryote)	Major classification of living cells that includes bacteria and archaea, but not eukaryotes. Prokaryotes are usually found as unicellular organisms that lack nuclei and membrane-bound organelles.	https://en.wikipedia.org/wiki/Prokaryote
Protein	Macromolecule consisting of one or more long covalently bound chains of amino acid residues. Protein molecules perform a vast array of functions within all organisms.	https://en.wikipedia.org/wiki/Protein
Protein chain	A covalently bound chain of specific amino acid residues that constitute a specific protein chain. Amino acid residues are joined by peptide bonds formed between amino groups and carboxyl groups of adjacent amino acids.	https://en.wikipedia.org/wiki/Protein
Protein sequence (amino acid sequence)	A covalently bound chain of specific amino acid residues that constitute a specific protein chain.	https://en.wikipedia.org/wiki/Protein_sequencing
Protein synthesis	The process whereby biological cells generate new protein chains by transcription and translation.	https://en.wikipedia.org/wiki/Protein_biosynthesis
Protein-coding	Sequences of DNA that determine the amino acid sequences of proteins.	https://en.wikipedia.org/wiki/Gene
Proton	A positively charged subatomic particle found in the nucleus of all atoms. Proton number defines the identity of a chemical element. A single proton is found in the nucleus of hydrogen, so ionized hydrogen atoms are equivalent to protons.	https://en.wikipedia.org/wiki/Proton
Proton gradient	An unequal distribution of protons (hydrogen ions), often across some sort of barrier such as a cell membrane, that represent potential	https://en.wikipedia.org/wiki/Electrochemical_gradient

(*Continued*)

Glossary Term	Descriptions or Definitions, Including Modified Wikipedia Excerpts	Relevant Link (Usually Wikipedia)
	energy. Proton gradients at undersea alkaline hydrothermal mounds (UAHM) are also hypothesized to have played an important role in the origin of life on Earth.	
Protozoan (protozoa)	Informal term for single-celled eukaryotes, either free-living or parasitic, which feed on organic matter such as other microorganisms or organic tissues and debris.	https://en.wikipedia.org/wiki/Protozoa
Pseudoscience	Beliefs and ideas that have the appearance of science, but are not actually based on genuine logic, empirical evidence, or the scientific method. Pseudoscience may be propagated either deliberately or unwittingly. Often, pseudoscience is spread by special interest groups that have political, economic, discriminatory, or religious agendas.	https://en.wikipedia.org/wiki/Pseudoscience
Q-cycle	A series of reactions describing how the sequential oxidation and reduction of electron carriers result in the movement of protons across a lipid bilayer.	https://en.wikipedia.org/wiki/Q_cycle
Quantitative inheritance	A branch of population genetics that deals with phenotypes that vary continuously (in characters such as height or mass)—as opposed to discretely identifiable phenotypes and gene-products (such as eye-color, or the presence of a particular biochemical).	https://en.wikipedia.org/wiki/Quantitative_genetics
Quaternary structure	The 3D shape formed when two or more folded protein chains bind together. Many functional proteins have multiple protein chains.	https://en.wikipedia.org/wiki/Protein_quaternary_structure

(*Continued*)

Glossary Term	Descriptions or Definitions, Including Modified Wikipedia Excerpts	Relevant Link (Usually Wikipedia)
Quinone	A class of organic compounds that are formally derived from aromatic compounds. Occurs by converting –CH= groups into –C(=O)= groups.	https://en.wikipedia.org/wiki/Quinone
Quorum sensing (bacteria)	The ability to detect and to respond to cell population density by gene regulation. As one example, quorum sensing (QS) enables bacteria to restrict the expression of specific genes to the high cell densities at which the resulting phenotypes will be most beneficial. Many species of bacteria use quorum sensing to coordinate gene expression according to the density of their local population.	https://en.wikipedia.org/wiki/Quorum_sensing
QWERTY keyboard layout	A widely-used standard arrangement of characters on both mechanical and virtual or digital keyboards. The name comes from the order of the first six keys on the top left letter row of the keyboard.	https://en.wikipedia.org/wiki/QWERTY
Random fertilization	The random union of one of many possible egg and sperm cells during sexual reproduction.	https://en.wikipedia.org/wiki/Quantitative_genetics#Random_fertilization
Random mutation	A heritable change in a DNA sequence where the different possible changes have equal probabilities.	https://en.wikipedia.org/wiki/Mutation
Random shuffling of genomic sequences (genetic recombination)	In genetic recombination, random crossover events that take place during meiosis that result in recombinant chromosomes that have portions of maternal and paternal sequences. Although crossover events are relatively random with respect to the loci involved, repetitive sequences are often hotspots for recombination.	https://en.wikipedia.org/wiki/Genetic_recombination

(*Continued*)

Glossary Term	Descriptions or Definitions, Including Modified Wikipedia Excerpts	Relevant Link (Usually Wikipedia)
Random variation (genetics)	Genotypic and phenotypic variation in which the variants have equal probabilities of occurrence by random chance alone.	https://en.wikipedia.org/wiki/Genetic_variation
Raw material (chemistry or biology)	Basic material, including reactants or energy sources, that can be transformed by natural processes into products or other material.	https://en.wikipedia.org/wiki/Raw_material
Reactant (reagent)	A chemical substance that undergoes a chemical reaction and is transformed into product substances.	https://en.wikipedia.org/wiki/Reagent
Receptor (biochemistry)	A protein molecule that receives chemical signals from outside a cell. When such chemical signals bind to a receptor, they cause some form of cellular/tissue response, e.g. a change in the electrical activity of a cell.	https://en.wikipedia.org/wiki/Receptor_(biochemistry)
Recessive	A Mendelian genetic relationship between alleles of a gene, in which the phenotype of one allele is masked or hidden in the presence of another allele, which is dominant.	https://en.wikipedia.org/wiki/Dominance_(genetics)
Recognition sequence	A DNA sequence motif that binds to another molecule, such as a restriction enzyme, in a specific manner.	https://en.wikipedia.org/wiki/Recognition_sequence
Recombinant DNA	DNA molecules formed by laboratory methods of genetic recombination. The technology that combines genetic material from multiple sources, creating artificial sequences not usually found in nature.	https://en.wikipedia.org/wiki/Recombinant_DNA
Recombination (meiosis)	Exchanges between sister chromatids of homologous chromosomes that result in recombinant chromosomes	https://en.wikipedia.org/wiki/Recombination

(*Continued*)

Glossary Term	Descriptions or Definitions, Including Modified Wikipedia Excerpts	Relevant Link (Usually Wikipedia)
	containing some segments inherited from the father and some from the mother. Occurs in a somewhat random fashion during meiosis, which produces sex cells.	
Recombination-activating gene	Genes that encode enzymes that play an important role in the rearrangement and recombination of the genes of immunoglobulin and T cell receptor molecules, however there is no evidence to suggest the developing T cells can undergo receptor editing in the same way that B cells do.	https://en.wikipedia.org/wiki/Recombination-activating_gene
Redox (reaction)	A chemical reaction involving exchanges of electrons between atoms or molecules, in which one substance is reduced (gains electrons) while the other is oxidized (loses electrons).	https://en.wikipedia.org/wiki/Recombination-activating_gene
Reduction	A chemical process that adds electrons (or adds hydrogen atoms and electrons) to atoms, molecules, or ions, thereby increasing the negative charge or reducing the positive charge and/or decreasing the oxidation number of that substance.	https://en.wikipedia.org/wiki/Reduction
Reduction potential	A measure of the tendency of a chemical species to acquire electrons and thereby be reduced. The more positive the potential, the greater its tendency is to be reduced.	https://en.wikipedia.org/wiki/Reduction_potential
Regional specification	The process whereby regions of the embryo respond to signals and change their patterns of gene expression in coordinated ways, in response to signals such as morphogens and/or by means of other molecular or cellular interactions during development.	(not applicable)

(*Continued*)

Glossary Term	Descriptions or Definitions, Including Modified Wikipedia Excerpts	Relevant Link (Usually Wikipedia)
Regulative embryo (cleavage, indeterminate)	Embryos that cleave into cells (blastomeres) that remain flexible in their developmental fates, where cellular interactions help to coordinate the range of different types of specialized cells produced by the daughter cells produced by early divisions.	https://en.wikipedia.org/wiki/Cleavage_(embryo)
Regulatory sequences	A segment of a nucleic acid molecule which is capable of increasing or decreasing the expression of specific genes within an organism. It is an essential feature of all living organisms and viruses.	https://en.wikipedia.org/wiki/Regulatory_sequence
Repeated sequence (genomic DNA)	Genomic DNA sequences that occur in multiple copies. Includes simple or complex tandem repeats or nonadjacent repeated elements.	https://en.wikipedia.org/wiki/Repeated_sequence_(DNA)
Repetitive DNA sequences	A broad category describing a variety of different types of repetitive genomic sequences.	https://en.wikipedia.org/wiki/Category:Repetitive_DNA_sequences
Replicate (biology)	An exact copy of a biological entity, a laboratory experiment, or samples used in an experiment.	https://en.wikipedia.org/wiki/Replicate_(biology)
Replication	Generally, reproduction of an exact copy of an entity, which may include a gene, a cell, a genome, or other entities.	https://en.wikipedia.org/wiki/Replication
Replication slippage (slipped-strand mispairing)	A form of mutation that leads to expansion or contraction of simple repetitive sequences such as dinucleotide or trinucleotide repeats. Occurs during DNA replication.	https://en.wikipedia.org/wiki/Replication_slippage
Reproductive isolation	A collection of evolutionary mechanisms, behaviors, and physiological processes that disrupt gene flow between populations.	https://en.wikipedia.org/wiki/Reproductive_isolation

(*Continued*)

Glossary Term	Descriptions or Definitions, Including Modified Wikipedia Excerpts	Relevant Link (Usually Wikipedia)
	A common mechanism by which populations evolve into distinct species.	
Res potentia (quantum physics)	Refers to a theory advanced by Ruth Kastner, Stuart Kauffman, and Michael Epperson, where potential events in the quantum realm have a real existence, but ordinary concepts of space and time do not apply.	https://arxiv.org/abs/1709.03595
Restriction enzyme	An enzyme that cleaves DNA into fragments. Depends on binding of the enzyme to specific nucleotide sequence motifs called recognition sites or restriction sites.	https://en.wikipedia.org/wiki/Restriction_enzyme
Retrovirus	A virus that produces reverse transcriptase and uses it to integrate RNA into DNA.	https://en.wikipedia.org/wiki/Retrovirus
Reverse transcriptase	Enzymes that synthesize complementary (base-paired) DNA strands from RNA templates. Discovered in retroviruses (such as HIV) and used in biotechnology.	https://en.wikipedia.org/wiki/Reverse_transcriptase
Ribosome	A complex molecular micro-machine found within all living cells, consisting of RNA and protein components. The site of biological protein synthesis.	https://en.wikipedia.org/wiki/Ribosome
RNA	Abbreviation for ribonucleic acid. Various types of RNA, such as messenger RNA, transfer RNA, and ribosomal RNA are involved in transcription and translation of genetic information, but there are many other types of RNA that have been more recently discovered, that play a variety of roles in the storage, transmission, and expression of genetic information.	https://en.wikipedia.org/wiki/RNA

(*Continued*)

Glossary Term	Descriptions or Definitions, Including Modified Wikipedia Excerpts	Relevant Link (Usually Wikipedia)
	RNA sequences have the sugar ribose as part of their backbone, and have nucleotides containing adenine, cytosine, guanine, and uracil nitrogenous bases.	
RNA polymerase	Enzymes involved in transcription, usually from DNA templates, that synthesize strands of RNA that are complementary to the DNA template.	https://en.wikipedia.org/wiki/RNA_polymerase
RNA synthesis	Synthesis of a complementary (base-paired) RNA strand from a DNA template by RNA polymerase enzymes.	https://en.wikipedia.org/wiki/RNA
RNA world (hypothesis)	Refers to a hypothetical stage in the origin of life in which the primary genetic material was RNA rather than DNA.	https://en.wikipedia.org/wiki/RNA_world
Rube Goldberg device (machine)	A Rube Goldberg machine is a machine intentionally designed to perform a simple task in an indirect and overcomplicated fashion. Often, these machines consist of a series of simple devices that are linked together to produce a domino effect, in which each device triggers the next one, and the original goal is achieved only after many steps. In biology, evolutionary processes often generate structures and functions that have the appearance of Rube Goldberg machines, but actually arise by blind natural processes.	https://en.wikipedia.org/wiki/Rube_Goldberg_machine
Salinity	The amount of salt in a solution. Salts are ionic substances such as sodium chloride or potassium chloride.	https://en.wikipedia.org/wiki/Salinity
Scanning electron microscope	A type of electron microscope that generates high-resolution images of the surfaces of tiny objects with focused beams of electrons.	https://en.wikipedia.org/wiki/Scanning_electron_microscope

(Continued)

Glossary Term	Descriptions or Definitions, Including Modified Wikipedia Excerpts	Relevant Link (Usually Wikipedia)
Schwann cell	Named after physiologist Theodor Schwann, cell types that are the principal glia of the peripheral nervous system (PNS). Glial cells function to support neurons and in the PNS, also include satellite cells, olfactory ensheathing cells, enteric glia and glia that reside at sensory nerve endings, such as the Pacinian corpuscle. The two types of Schwann cells are myelinating and nonmyelinating.	https://en.wikipedia.org/wiki/Schwann_cell
Scientific theory	An explanation of an aspect of the natural world that can be repeatedly tested. Scientific theories are often tested in accordance with the scientific method, using a predefined and reproducible protocol of observations and experiments.	https://en.wikipedia.org/wiki/Scientific_theory
Secondary messenger (system)	Cytoplasmic substances involved in transducing and transmitting signals from cell surface receptors in signal transduction pathways.	https://en.wikipedia.org/wiki/Second_messenger_system
Secondary structure (protein)	The 3D form of local segments of protein chains (but not the entire folded chain). The most common elements are alpha helixes and beta sheets.	https://en.wikipedia.org/wiki/Protein_secondary_structure
Secreted protein (secretory protein)	A protein that is normally transported outside the cell membrane by normal physiological processes. Usually refers to functional proteins, such as milk proteins or antibodies, rather than waste materials which are referred to as excreted materials.	https://en.wikipedia.org/wiki/Secretory_protein
Segmentation (biology)	The division of some animal and plant body plans into a series of repetitive segments, which are specified during development.	https://en.wikipedia.org/wiki/Segmentation_(biology)

(*Continued*)

Glossary Term	Descriptions or Definitions, Including Modified Wikipedia Excerpts	Relevant Link (Usually Wikipedia)
Selection (Natural Selection)	The natural preservation of variants in a population that prove useful within the context of survival and/or reproduction. A critical component of Natural Selection.	https://en.wikipedia.org/wiki/Selection
Selective advantage (Natural Selection)	Variants of structures and functions that are more useful than others in the struggle for existence and therefore tend to be preserved by Natural Selection.	https://en.wikipedia.org/wiki/Adaptation
Selective forces (evolutionary pressure)	Environmental factors (both living and nonliving, biotic and abiotic) that act on natural populations to preserve (naturally select) certain variants more than others, because those factors contribute more to survival and reproduction in a particular context.	https://en.wikipedia.org/wiki/Evolutionary_pressure
Self-accelerating process	Any process that increases in rate. For example, length increases in simple-repetitive sequences by slipped-strand mispairing, can be a self-accelerating process in genomic evolution.	https://en.wikipedia.org/wiki/Acceleration
Self-organization	A process where some form of overall order arises naturally from local interactions. Interactions between parts of an initially disordered system can result in emergent structures and functions.	https://en.wikipedia.org/wiki/Self-organization
Selfish DNA (selfish genetic element)	DNA sequences that replicate and often move around by relatively random and non-Mendelian processes. The teleological term "selfish" refers to the fact that the sequences replicate and insert themselves in genomes by molecular mechanisms that are arbitrary, due to their molecular structure. Transposable elements and retroviruses are examples of selfish DNA.	https://en.wikipedia.org/wiki/Selfish_genetic_element

(*Continued*)

Glossary Term	Descriptions or Definitions, Including Modified Wikipedia Excerpts	Relevant Link (Usually Wikipedia)
Semantics (linguistics)	The subfield that is devoted to the study of meaning, as inherent at the levels of words, phrases, sentences, and larger units of discourse (termed texts, or narratives). The study of semantics is also closely linked to the subjects of representation, reference, and denotation.	https://en.wikipedia.org/wiki/Semantics
Semi-permeable	A barrier that selectively lets specific substances or objects through. The ability to pass through is often determined by physical characteristics such as size and/or shape, or by chemical characteristics such as attractive or repulsive forces.	https://en.wikipedia.org/wiki/Permeability
Semi-permeable membrane	A type of biological or synthetic membrane. It will allow certain molecules or ions to pass through by diffusion.	https://en.wikipedia.org/wiki/Semipermeable_membrane
Serpentinization	A hydration and metamorphic transformation of ultramafic rock from the Earth's mantle. The mineral alteration is particularly important at the sea floor at tectonic plate boundaries.	https://en.wikipedia.org/wiki/Serpentinite
Sex cells (sperm or egg)	Sperm or egg cells which are produced by male or female germ line cells, respectively, during meiosis, in sexually reproducing organisms.	https://en.wikipedia.org/wiki/Gamete
Sex chromosome	A chromosome that determines the sex of an individual. Sex chromosomes are designated as X, Y, W, or Z, depending on the mechanism of sex determination.	https://en.wikipedia.org/wiki/Sex_chromosome

(*Continued*)

Glossary Term	Descriptions or Definitions, Including Modified Wikipedia Excerpts	Relevant Link (Usually Wikipedia)
Sex determination (system)	Determination of biological gender by genetic elements.	https://en.wikipedia.org/wiki/Sex-determination_system
Sex-linked (sex-linkage, X-linkage	Refers to genes that are located on sex chromosomes. In mammals, sex-linkage usually refers to genes linked to the X chromosome, but can also refer to sequences found on Y chromosomes, or on W or Z sex chromosomes in certain groups of animals such as amphibians.	https://en.wikipedia.org/wiki/Sex_linkage
Sexual reproduction	A form of reproduction in which sex cells (sperm and eggs) fuse together. The cells fuse during fertilization, resulting in a zygote that develops into a new individual.	https://en.wikipedia.org/wiki/Sexual_reproduction
Sexual selection	A mode of natural selection where members of one biological sex choose mates of the other sex to mate with (intersexual selection), and compete with members of the same sex for access to members of the opposite sex (intrasexual selection). These two forms of selection mean that some individuals have better reproductive success than others within a population, either from being more attractive or preferring more attractive partners to produce offspring.	https://en.wikipedia.org/wiki/Sexual_selection

(*Continued*)

Glossary Term	Descriptions or Definitions, Including Modified Wikipedia Excerpts	Relevant Link (Usually Wikipedia)
Side-chain	A chemical group that is attached to the backbone of a molecule. It is one factor determining a molecule's properties and reactivity. For example, the 20 types of amino acids found in living cells are distinguished by their side chains.	https://en.wikipedia.org/wiki/Side_chain
Signal peptide	Sometimes referred to as signal sequence, targeting signal, localization signal, localization sequence, transit peptide, leader sequence or leader peptide), signal peptides are short amino acid sequences (usually 16–30 amino acids long), present at the N-terminus of the majority of newly synthesized proteins, that are destined towards the secretory pathway. These proteins include those that reside either inside certain organelles (the endoplasmic reticulum, Golgi or endosomes), are secreted from the cell, or are inserted into most cellular membranes.	https://en.wikipedia.org/wiki/Signal_peptide
Signal transduction	The process by which a chemical or physical signal is transmitted, often from a transmembrane cell-surface receptor, through a cell as a series of molecular events, most commonly protein phosphorylation catalyzed by protein kinase enzymes but also by means of other intermediaries such as secondary messengers, which ultimately results in a cellular response.	https://en.wikipedia.org/wiki/Signal_transduction

(*Continued*)

Glossary Term	Descriptions or Definitions, Including Modified Wikipedia Excerpts	Relevant Link (Usually Wikipedia)
Signal transduction pathway	A pathway by which a chemical or physical signal is transmitted, often from a transmembrane cell-surface receptor, through a cell as a series of molecular events, most commonly protein phosphorylation catalyzed by protein kinase enzymes but also by means of other intermediaries such as secondary messengers, which ultimately results in a cellular response.	https://en.wikipedia.org/wiki/Cell_signaling - Signaling_pathways
Simple-repetitive DNA sequence (repeated sequence, simple sequence)	A sequence containing large numbers of short tandem repeats ranging in length from single nucleotides to several nucleotides.	https://en.wikipedia.org/wiki/Repeated_sequence_(DNA)
Single-stranded (not base-paired)	Strands of DNA or RNA that are not base-paired with other strands (and do not have substantial internal base-pairing).	https://en.wikipedia.org/wiki/Base_pair
Skeletal muscle	One of three major muscle types, apart from cardiac muscle and smooth muscle. Skeletal muscle refers to multiple bundles (fascicles) of cells joined together called muscle fibers. Muscle fibers are in turn composed of myofibrils. The myofibrils are composed of actin and myosin filaments, repeated units are known as sarcomeres, which are the basic functional units of the muscle fiber.	https://en.wikipedia.org/wiki/Skeletal_muscle
Slipped-strand mispairing	A mutation process in which simple sequence repeat units are added or deleted, usually due to mispairing of tandem repeat units during DNA replication. A major evolutionary source of simple-repetitive DNA sequences, particularly in eukaryotic genomes.	https://en.wikipedia.org/wiki/Slipped_strand_mispairing

(*Continued*)

Glossary Term	Descriptions or Definitions, Including Modified Wikipedia Excerpts	Relevant Link (Usually Wikipedia)
Social sciences	Sciences that are concerned with society and the relationships among individuals within a society. Examples of social sciences include social psychology, sociology, history, and political science.	https://en.wikipedia.org/wiki/Social_science
Sociobiology	A field of biology that aims to examine and explain social behavior of animals in terms of evolution. When applied to humans, great care must be taken to rely on empirical scientific data when drawing conclusions.	https://en.wikipedia.org/wiki/Sociobiology
Solution	A homogeneous mixture in which a solvent surrounds solute particle at the molecular level.	https://en.wikipedia.org/wiki/Solution
Somatic hypermutation (SHM)	A cellular mechanism by which the immune system adapts to the new foreign elements that confront it. A major component of the process of affinity maturation, SHM diversifies B cell receptors used to recognize foreign elements (antigens) and allows the immune system to adapt its response to new threats during the lifetime of an organism. SHM involves a programmed process of mutation affecting the variable regions of immunoglobulin genes. Unlike germline mutation, SHM affects only an organism's individual immune cells, and the mutations are not transmitted to the organism's offspring.	https://en.wikipedia.org/wiki/Somatic_hypermutation

(*Continued*)

Glossary Term	Descriptions or Definitions, Including Modified Wikipedia Excerpts	Relevant Link (Usually Wikipedia)
Sox (gene family)	A family of transcription factors with a highly-conserved DNA binding domain called the HMG-box. Homologues have been identified in insects, nematodes, amphibians, reptiles, birds and a range of mammals. HMG boxes called SRY resides on the Y-chromosome and many other different aspects of development in various species of eukaryotic organisms.	https://en.wikipedia.org/wiki/SOX_gene_family
Sperm cell	The male reproductive cell (sex cell) that develops after a diploid cell from the male germ line undergoes meiosis to form four haploid daughter cells.	https://en.wikipedia.org/wiki/Sperm
Sponge (biology)	Members of the phylum Porifera which are multicellular organisms that have bodies full of pores and channels allowing water to circulate through them, consisting of jelly-like mesohyl sandwiched between two thin layers of cells.	https://en.wikipedia.org/wiki/Sponge
Spontaneous mutation	A mutation that occurs by chance in a genome in a natural setting.	https://en.wikipedia.org/wiki/Mutation
Spontaneous reaction (chemical)	The time-evolution of a chemical system in which it releases free energy. It moves to a lower, more thermodynamically stable energy state.	https://en.wikipedia.org/wiki/Spontaneous_process
Sporulation (spore)	Formation of spores, which are dormant forms of bacteria, or units of sexual or asexual reproduction in eukaryotes, that may be adapted for dispersal and for survival, often for extended periods of time, in unfavorable conditions.	https://en.wikipedia.org/wiki/Spore

(*Continued*)

Glossary Term	Descriptions or Definitions, Including Modified Wikipedia Excerpts	Relevant Link (Usually Wikipedia)
Slipped-strand mispairing (SSM)	Also known as replication slippage, SSM is a mutation process which occurs during DNA replication. It involves denaturation and displacement of the DNA strands, resulting in mispairing of the complementary bases. When it synergizes with additional mutational mechanisms such as point mutations and unequal crossing-over, offers a comprehensive explanation for the origin and evolution of repetitive DNA sequences.	https://en.wikipedia.org/wiki/Slipped_strand_mispairing
Shape-specific molecular interactions and binding events (SSM-IBE)	Descriptive term introduced in *Rethinking Evolution* referring to highly-specific molecular binding events and interactions between biologically important molecules such as enzymes and their reactants (substrates), antibodies and their specific epitopes of antigens, cell-surface receptors and their signals, and DNA sequences and transcription factors.	(not applicable)
Stability (chemical stability)	The tendency of a chemical substance to remain the same rather than undergoing chemical reactions.	https://en.wikipedia.org/wiki/Chemical_stability
Stem cell	A cell that can differentiate into other types of cells and can divide to produce more of the same type of stem cells. They are found in multicellular organisms.	https://en.wikipedia.org/wiki/Stem_cell
Strong chemical bond (covalent bond)	A chemical bond that involves sharing of one or more electron pairs in hybrid orbitals, binding atoms together by an electromagnetic force. Strong chemical bonds often require chemical reactions to break them.	https://en.wikipedia.org/wiki/Chemical_bond

(*Continued*)

Glossary Term	Descriptions or Definitions, Including Modified Wikipedia Excerpts	Relevant Link (Usually Wikipedia)
Subfunctionalization	One of the possible outcomes of functional divergence that occurs after a gene duplication event, in which pairs of genes that originate from duplication, or paralogs, take on separate functions. Subfunctionalization is a neutral mutation process; meaning that no new adaptations are formed.	https://en.wikipedia.org/wiki/Subfunctionalization
Submicroscopic realm	A general term with two general meanings: (1) Describes small objects, such as cells and organelles, that are too small to be resolved with a light microscope. (2) Applies to objects at the atomic or subatomic size scale, which are governed by the laws of quantum mechanics rather than classical physics.	https://en.wikipedia.org/wiki/Microscopic_scale
Substrate-level phosphorylation	A metabolic reaction that results in the formation of ATP or GTP by the direct transfer of a phosphoryl (PO3) group to ADP or GDP from another phosphorylated compound. Unlike oxidative phosphorylation, oxidation and phosphorylation are not coupled in the process of substrate-level phosphorylation.	https://en.wikipedia.org/wiki/Substrate-level_phosphorylation
Subunit (biology)	Part of a larger entity, called a unit. Subunits usually have structural and functional integrity, but can interact with other subunits to form higher levels of complexity which have emergent properties.	https://en.wikipedia.org/wiki/Subunit
Sugar	The generic name for soluble carbohydrates. Simple sugars are called monosaccharides and include glucose (also known as dextrose),	https://en.wikipedia.org/wiki/Sugar

(*Continued*)

Glossary Term	Descriptions or Definitions, Including Modified Wikipedia Excerpts	Relevant Link (Usually Wikipedia)
	fructose, and galactose, as well as five-carbon sugars such as ribose and deoxyribose found in nucleic acids.	
Symbiosis	Any type of a close and long-term biological interaction between two organisms. This can benefit both organisms, harm one and benefit the other, or benefit one and have no effect on the other.	https://en.wikipedia.org/wiki/Symbiosis
Synthesis (organic)	The chemical synthesis of organic compounds from chemical precursors.	https://en.wikipedia.org/wiki/Synthesis
T cell	One of the subtypes of a white blood cell in a vertebrate's immune system.	https://en.wikipedia.org/wiki/Lymphocyte
T-cell receptor	A molecule found on the surface of T cells, or T lymphocytes, that is responsible for recognizing fragments of antigen as peptides bound to major histocompatibility complex (MHC) molecules. T-cell receptor genes are diverse members of the immunoglobulin superfamily.	https://en.wikipedia.org/wiki/T-cell_receptor
Tandem array	A gene cluster created by mechanisms that generate new genetic material. They serve to encode large numbers of genes at a time.	https://en.wikipedia.org/wiki/Tandemly_arrayed_genes
Teleological	An argument for the existence of God. The argument for an intelligent creator is based on perceived evidence of deliberate design in the natural world.	https://en.wikipedia.org/wiki/Teleological_argument
Temperature-sensitive (mutation)	A mutation, usually in a protein-coding gene, whose phenotype is expressed at certain temperatures and not others.	https://en.wikipedia.org/wiki/Temperature-sensitive_mutant

(*Continued*)

Glossary Term	Descriptions or Definitions, Including Modified Wikipedia Excerpts	Relevant Link (Usually Wikipedia)
Template (DNA or RNA)	During DNA replication or transcription, a single strand of DNA serves as a template for the synthesis of a complementary (base-paired) strand of DNA or RNA by DNA or RNA polymerase enzymes, respectively. During reverse transcription, a strand of RNA serves as a template for synthesis of a complementary DNA strand by reverse transcriptase.	(not applicable)
Template-based replication	During DNA replication or transcription, a single strand of DNA serves as a template for the synthesis of a complementary (base-paired) strand of DNA or RNA by DNA or RNA polymerase enzymes, respectively. During reverse transcription, a strand of RNA serves as a template for synthesis of a complementary DNA strand by reverse transcriptase.	(not applicable)
Tertiary structure (protein)	The 3D, usually folded shape of a protein. It will have a single amino acid primary sequence as well as secondary structure.	https://en.wikipedia.org/wiki/Protein_tertiary_structure
Tether (molecular)	Attach a molecule or molecular structure to a rigid support by means of an intermediate substance.	https://en.wikipedia.org/wiki/Tether_(disambiguation)
Tetrapod	The superclass Tetrapoda contains the four-limbed vertebrates known as tetrapods. It includes extant and extinct amphibians, reptiles (including dinosaurs and thus birds) and mammals (including primates, and all hominid subgroups including humans), as well as earlier extinct groups.	https://en.wikipedia.org/wiki/Tetrapod

(*Continued*)

Glossary Term	Descriptions or Definitions, Including Modified Wikipedia Excerpts	Relevant Link (Usually Wikipedia)
TGF-beta (superfamily)	A multifunctional member of the transforming growth factor superfamily. The superfamily has numerous functions in development and signaling by white blood cell lineages.	https://en.wikipedia.org/wiki/Transforming_growth_factor_beta
Theory of Natural Selection	Theory based on the differential survival and reproduction of individuals due to differences in phenotype. It is a key mechanism of evolution, the change due to heritable traits characteristic of a population over generations.	https://en.wikipedia.org/wiki/Natural_selection
Thymine	One of the four nitrogenous bases found in the nucleotides of DNA.	https://en.wikipedia.org/wiki/Thymine
Tissue (biology)	A cellular organizational level between cells and a complete organ. A tissue is an ensemble of similar cells and their extracellular matrix from the same origin that together carry out a specific function. Organs are then formed by the functional grouping together of multiple tissues.	https://en.wikipedia.org/wiki/Tissue_(biology)
Toolkit (developmental or genomic)	Metaphorical description of modular, reusable elements that can be modified, fine-tuned, integrated and reused, under Natural Selection.	(not applicable)
Total genomic DNA	The complete set of DNA sequences (i.e. the genome) found in the cells of a species of organism, or cells from individuals of that species.	https://en.wikipedia.org/wiki/Genomic_DNA
Transactional interpretation	Takes the psi and psi* wave functions of the standard quantum formalism to be retarded (forward in time) and advanced (backward in time) waves that form a quantum interaction as a Wheeler–Feynman handshake or transaction.	https://en.wikipedia.org/wiki/Transactional_interpretation

(*Continued*)

Glossary Term	Descriptions or Definitions, Including Modified Wikipedia Excerpts	Relevant Link (Usually Wikipedia)
Transcription	Synthesis of an RNA chain from a DNA template. One type of RNA gets processed during production of messenger RNA from protein-coding genes. There are several other types of RNA molecules that are also the products of transcription, including, but not restricted to, ribosomal RNAs and transfer RNAs.	https://en.wikipedia.org/wiki/Transcription_(biology)
Transcription factor	A protein that controls the rat of transcription. Controls the rate of transcription of RNA from DNA by binding to specific DNA sequences and/or other proteins attached to a genetic locus on a chromosome.	https://en.wikipedia.org/wiki/Transcription_factor
Transfer RNA	A specialized folded RNA molecule that carries a specific amino acid to a ribosome. At the ribosome, the anticodon of the tRNA base-pairs with a complementary codon of mRNA, which places the amino acid into the correct sequence in a growing protein chain during protein synthesis (translation).	https://en.wikipedia.org/wiki/Transfer_RNA
Translation	The process in which ribosomes in the cytoplasm or endoplasmic reticulum synthesize proteins. This occurs after transcription of DNA and processing of messenger RNA in the cell's nucleus.	https://en.wikipedia.org/wiki/Translation_(biology)
Transmembrane receptor	Receptors that are embedded in the cell membrane (plasma membrane). They act in cell signaling by receiving (binding to) extracellular molecules. They are specialized integral membrane proteins that allow communication between the cell and the extracellular	https://en.wikipedia.org/wiki/Category:Transmembrane_receptors

(*Continued*)

Glossary Term	Descriptions or Definitions, Including Modified Wikipedia Excerpts	Relevant Link (Usually Wikipedia)
	space. The extracellular molecules may be hormones, neurotransmitters, cytokines, growth factors, cell adhesion molecules, or nutrients; they react with the receptor to induce changes in the metabolism and activity of a cell. In the process of signal transduction, ligand binding affects a cascading chemical change through the cell membrane.	
Transmission (genetic)	The transfer of genetic information from genes to another generation.	https://en.wikipedia.org/wiki/Transmission_(genetics)
Transmission electron microscopy	A microscopy technique used for high magnifications that exceed the resolution of images formed by light. A beam of electrons is transmitted through a specimen to form an image. The specimen is most often an ultrathin section less than 100 nm thick or a suspension on a grid.	https://en.wikipedia.org/wiki/Transmission_electron_microscopy
Transposable element	A DNA sequence that can change its position within a genome. Sometimes it creates or reverses mutations and alters the cell's genetic identity and genome size.	https://en.wikipedia.org/wiki/Transposable_element
Tree of life (biology)	The tree of life or universal tree of life is a metaphor, model, and research tool. It is used to explore the evolution of life and describe the relationships between organisms, both living and extinct. Was described in a famous passage in Charles Darwin's *On the Origin of Species* (1859).	https://en.wikipedia.org/wiki/Tree_of_life_(biology)

(*Continued*)

Glossary Term	Descriptions or Definitions, Including Modified Wikipedia Excerpts	Relevant Link (Usually Wikipedia)
Transfer RNA (tRNA)	Folded, single-stranded RNA molecule with a clover-leaf shape that carries a specific amino acid to the ribosome, and base-pairs with a specific messenger RNA codon, during protein synthesis (translation).	https://en.wikipedia.org/wiki/Transfer_RNA
UAHM (Lost City)	Abbreviation for undersea alkaline hydrothermal mound. An example, called the Lost City hydrothermal field, was discovered in the Mid-Atlantic ocean.	https://en.wikipedia.org/wiki/Lost_City_Hydrothermal_Field
Ultramafic (rock)	Igneous and meta-igneous rocks with a very low silica content. The Earth's mantle is composed of ultramafic rocks.	https://en.wikipedia.org/wiki/Ultramafic_rock
Ultrastructure	The internal architecture of cells. It is visible at higher magnifications than found on a standard optical light microscope, such as those provided by the transmission electron microscope.	https://en.wikipedia.org/wiki/Ultrastructure
Undersea alkaline hydrothermal mound (see alkaline hydrothermal vent)	An undersea projection above the seafloor formed by the porous precipitates that form from adjacent alkaline hydrothermal vents.	https://en.wikipedia.org/wiki/Hydrothermal_vent
Unequal crossing-over	A type of crossover that results in duplication or deletion of segments of genomic DNA. A major mechanism involved in gene duplication.	https://en.wikipedia.org/wiki/Unequal_crossing_over
Unicellular (organism)	A prokaryotic or eukaryotic organism that usually carries out its life cycle as separate individual cells.	https://en.wikipedia.org/wiki/Unicellular_organism
Unit	An entity that represents a cohesive whole, often composed of interacting subunits.	https://en.wiktionary.org/wiki/unit
Universe	All of space and time and their contents, including planets, stars, galaxies, and all other forms of matter and energy.	https://en.wikipedia.org/wiki/Universe

(*Continued*)

Glossary Term	Descriptions or Definitions, Including Modified Wikipedia Excerpts	Relevant Link (Usually Wikipedia)
Updated evolutionary synthesis (UES)	A term introduced in Rethinking Evolution to describe a 21st century synthesis of evolutionary theory that recognizes the significance of Natural Selection and also several new complementary principles that are well-supported by wide-ranging empirical discoveries in molecular, cellular and developmental biology.	(not applicable)
Uracil	One of the four nitrogenous bases found in nucleotides of RNA.	https://en.wikipedia.org/wiki/Uracil
UV (radiation)	Electromagnetic radiation consisting of photons that have higher energy than violet colored light and fall outside of the spectrum of light visible to humans. Certain wavelengths of UV light can mutate DNA or damage proteins.	https://en.wikipedia.org/wiki/Ultraviolet
V(D)J recombination	The unique mechanism of genetic recombination that occurs only in developing lymphocytes during the early stages of T and B cell maturation. It involves somatic recombination, and results in the highly diverse repertoire of antibodies/immunoglobulins (Igs) and T-cell receptors (TCRs) produced by individual B cells and T cells, respectively.	https://en.wikipedia.org/wiki/V(D)J_recombination
Variation	Means that biological systems are different over space. Occurs within and among populations.	https://en.wikipedia.org/wiki/Genetic_variation
Venting (hydrothermal, undersea)	Release of geothermally heated water from an oceanic fissure in the Earth. In the context of the Origin of Life, this continuous flow can release waste products from organic chemical reactions.	https://en.wikipedia.org/wiki/Hydrothermal_vent

(*Continued*)

Glossary Term	Descriptions or Definitions, Including Modified Wikipedia Excerpts	Relevant Link (Usually Wikipedia)
Vertebrate	A species of animal with a backbone, classified within the vertebrate subphylum of chordates.	https://en.wikipedia.org/wiki/Vertebrate
Viable offspring	Offspring that are able to survive up to reproductive age. To continue an evolutionary lineage, they must be able to breed and produce another generation of viable offspring.	https://en.wikipedia.org/wiki/Genetic_viability
Virulent (pathogen)	An adjective describing a pathogen's or microbe's ability to infect and/or damage an organism's host cells.	https://en.wikipedia.org/wiki/Virulence
Virus	A small infectious agent that replicates only inside living cells of other organisms. Viruses usually have a high degree of host specificity but are widespread in prokaryotes (e.g. bacteriophages) and eukaryotes (e.g. influenza, HIV, etc.).	https://en.wikipedia.org/wiki/Virus
Waste products (metabolic waste)	In metabolism, refers to unwanted or unusable materials. Often, such materials are harmful to living cells and must be excreted.	https://en.wikipedia.org/wiki/Metabolic_waste
Water molecules	Molecules consisting of two covalently bonded hydrogen atoms and one oxygen atom (H_2O). Water molecules are vital for all forms of life, and are the main constituent of Earth's streams, lakes, and oceans.	https://en.wikipedia.org/wiki/Water
Wnt (signaling pathway)	A group of signal transduction pathways which begin with proteins that pass signals into a cell through cell-surface receptors. Play a variety of roles in development.	https://en.wikipedia.org/wiki/Wnt_signaling_pathway
Wood-Ljungdahl (acetyl-CoA) pathway	A set of biochemical reactions used by some bacteria and archaea. Enables organisms to use hydrogen as an electron donor and carbon dioxide as	https://en.wikipedia.org/wiki/Wood%E2%80%93Ljungdahl_pathway

(*Continued*)

Glossary Term	Descriptions or Definitions, Including Modified Wikipedia Excerpts	Relevant Link (Usually Wikipedia)
	an electron acceptor, providing raw materials for organic biosynthesis and energy transformation.	
X-ray diffraction	A technique used for determining the atomic and molecular structure of a crystal. Can produce a 3D picture of the density of electrons within a crystal.	https://en.wikipedia.org/wiki/X-ray_crystallography
X-rays	A form of high-energy electromagnetic radiation. Can break DNA and cause mutations.	https://en.wikipedia.org/wiki/X-ray
Yin-yang	Describes how seemingly opposite or contrary forces may be complementary. Each may give rise to the other.	https://en.wikipedia.org/wiki/Yin_and_yang
Zebrafish (*Danio rerio*)	A freshwater fish belonging to the minnow family. The zebrafish is an important and widely used vertebrate model organism in scientific research, including developmental biology and evo-devo.	https://en.wikipedia.org/wiki/Zebrafish
Zeitgeist	An invisible agent or force dominating the characteristics of a given epoch in world history.	https://en.wikipedia.org/wiki/Zeitgeist
Zygote (fertilized egg)	A fertilized egg resulting from the fusion of a sperm cell and an egg cell during sexual reproduction.	https://en.wikipedia.org/wiki/Zygote

About the Cover Photograph

In the cover photo, nature photographer Andreas Kay has captured one of the most striking examples of mimicry in nature. If you thought you were looking at a photograph of a venomous snake, you are not alone. A closer look reveals that this denizen of the Amazonian rainforest has legs. It is using them to hang upside down as it clings to a stem. The large black eyes, complete with light reflections, are not actually eyes at all, and this creature's vision is limited to distinguishing dark from light.

The snake mimic is actually a caterpillar, the larval form of a moth, (*Hemeroplanes triptolemus*). Its extraordinary resemblance to the triangular head of a viper is achieved by expanding its first body segments when disturbed. For more about this natural wonder, see the popular nature article by Sarah Keartes.[a]

Long before much was known about the inner-workings of cells, **Charles Darwin**'s classical theory could explain the origins of **mimicry** at a high level. **Adaptations** that prove useful in the struggle for existence arise by a blind process in which useful variations gradually accumulate by means of Natural Selection.

But today, biologists understand far more about the biological principles that make **Natural Selection** possible. *Rethinking Evolution* brings these recent discoveries together as an **Updated Evolutionary Synthesis** (**UES**). The photograph by Andreas Kay is reprinted with permission.

---

[a]https://www.earthtouchnews.com/wtf/wtf/this-is-not-a-snake-its-one-of-the-best-mimics-in-nature/

Index

A

ability to evolve, 330
abiogenesis, 84–86, 98
abiotic, 321
accessory pigments, 144
acetogenesis, 102
acetogens, 97, 116
acetylation, 193
acetyl-CoA, 102
A, C, G, and T, 324
acidity, 88
actin, 209, 212, 214
activation energy, 59, 90, 93–94, 244
active site, 242–243
adaptations, viii, xi, 3, 6, 13, 25, 32, 37, 51, 56, 145, 150, 159–160, 172, 185, 196–197, 207, 236, 238, 250, 256–257, 259–260, 269–270, 276, 278–279, 303, 310, 313, 316, 322, 327, 329–330, 457
adaptive immune system, 313, 331
adenosine triphosphate, 59
aggregates, 199
AIS, 314
algae, 128–129
alkaline hydrothermal vent, 40, 87, 100, 102
alkalinity, 88
allele, 165, 167, 169–171, 173, 183, 185, 191, 227, 322
allele frequencies, 38, 175, 295–296, 323
alpha helices, 56
alpha-proteobacteria, 137, 140
amino acid sequences, 43, 53, 55, 57, 60, 66, 86, 107, 115, 170, 173, 177, 183–184, 225, 233, 242, 249, 252, 278, 303, 305, 324–325, 329
amino-acyl tRNA synthases (aaRS), 66, 241–242, 251, 325
amino groups, 83
amplified, 188
animal cell, 104, 207, 209, 212
animal development, 39, 209, 216–217, 226, 293
animal diversity, 4
animals, 207
animals, plants, and fungi, 327
anthropomorphic, 216
antibiotic resistance, 148
antibody, 313, 316
anticodon, 66–68, 242, 249
anticodon base-pairs, 325
antigen, 170, 185, 313–316

antiporter, 106
apoptosis, 213
aqueous, 245
aqueous solutions, 88
archaeal, 97, 103–105, 131, 133, 138–140, 197, 326
archaean, 57, 148
argument from design, 157
arms race, 146, 199
astrobiology, 86, 224
atomic nuclei, 17
ATP, 59–60, 74, 82, 102, 104–108, 118, 140, 211–212, 214
autonomous specification, 217
autosome, 169–170, 183
autotrophs, 202

B

bacteria, 43, 97, 103, 131, 138, 148, 197, 314, 326
bacteriophage, 43, 46
Baldwin effect, 77, 215
base-pair, 66, 242, 249, 251, 324
basic research, 46
BBC, 7
B cells, 214, 314
beta-pleated sheets, 56
Bicoid, 281, 287
binary bits, 235
bind and interact, 309
binocular vision, 228
biochemical pathway, 57, 119, 244, 328
biofilms, 148, 271
bioinformatics, 47, 143–144, 187
biological complexity, 76, 123, 155, 205
biological evolution, viii, x–xi, 3, 6, 13, 15, 17, 21, 50, 53, 236, 276, 318, 331
biological membranes, 20, 134
biological organization, 25, 132
biology researchers, 54
biomass, 197
biosynthetic pathways, 116, 183
biotechnology, 57, 186–188
biotic, 321
Bithorax Complex (BX-C), 284
black box, 216, 258
blastocyst, 218
blastomeres, 215
blastula, 218
blind and unconscious natural result, 258
blind watchmaker, 60, 62, 236–238, 250, 255–256, 318
body plan, 39, 47, 328
Boveri–Sutton chromosomal theory of inheritance, 163–164
bricolage, 257
buoyant-density, 44
butterfly wings, 292

C

Cambrian explosion, 75, 150, 327
cancer, 213, 314–315
capacity to interact, 252
capillary action, 20
capture of energy and nutrients, 145, 196, 206
carbohydrate, 44, 53, 82, 95, 102, 251, 323
carbon-based life forms, 114–115
carbon fixation, 197
carnivorous plant, 168, 211
catalysts, 90, 92–94, 96, 109, 113–114, 117
caterpillar, 457
CD4, 314

CD8, 314
cell adhesion, 214
cell biologists, 107, 214
cell biology, 51
cell–cell communication, 316
cell cycle, 214
cell differentiation, 195, 210
cell division, 43, 222, 281
cell fractionation, 43
cell-free preparations, 43
cell membrane, 61, 82, 212
cell motility, 59, 213, 283
cell physiologists, 214
cell signaling, 60
cell-surface receptor, 62, 213, 278, 283, 312
cell theory, 218
cellular adhesion, 283
cellular and molecular determinants, 292
cellular differentiation, 208, 220
cellular metabolism, 82
cellular slime mold, 208–209
central dogma, 180
central dogma of molecular biology, 181
central dogma of molecular genetics, 86
chaperonins, 56
charged, 88
chemical bond energy, 94
chemical energy, 91
chemical mutagen, 171, 181, 323
chemiosmosis, 99
chemoautotrophic, 112
chimpanzees, 39, 46, 200
chloroplasts, 59, 103–104, 107, 134, 136, 211, 326
choanoflagellate, 196, 200, 203–205, 208, 327
chromosomal rearrangements, 193
chromosomal theory of inheritance, 154, 161, 163
chromosome, 38, 57, 163, 166, 171, 263, 308, 311
classical Darwinism, 262
cleavage, 206, 218, 220, 274, 281
cloning, 46
coding metaphor, 233
co-dominance, 169, 184
codon, 43, 233
coevolution, 123–124, 126, 132, 145–147, 210, 312, 318, 329
colonial, 201
combinatorial phenotype, 255
commensalistic, 133
complementary 3D shape, 243
complementary DNA (cDNA), 46, 187–188
complex adaptations, 267, 313
complex adaptive system, 28
complex biological organization, 207
complex cellular structures, 303
complex molecular binding events, 323
complex molecular organization, 240
complex multicellular structures and functions, 219
complex organic chemicals, 103
complex organic materials, 96
complex organization, viii, 5–6, 9, 14, 49, 63, 109, 121, 189, 201, 207, 218, 226, 230, 237–238, 245, 252, 269–270, 274, 297, 299, 327
complex phenotypes, 259, 312
complex sequences, 72
complexity, 23–24, 32, 41, 52, 55, 63, 65, 68, 72, 74–75, 81, 86, 107, 111

composite biological organization, 131
composite cells, 123, 131, 133
composite genomes, 123, 131
composite organisms, 123, 131, 141, 151
computers, 221
concentrated, 81, 83
conceptual framework, x, 10, 15, 27, 28, 31–32, 38, 41–42, 48, 52, 80, 82, 115–116, 118, 134, 151, 153, 162, 185, 189, 190–192, 216, 223–224, 238, 240, 249, 295, 299, 318
conditional specification, 217
consciousness, 215
consilience, 52
consumers, 144, 200, 202
containment, 90, 96, 102, 109, 113–114, 117
control of gene expression, 39
co-option, 272, 330
core metabolic pathways, 97–98
core metabolism, 118
covalent, 94
covalent bond, 17, 93
creationist attacks, 277
creationists, 5, 107
creative forces, 275
creative forces of natural selection, 276
CRISPR, 187–188, 281, 294–295
cross-hybridization, 305
crossing-over, 165, 171
cross over events, 193
cyanobacteria, 104, 106, 137, 141, 143, 199, 326
cysteine, 56
cytologists, 311
cytoplasm, 54, 66, 91, 163, 171, 213, 324
cytoplasmic determinants, 220, 232, 274, 280
cytoplasmic DNA, 281
cytoskeletal elements, 212
cytoskeleton, 209

D

Darwin, Charles, 112, 121, 153, 158, 457
databases, 187
decomposers, 144
deep history, 326
deep homology, 249, 253, 270, 272, 281, 285–286, 289–291, 293, 327
deletion, 181, 183–184, 193, 278–279, 304, 306–307
denaturation, 57
deoxyribonucleic acid, 173
desiccation, 56
development, 60, 63, 71–72, 75–76, 78, 166, 176, 185, 195–196, 205, 208, 209
developmental biology, viii, x, xix, 6, 11, 47, 51, 186, 269, 279, 283, 292, 298–299
developmental determinants, 270
developmental neurobiology, 224
developmental toolkit, 76, 195, 253, 258, 262, 268, 270–271, 273, 280, 285, 286, 291, 303–304, 329, 330
development of pattern and form, 287
differentiation, 210, 213, 281
digital computer code, 225
digital ontology, 230
digital physics, 230

diploid, 164–165, 183, 322
dipole, 19
dipole–dipole attractions, 19–20
disequilibrium, 93–94
disorder, 95
dispersal, 146
dispersal vectors, 147
diversity, 327
DNA, 148, 303
DNA polymerase, 57, 64, 306–307
DNA replication, 57, 165, 302, 306, 324
DNA/RNA, 234, 294
DNA–RNA hybrids, 324
DNA/RNA templates, 82
DNA sequence evolution, 308
DNA sequence variation, ix, 265, 303
DNA sequences, 187, 225, 311, 314
DNA synthesis, 164
doctrine of Malthus, 156
dogs, 126
dominance, 169, 174, 312
double-helix, 180
double-stranded, 65, 323–324
downstream, 183–184
Drosophila, 43, 47, 167–169, 171–172, 192, 281, 284–285, 288
Drosophila embryogenesis, 287
duplicated genes, 302, 311, 313

E

E. coli, 58, 133, 146, 187, 205
ecological impact, 193
ecological *niche*, 330
ecological phenotype, 50, 76, 268–270, 272–273, 278–280, 283, 294, 298, 303, 312, 329–330
ecological relationship, 133, 145, 168, 329
ecological symbionts, 146
ecosystems, 24, 25, 133, 144, 196
ectosymbiotic, 151
efficacy, viii, 3, 6, 11, 107, 116, 160, 238, 256, 271–272, 276, 283, 316–317, 327, 330
egg cell, 164–166, 218, 263, 277, 322, 324
electron bifurcation, 116
electron clouds, 19, 59
electron microscope, 135
electron transfer, 100
electrons, 17, 19, 148
embryo, 163, 176, 195–196, 201, 203, 208–209, 287
embryogenesis, 209, 212, 218, 226, 229, 282, 285
embryonic development, 195, 203, 216, 221–222, 272, 276, 281, 292, 328
embryonic mutations, 285
emergence, 13, 18, 21, 312, 327
emergent, 235
emergent evolutionary potential (EEP), viii, 10, 13–14, 17–18, 21, 24–25, 40, 50, 53, 60, 63, 109, 111, 115, 123, 133, 144, 146–147, 149–151, 215, 230, 247, 249–250, 257, 269–272, 283, 304, 309, 318, 327, 331
emergent innovations, 116
emergent interactions, 118, 278, 303
emergent organization, 230
emergent potential, 328
emergent properties, 326
emergentism, 215
empirical, vii

empirical advances, 40, 298
empirical data, 6, 8, 37, 139, 161, 222, 318
empirical discovery, 223
empirical evidence, xi, 7–9, 52, 80, 101, 103, 117, 119, 132, 136, 150, 155, 160, 247, 253, 305
empirical knowledge, 162, 229
empirical methodology, 154
empirical observations, 159
empirical research, xii, 33
empirical technique, 143, 179, 274, 281, 309, 314
empirical tools, 160
ENCODE, 72
encyclopedia of DNA elements, 72
endergonic, 114
endosymbionts, 134
endosymbiosis, 131, 136–137, 139, 145, 201–202
endosymbiotic, 103–104, 107, 133–134, 136–137, 140, 151, 326
endosymbiotic theory, 136, 141
energy and metabolism, 97
energy and nutrients, viii, 4, 22, 82, 112, 147, 197, 199–202, 207, 210, 212–213, 250, 252, 329
energy and raw materials, 114, 146
energy capture, 149
energy dissipation, 93–94
energy metabolism, 102
entelechy, 217
enthalpy, 94
entity, 23, 86, 132, 145, 278
entropy, 95, 246
enzymatic activity, 59, 62, 67, 325
enzymes, 20, 43, 46, 57–59, 63, 65–67, 74, 82, 96, 107–108, 145, 149, 168, 176–177, 180, 187, 196, 210, 214, 227, 241, 243–244, 278, 306, 323, 328
enzyme-substrate, 145
epigenetic, 208, 220, 229–230, 232
epistasis, 169
epistemological, 149
equilibrium, 89, 93–94, 102
euglenid, 201
eukaryotes, 91, 117, 136, 138, 177, 183, 199–200, 207, 220, 271, 325
eukaryotic, 56, 72, 79–80, 82, 97, 102–105, 119, 131, 134, 136–137, 140, 148, 176–177, 196, 201–204, 210–211, 219–220, 249, 303, 306, 324, 326
eukaryotic nucleus, 140
eukaryotic RNA transcription, 64
evolutionary biology, 189
evolutionary change, 304
evolutionary conservation, 305
evolutionary developmental biology (evo-devo), x, 6, 10, 40, 47, 75, 196, 208, 226, 262, 272–273, 275, 279, 282–285, 287, 289–290, 292, 294, 298, 299, 305, 328, 330
evolutionary innovations, 24, 49, 200–201, 212, 329–330
evolutionary opportunism, 141
evolutionary potential, 247
evolutionary psychology (evo-psych), xi, 40, 231, 331
evolutionary success, 252
evolutionary theory, vii–viii, ix, xi, 6–7, 11, 13, 15, 22–23, 28, 33–34, 38, 40, 47–48, 84–85, 107, 113, 116, 119, 134, 137, 140, 149–150,

154, 174–175, 185–186, 189–190, 207, 239, 250, 264, 267, 280, 285, 289–290, 296, 299, 302, 309
evolutionary tinkering, 253
evolution of complexity, 245
evolvability, 9, 40, 111, 116, 150, 196, 199, 246, 249, 263, 271–272, 279, 329–330
exaptation, 131, 197, 250, 258, 272, 279, 330
exergonic, 114
exons, 177, 233
experimental embryologists, 214
extant, 272
extinct, 29, 147, 201, 263, 272
extinction, 272
extracellular matrix, 207
extremophiles, 56, 57
eyespot apparatus, 201
eyespot patterns, 295

F

facilitated variation, x, 10, 40, 49, 331
falsifiability, 149
families of hypotheses, 80
fate maps, 282
fatty-acid, 82, 90, 95, 251
feathers, 293
fertilization, 218, 236, 263, 274, 277–278
fertilized egg, 165, 167, 203, 209, 218, 222, 229, 235, 268, 280, 282
fine-tuning, 304
fitness, 270
flagellum, 200
flexible, adaptive process, 228
flow cytometry, 314
flower development, 209
flowering plants, 147, 210
food web, 133, 144, 211
fossilized tetrapods, 293
fossil record, 25, 30, 136, 154, 204, 326
four levels of protein structure, 55
frameshift mutation, 183–184
frameshifts, 193, 304
fruitfly, 154, 167–168, 185, 218, 284–285, 292–293, 305
functional groups, 207, 248
functional units, 23
fundamental research, 178
fungi, 127, 129, 142, 144, 176, 201, 206

G

gap gene, 287–288
GAs, 260
gastrulation, 218, 221, 274, 282–283
gene, 190–193
gene duplication, 46, 193, 263, 289, 301–302, 304, 308–313, 317, 331
gene expression, 47, 61, 213, 222, 282, 325
gene families, 204–205, 209, 327
gene flow, 175, 323
gene pool, 133, 175, 188, 261–262, 264, 271, 278, 289, 292, 323, 329
gene products, 174, 185, 267, 293
gene regulatory networks, 328
generalists, 141
generative effects, 193
generative phenotype, 76, 268–274, 278, 280–281, 283, 292, 298, 312, 330
generative toolkit, 292

genes, 121, 174, 177, 219
genetic algorithm (GA), 255, 259–260, 264, 273
genetic apparatus, 63, 66, 81, 111, 116, 119, 186, 239–241, 249–253, 325
genetic barcode, 281
genetic blueprint, 223–224, 237
genetic code, 17, 39, 43, 66, 68–69, 79, 81–82, 86–87, 95–98, 102, 109, 111, 115–116, 118–119, 174, 177, 179, 181–182, 189, 225, 229, 233–234, 237–238, 240–242, 246–247, 249, 267, 297, 325
genetic crosses, 172
genetic determinants, viii, 6, 39, 53, 63, 65, 196, 221–222, 226, 233, 238, 250, 255, 257, 267, 269, 270, 272, 276–277, 280, 287, 292, 294, 298, 324, 330
genetic diversification, 264
genetic drift, 175, 323
genetic material, 38, 43–44, 53, 65, 132, 173, 177–180, 241, 252, 323, 326
genetic mutations, 289
genetic transformation, 178
genetic variation, 70, 192, 278, 303, 313
genetics, 216
genome, 189, 199, 204
genome evolution, 50, 70–71, 74, 262, 301, 304
genomic DNA, 44, 72, 141, 196, 221, 224, 232, 253, 263, 278, 289, 294, 303–305, 307–308
genomic evolution, 302, 308, 326
genomic scrapyard, 233, 258, 263, 270–272
genomic sequence evolution, 301
genomic toolkit, 262
genomic variation, 279, 301, 331
genotype and phenotype, 76, 167, 267, 302
genotypes, 323
genotypic gestalt, 298
geochemical energy, 96
geology, 29–30, 34, 79, 98, 154, 156, 159
germ layers, 268, 282
germ line, 164–166, 171, 283, 324
gestalt, 15, 21
goal-seeking, 277
G protein-coupled receptors (GPCR), 312
gradient, 91, 93–94, 220, 281
gradual accumulation, 23, 25, 76, 107, 308, 316, 329
gradualism, 10, 153, 156
green algae, 143, 199, 202, 206
group selection, 271, 331
growth and development, 210

H

H^+, 88–89
hairs, 293
Hardy–Weinberg equilibrium, 323
heat, 17, 20, 57, 81, 83, 91–92, 93–94, 96, 145, 227
Hedgehog, 205, 217, 293
Hemeroplanes triptolemus, 457
hereditary factors, 153, 161–164, 166, 168–169, 171, 174, 276
hereditary variation, 280
heredity, 25, 38, 97, 119, 277, 322
heritable, 321
heterogeneous, 44

heterotrophs, 201–202
heterozygous, 169, 183–184, 323
high-energy compounds, 97, 102, 118
higher-level organized structures and functions, 328
higher levels of emergent structure and function, 330
holon, 22
homeobox, 281
homeodomain, 281
homologous chromosomes, 165, 169, 263, 322
homologous DNA, 293
homologous pair, 165, 169, 322
homologs, 205
homology, 305
homo sapiens, 326
homozygous, 169, 183, 323
horizontal gene transfer, 133, 148, 197, 270–271, 329
housekeeping genes, 210, 279
Hox gene complexes, 286
Hox genes, 284–285, 291, 328
human, 285
human evolution, 200, 331
human genome, 47, 72, 74, 303
hybrid orbital, 17
hydrogen atom, 17, 19, 88
hydrogen bonding, 19–20, 64, 324
hydroxide, 89
hypothesis, 52, 80, 82, 84, 204, 307–308

I

immune system, 62, 170, 185, 302, 315, 317–318, 328
immunoglobulin gene superfamily, 314
immunoglobulin superfamily (ISF), 313, 315
implicate and generative order, 267
implicit informational content, 95–96, 229–230, 232, 237
incomplete dominance, 169, 183–184
incremental accumulation of random events, 151
incremental change, 78, 268, 296, 297–299, 311, 327, 329
induced fit, 243–244
infinitesimal changes, 156, 301
infinitesimal variation, 23–24, 41, 75, 108, 160
influenza, 178
information content, 96
innate immune system, 205
inner-workings of cells, v, 6, 19, 23, 35, 42, 48, 52, 54, 78, 133, 151, 168, 174, 177, 189, 215, 222, 229, 245, 276, 331
inner-workings of interacting cells, 280
inner-workings of living cells, 160, 162
inorganic catalysts, 102, 247, 251
inorganic catalytic cofactors, 251
inorganic products, 247
insertion, 181, 183–184, 193, 304, 306–307, 317
in situ hybridization, 216, 288
in situ imaging, 285
in situ visualization, 285
instincts, 77
integral membrane proteins, 278
intelligence play, 277
intelligent design, 6, 41, 61–62, 234, 237, 241, 250
intercellular signaling, 220
intermediate products, 117

intermolecular binding forces, 19
introns, 64, 233
ion-dipole, 20
ionic compounds, 89, 91
ion–ion, 89
ionization, 88
ionizing radiation, 56, 323
ions, 88–89, 91, 106
iron monosulfide, 90, 92
irreducible complexity, 23, 107, 250

J

junk DNA, 46, 72, 263

L

lateral gene transfer, 131, 133, 329
law of dominance, 162, 169
law of independent assortment, 162
law of segregation, 162
laws of heredity, 153, 161, 164
lay the groundwork, 9, 49, 230, 239, 249–250, 252, 302, 317
levels of complexity, 329
levels of organization, 123, 149, 185, 193, 207–208, 222, 229–230, 235, 329
lichen, 131, 137, 141–144, 151, 156
life-history strategies, 202
ligase, 187
light-dependent process, 228
limiting resources, 198
linkage groups, 169, 171
lipid bilayers, 20
lipids, 53, 90–91, 95, 102, 107, 251, 323
lock and key, 243
lone pairs, 19
lost city, 100–101

M

macroevolution, 295–296, 299
macroevolutionary adaptations, 297
macromolecules, 43, 52–53, 74–75, 82, 84, 95–97, 102, 111, 168, 185, 195, 199–200, 210, 222, 234, 243, 245, 257
macroscopic realm, viii, 14–15, 17, 21
MADS-box, 208
MADS-box transcription factor, 209
maternal effects, 281
maternal genes, 232
meiosis, 164–165, 171, 263, 308
membrane, 44, 54, 60, 81, 92–94, 96, 100, 104, 114, 244, 278, 288
membrane-bound, 60, 62, 106–107, 140, 328
memory cells, 314
Mendelian genetics, 27, 38, 43, 153–154, 161, 163, 166–167, 172–174, 182, 189, 192, 279, 284–285
Mendelian inheritance, 163
Mendel's laws, 162
meristematic, 209
messenger RNA (mRNA), 39, 43, 58, 64, 67–68, 171, 182, 186–187, 225, 233, 241–242, 249, 252, 289, 324–325
metabolic evolution, 117
metabolic evolutionary advances, 117
metabolic innovations, 119, 197
metabolic pathways, 58, 100, 103, 114–118, 140, 149, 196, 210, 223, 326
metabolic scrapyard, 114
metabolism, 57, 149
metabolomics, 116
metagenome, 133
metals, 88, 96, 98

metamorphosis, 209
metaphorical blind watchmaking, 270
metaphorical building blocks, 290
metaphorical language, 238
metaphorical "scrapyard" of reusable elements, 329
metaphorical "toolkit" genes, 328
methanogenesis, 102
methanogens, 97, 116
methylation, 193
mice, 172
microbial consortia, 148
microevolution, 295–297, 299
micro-machine, 54, 108
migration, 323
Miller–Urey hypothesis, 84
mimicry, 457
mineral catalysts, 90, 92
mineral membranes, 91
mineral or organic catalysts, 248
mitochondria, 59, 103–104, 107, 118, 134, 136–137, 139–140, 211–212, 232, 326
mitochondrial proton gradients, 140
mitochondrion, 105, 139–140
mitosis, 164–165, 171
model system, 292–293, 305
modern synthesis, vii, viii, x, 1, 10, 25, 27, 40–41, 49, 68–71, 74–75, 77, 82, 173–177, 189–192, 195, 262, 265, 271, 273, 277, 279, 287–289, 292, 296–297, 299, 301, 303, 316, 322
modular, 50, 51, 112, 268, 278, 290, 294, 303
modular structures and functions, 330
module, 7, 49, 76, 114, 119, 293, 303
molarity, 89
molecular and cellular interactions, 330
molecular binding events, 235
molecular biology, 139, 180–181, 191, 285
molecular cloning, 47
molecular evolution, 65, 115, 252, 289, 314, 316, 318
molecular genetics, 51, 86–87, 115, 168, 173–174, 177, 182, 185, 189, 191, 195, 216, 234, 282, 287, 328
molecular genetics of development, 280
molecular genetics of fruitfly development, 284
molecular genetics of vertebrate development, 293
molecular living fossil, 317
molecular phylogeny, 326
molecular proliferation, 248, 252
molecular recognition, 243
monoclonal antibodies, 314
monomers, 186
morphogenesis, 208–209, 213, 287
morphogenic fields, 220
morphogens, 220, 274, 281, 285, 287, 298, 328
morphology, 199, 295, 297–298
mosaic, 217
mosaic embryos, 217
motility, 212
multicellular, 118, 131, 196, 218, 222
multicellular eukaryotes, 197
multicellular organisms, 207
multicellularity, 193, 195–196, 199–200, 202, 206, 210
multigene families, 208, 279, 302, 309–310, 312, 315–316, 331

mutation, 47, 182, 227, 277–278, 289, 323
mutualism, 132, 141, 145–146, 148, 321, 329
mutualistic, 133, 210–211
mycorrhizae, 130
myelin sheaths, 214
myosin, 214

N

natural evolutionary systems, 255
natural history, 7
naturalism, 18, 33–35, 274
natural science, ix, xi, xxi, 33–34, 156
natural selection, x, xix, 3, 6, 9–10, 15, 22–23, 25, 27–28, 32–35, 37, 50–51, 57–58, 60, 63, 68, 71, 74–76, 81–82, 86, 97–98, 107–109, 111–112, 115–116, 118–119, 132, 145–146, 149, 153, 155, 157–160, 163, 172, 174–175, 199, 205, 212, 234, 236, 238, 241, 247, 250–253, 256–261, 265, 268–269, 271–273, 275–279, 283, 289, 296–297, 299, 304, 311–312, 314, 316–318, 322–323, 327, 331, 457
natural theological, 29, 33–34, 158–159, 237
negative logarithm, 89
neo-Darwinian, 82, 190, 303
neo-Darwinian theory, 267
neo-Darwinism, 68, 153, 189–191
neofunctionalization, 311
neurulation, 219
niche, 3–4, 22, 50, 146, 197, 201–202, 206, 211, 236, 269–270, 272, 280, 296, 321–322
nitrogenous base, 66, 95, 324
nonantigenic, 185
noncoding DNA, 46, 233
nonmetals, 88
nonrandom, 78, 278–279, 289, 304, 313, 327
non-random linkage, 264
nonrepetitive, 72
nonsense mutation, 184
nonspontaneous, 74, 93–94
nuances, 7–8, 28, 71, 77–78, 219, 264, 271, 325
nucleated cells, 66, 112
nucleosomes, 325
nucleic acid, 53, 86–87, 95, 97, 102, 180–181, 281
nucleic acid hybridization, 44
nucleotide sequence, 225–226, 229, 249
nucleus, 62, 66, 173, 213, 324

O

odorant receptors, 312
OH^-, 89
olfactory, 312
olfactory receptor, 302, 312
olfactory receptor family, 312
one gene, one enzyme, 177
one gene, one enzyme hypothesis, 176
one gene, one protein-chain, 177
operator, 58
operon, 58, 196
opportunism, 258
opportunist, 145
opportunities for change, 37
optix, 295
order, 95
organelles, 91, 103, 105, 107, 134, 139, 151

organic compounds, 39, 83–84, 90, 92, 111, 113–114, 116, 247–249, 251, 326
organic molecules, 93, 113, 245, 252
organic products, 114–115, 118
organizing power, 318
organogenesis, 219, 274
origin of life, 28, 79–80, 83, 87, 91, 94, 96–98, 102, 109, 111–113, 118, 223, 247, 249, 250, 267, 325
On the Origin of Species, 123, 322
origins of complexity, 111
oxidation, 91
oxidation–reduction reactions, 92
oxidizing agents, 149
oxygen atom, 18–19, 51
oxytocin, 126

P

pair-rule genes, 287
paradigm shifts, 28, 37
parameters, 259
parasites, 146, 199
parasitic, 133
parasitism, 146
pathogenic, 317
pathogens, 148, 179, 205, 214, 314–315
pattern formation, 39
patterns of gene expression, 285
Pax, 205, 291
peer-reviewed, viii, x, xx, 40, 81, 97–98, 100–101, 103, 181, 238, 247
pH, 56, 88–89, 113, 117
pH gradient, 100
phenotype, 175, 197, 227, 280, 323
phenotypic gestalt, 50, 78, 258, 261, 267–269, 272–273, 292–294, 312
phospholipids, 20
photosynthesis, 168, 197–198, 201, 203, 211
photosynthesizers, 144
phylogenetic, 141
phylogenomics, 204
plant evolution, 311
plants, 127, 129
plasmid, 148, 187
plastid, 137, 139
pleiotropic, 169
point mutations, 73, 193, 261–262, 264, 301, 303, 307, 310–311, 316–317, 331
polar bonds, 19
pollination vectors, 147
polygenic effects, 78
polymerase chain reaction (PCR), 39, 47, 57, 187, 188
polymers, 186, 234, 251, 324, 328
polyribonucleotide sequences, 251
polytene chromosomes, 171
population genetics, 38, 174–175, 189, 216, 262
porosity, 114, 248
potential biological evolution, 18
potential energy, 17–18, 59, 91–92, 94–95, 102, 107–108, 247
prebiotic soup, 79–84, 115, 325
prebiotic soup hypothesis, 39
precipitation, 89, 91
predators and prey, 145–146, 168, 196, 211, 312, 321, 329
preformationism, 231–232
pressure, 92, 117
primary producers, 144
producers, 202, 321

producers and consumers, 168, 196, 329
proflavin, 181
programmed cell death, 213
prokaryotes, 97, 103, 112, 117, 133, 136, 139, 147, 176, 199, 202, 205, 207, 271, 302
prokaryotic, 56–58, 79–80, 82, 102–105, 116–117, 119, 131, 133, 136, 138, 140, 147, 149, 183, 195–198, 203, 210, 249, 326
protected natural mineral compartments, 326
protein chains, 43–45, 53, 55–58, 62–63, 66, 95–96, 108, 170, 174, 177, 180–185, 193, 213, 233, 245–246, 253, 277–278, 305, 313, 324–325, 329
protein-coding, 46, 182–183, 208, 225–226, 234, 262–264, 270, 297, 305–306, 308, 310, 316, 324, 329
protein sequence, 17, 39, 46, 66, 174, 188, 234, 279, 293, 314–315
protein synthesis, 173, 177
proto-cells, 81, 82
proton fluxes, 248
proton gradients, 9, 91–94, 96, 103–105, 107–108, 112, 117, 134, 140, 211–212, 247, 326
proximity, 143
pseudoscience, 41, 151, 237
pseudoscientific, 107

Q

quantitative inheritance, 169
quantum physics, viii, 14, 16
quorum sensing, 148, 271
QWERTY effect, 38, 98

R

RAG1, 317
RAG2, 317
random fertilization, 277, 322
random mutations, ix, 9, 49, 68–69, 71, 74, 192, 290
random point mutations, 304
random shuffling of genomic sequences, 277
random variation, 10, 41, 76–77, 82, 116, 259, 277, 279, 322
raw materials, 109
raw materials and energy, 113
rearrangements, 304
recessive, 169–171, 183, 185, 232, 264
recognition sequences, 46
recombinant DNA, 39, 46, 177, 186–189, 307
recombination, 171, 261, 263
recombination-activating genes, 316
redox potential, 93, 96
redox reactions, 92, 111, 247
reducing agents, 149
reduction, 91
reduction potential, 92
regional specification, 208
regulation of gene expression, 283, 292, 298, 329
regulation of genes, 304
regulation of protein synthesis, 57
regulative, 217
regulative development, 217
regulative embryos, 217
regulatory DNA sequences, 226, 270
regulatory functions, 311
regulatory genes, 329
regulatory mutations, 46

regulatory sequences, 40, 50, 53, 208, 263, 277, 309
release of waste products, 113, 117
repeated elements, 308
repetitive sequences, 301
replacing outdated metaphors with molecular interactions, 240
replication, 97
replication slippage, 47, 302, 304–305, 307
replicator, 85, 97
reproductive isolation, 175
res potentia, 14–16, 215, 230
restriction enzymes, 46, 187
retroviruses, 46
reusable modules, 124, 130
reusable "toolkit" genes, 330
reverse transcriptase, 46
ribosomal RNA, 138
ribosome, 66–67, 173, 181–184, 233, 241–242
RNA, 293
RNA polymerase enzymes, 64
RNA template, 188
RNA transcription, 73
RNA world, 65, 115–116, 119
Rube Goldberg device, 75, 257, 261, 270, 284

S

3D shapes, 53, 56, 59, 234, 242, 251, 328
salinity, 56
scales, 293
scanning electron microscope, 134
science communicator, 226
scientific method, 274
scientific theory, viii, 35
secondary messengers, 63
secreted protein, 317
segmentation, 284, 286–287, 328
selective forces, 57
selective pressure, 61, 247, 250, 264, 270–271, 290
self-accelerating, x, 73, 304, 307, 309
selfish, 303
selfish DNA, 233
self-organization, 10, 40, 56, 111
self-replication, 86
semantic limitations, 240–241
semi-permeable membranes, 90–91, 100
sequence alignment, 46, 187
sequence of amino acids, 177
serendipity, 305
serpentinization, 98, 113, 117, 247
sex cells, 171, 277, 308, 311
sex chromosomes, 169, 171, 183
sex determination, 305
sex-linked, 183
sexual reproduction, 147, 154, 163, 171, 174–175, 209, 218, 261, 277
sexual selection, 7, 318
shape and form, 280, 323
shape, form, and capabilities, 289
shape-specific molecular interaction and binding events (SSM-IBE), 239, 245–247, 249, 253, 257, 269, 283, 313, 318, 328
Siamese cat, 227–229, 237
signaling pathways, 309
signal peptide, 107, 317
signal transduction cascades, 63
signal transduction pathways, 60, 75, 213, 257, 279, 328
simple repeats, 307–308
simple repetitive DNA sequences, 305

simple repetitive sequences, 302, 305–308, 331
single-celled eukaryotes, 201
singularity, 319
slipped-strand mispairing (SSM), x, 47, 193, 302, 305, 307, 331
snake mimic, 457
software, 221
somatic hypermutation, 302, 313, 316
Sox, 205
speciation, 38, 173, 175, 261, 292, 323
sperm, 322
sponges, 204
spontaneous, 56, 74, 93, 323
spontaneous mutations, 171
spontaneous reactions, 93
sporulation, 209
stem cells, 210
stromatolites, 104, 106
strong chemical bonds, 93
structure and function, 3, 5, 9, 21–24, 28, 41, 47, 51, 65, 71, 74–75, 78, 107, 113, 160, 177, 207, 239, 250–251, 253, 267, 312, 329, 323
struggle for existence, xix, 3, 9, 156, 250, 252, 258, 260, 278, 309, 311
subfunctionalization, 311, 317
submicroscopic realm, 14
substrates, 309
subunits, 22–24
success criterion, 252
supernatural design, 277
supernatural explanations, 217
survival and reproduction, 109, 193
symbiosis, 123, 133, 136, 140, 145
symbiotic, 58, 124, 126–127, 144, 205, 326, 329
synergistic, 304
synergy, 308
synthesis, 63

T

tandem repeats, 73, 304–305, 331
T-cell receptor, 214, 313, 314, 316
T cells, 314
teleological, 3
temperature-sensitive, 227
temperature-sensitive mutation in pigmentation, 228
template-based replication, 97
terrestrial plants, 202
tether, 114, 251
TGF-beta, 205
theorists, v, 34, 141, 150, 230, 262, 273, 319
three domains of life, 138
tinkerer, 145
tinkering, 257, 270
tissues and organs, 207, 280
toolbox, 196
toolkit, 196, 270, 298
total genomic DNA, 188
toxins, 214, 314
transactional interpretation, 14
transcription and translation, 281
transcription factors, 47, 63–64, 181, 192, 196, 205, 208–209, 217, 219–220, 233, 274, 281, 287, 291, 295, 298, 325, 328
transfer RNAs (tRNAs), 66–67, 173, 181, 241, 324
translation, 58, 66, 70, 241
transmembrane, 62, 312
transmission electron microscope, 134
transposable elements (TE), 39, 73–74, 303, 316

transposon, 317
tree of life, x, 147
trial-and-error, 305
tRNA-ligase, 241

U

ultramafic, 98
ultrastructure, 134
undersea alkaline hydrothermal mounds (UAHM), 28, 57, 79–81, 87–90, 94, 96–98, 100–103, 105–106, 109, 111–119, 140, 247, 325–326
unequal crossing-over, 301, 304, 308
units, 22, 24
updated evolutionary synthesis (UES), vii–ix, xi, xix–xx, 1, 6, 8–11, 27, 40–41, 50–52, 76, 78, 107, 109, 132, 193, 215, 221, 233, 240, 255, 271, 273–275, 279, 301, 304, 317–318, 321, 457
UV radiation, 56

V

variation and randomness, 277
variation and selection, xx, 75, 258, 264, 296, 299, 322
V(D)J recombination, 302, 313, 316, 318
venting of wastes, 109
venting waste products, 96
vertebrate skin appendages, 293
vertical inheritance, 151
viral genomes, 46
viruses, 314
vitalism, 217

W

Wallace, Alfred Russel, 112
warm little pond, 81, 83, 112, 325
ways of life, 4, 50, 159, 196, 201, 206, 268, 280, 296, 321–322
Wnt, 220, 293
WntA, 295
Wood-Ljungdahl, 102
wound healing, 210

X

X-linked genes, 171
X-ray diffraction, 179

Y

yeasts, 202

Z

zebrafish, 172, 292–293
zygote, 203, 218, 222

www.ingramcontent.com/pod-product-compliance
Lightning Source LLC
Chambersburg PA
CBHW061801200525
27002CB00004B/27